CULTURE, PLACE, AND NATURE
Studies in Anthropology and Environment
K. Sivaramakrishnan, Series Editor

Centered in anthropology, the Culture, Place, and Nature series encompasses new interdisciplinary social science research on environmental issues, focusing on the intersection of culture, ecology, and politics in global, national, and local contexts. Contributors to the series view environmental knowledge and issues from the multiple and often conflicting perspectives of various cultural systems.

Ordering the Myriad Things

FROM TRADITIONAL KNOWLEDGE TO
SCIENTIFIC BOTANY IN CHINA

Nicholas K. Menzies

UNIVERSITY OF WASHINGTON

Seattle

Ordering the Myriad Things was made possible in part by a grant from the Chiang Ching-kuo Foundation for International Scholarly Exchange.

Copyright © 2021 by the University of Washington Press

Design by Katrina Noble
Composed in Warnock, typeface designed by Robert Slimbach

25 24 23 22 21   5 4 3 2 1

Printed and bound in the United States of America

All rights reserved. No part of this publication may be reproduced or transmitted in any form or by any means, electronic or mechanical, including photocopy, recording, or any information storage or retrieval system, without permission in writing from the publisher.

UNIVERSITY OF WASHINGTON PRESS
uwapress.uw.edu

LIBRARY OF CONGRESS CATALOGING-IN-PUBLICATION DATA
Names: Menzies, Nicholas K., author.
Title: Ordering the myriad things : from traditional knowledge to scientific botany in China / Nicholas K. Menzies.
Description: Seattle : University of Washington Press, [2021] | Series: Culture, place, and nature : studies in anthropology and environment | Includes bibliographical references and index.
Identifiers: LCCN 2021009138 (print) | LCCN 2021009139 (ebook) | ISBN 9780295749457 (hardcover) | ISBN 9780295749464 (paperback) | ISBN 9780295749471 (ebook)
Subjects: LCSH: Botany—China—History. | Plants—Identification. | Plants—Classification.
Classification: LCC QK21.C6 M46 2021 (print) | LCC QK21.C6 (ebook) | DDC 581.951—dc23
LC record available at https://lccn.loc.gov/2021009138
LC ebook record available at https://lccn.loc.gov/2021009139

The paper used in this publication is acid free and meets the minimum requirements of American National Standard for Information Sciences— Permanence of Paper for Printed Library Materials, ANSI Z39.48–1984.∞

In China, as descendants of enlightened spirits living in a flourishing region, we have assigned correct names to the myriad things from the earliest times in the past. So if we now cut ourselves off and reject the old, if we discard and throw out and do not avail ourselves of the established nomenclature, there is absolutely no doubt that we would be putting ourselves at the same level as the uncivilized tribes of the south.

—ZHONG GUANGUANG, *ON THE IMPORTANCE OF CHINESE NAMES FOR PLANTS WITH SUGGESTIONS FOR A PROPER SYSTEM OF NOMENCLATURE*

Trees help us to calibrate the speed of current environmental change: they provide a context more in tune with the tempo of the Earth. They slow us down, they teach us the virtue of patience, and they remind us to think about all that has gone before and what is to come; the legacy we received and the legacy that we will leave.

—PETER CRANE, *GINKGO: THE TREE THAT TIME FORGOT*

CONTENTS

Foreword by K. Sivaramakrishnan ix

Acknowledgments xiii

Timeline for Botany in China xvii

Introduction 3

CHAPTER ONE
How the Southern Mountain Tea Flower Became
Camellia reticulata 9

CHAPTER TWO
The Historical Context of an Epistemic Transition 27

CHAPTER THREE
Nature, the Myriad Things, and Their Investigation 42

CHAPTER FOUR
A New Language to Name and Describe Plants 56

CHAPTER FIVE
Observing Nature, Practicing Science 75

CHAPTER SIX
The Inventory of Nature 100

CHAPTER SEVEN
Botanical Illustration 117

CHAPTER EIGHT
Spaces for Communicating and Informing 140

CHAPTER NINE
Museums, Exhibitions, and Botanical Gardens   158

CHAPTER TEN
Metasequoia glyptostroboides, the Dawn Redwood   173

Glossary   183

Notes   197

References   217

Index   271

FOREWORD

Travelers, amateur explorers and collectors, and people prospecting for vegetative and floral treasures that might present commercial opportunities had traversed the globe by the late nineteenth century. They came mostly from Europe to Asia, South America, and sub-Saharan Africa. And through their efforts, a science of botany emerged, including fields like economic botany, that aligned the identification and classification of plants with lucrative trade in plant materials and products beyond the staples of grains, oilseeds, and fibers.

This scientific botany developed in Europe as part of what historian of science Lorraine Daston (2016) has called the first two scientific revolutions of the seventeenth and eighteenth centuries, which brought science into existence as a confrontation with modernity, giving it forms of knowledge production that included field observation and laboratory experiments. An empire of science emerged alongside modern European expansion into the tropical world, and botany participated in the consolidation of both trade and science as modalities of imperial reach into distant regions.

As another historian of science, Londa Schiebinger (2004), evocatively describes it, bioprospectors returned with medicinal plants, exotic aromas and delicacies, and the core substances for industrial revolution for their royal sponsors and national economies. They created nothing less than a global culture of botany. Systematics and the description of plant properties and attributes had also entered the scientific repertoire by the end of the eighteenth century. Carl Linnaeus, for instance, had already produced *Species plantarum* in 1753 (see Hedberg 2005), and Erasmus Darwin (1791) had celebrated in verse the wonderful gifts of the botanical garden by the end of the eighteenth century! Science and adventure combined in service of empire, and botany, as anthropologist Lucile H. Brockway (1979) so ably showed, was a key mode of colonial knowledge (also see Schiebinger and

Swan 2007). In some ways, the emphasis on field exploration, observation, and developing a language of classification brought scientific botany into line with the modalities developed more generally by colonialism for what anthropologist Bernard Cohn (1996) so persuasively argued were its forms of knowledge.

Nicholas Menzies is aware that much of this is known to scholarship in the field of environmental studies. Pertinent to this book, however, is that this imperium of plant science also came to be known to Chinese plant scientists in the late nineteenth century. They initiated debates in the Chinese scientific academy about the relationship between botany, environmental knowledge—especially of plants, soils, hydrology, and variation—and the building of a national heritage of plant sciences research. Along the way, he addresses big questions about how knowledge about an animate nonhuman world was transformed during the nineteenth century in China. Budding Chinese scientists trained in Europe and Japan rethought foundational analytical terms like essences and relations, reflecting on life cycles of growth and decay and how individuals shaped communities in the plant world.

Cognizant of the way foreign collectors had already located Chinese plants in world herbaria and research gardens in the European metropole, Chinese field science was born to replicate the methods of its counterpart in the Occident. And thus, as Menzies shows, Chinese botany came to be between the mid-nineteenth and mid-twentieth centuries through a dialogue between traditional forms of plant knowledge in China and the construction of modern scientific botany as a Chinese enterprise, a process as much of the making of Chinese plant scientists as it was of the making of Chinese plant science.

Ordering the Myriad Things, then, sets out to answer basic questions about how a technical vocabulary and a set of scientific practices emerged as the instruments of Chinese botany, and how this new field of knowledge owed a debt to, but also differentiated itself from, earlier ways of knowing the world of plants, as situated in a realm of varied forms of life, human and nonhuman, material and spiritual. In a striking example, the case of the southern mountain tea flower with which the book opens becomes illustrative of the modes of knowledge and codification that encompass all plants in the Chinese landscape. As the dawn redwood case that comes at the end shows, by the twentieth century, knowledge-making protocols about plants and the plants themselves were being transported across continents, even as Chinese botany was consolidated as a distinct field.

Menzies locates these scientific emergences in a wider cultural and political history of imperial and republican China. Contact with European naturalists and their systems of plant classification had endured for several centuries before the political events at the end of the nineteenth century accelerated institutional investments in modern scientific botany in China, now seen as integral to geopolitical ambition. Menzies tracks science and political change in their entanglements, ending with the transition to revolutionary government in the middle of the twentieth century. Drawing on painstaking research, written in spare and accessible prose, and suitably illustrated, this book tells a fascinating story of scientific revolution in the midst of tremendous political upheaval and change.

The range and complexity of the material covered in this work are noteworthy. The nationalist impulses of the first generation of Chinese botanists are examined in relation to their intellectual debt to and appreciation for their teachers in Europe and Japan. Topics like fieldwork, visual representation, translation, and the popularization of science in China in late imperial and republican times are also insightfully brought into view. Menzies has delivered a study that will surely tap into the rising interest in both the history of science in Asia and the anthropology of human-plant relations, while enriching the study of East Asia, the history of plant sciences, and the broader intertwined history of imperial exploration and modern science that has been documented in other parts of the world.

K. SIVARAMAKRISHNAN
Yale University

ACKNOWLEDGMENTS

I first realized how fluid boundaries between categories can be when I was beginning to learn Chinese and encountered the word *qing* 青. Dictionaries translate *qing* as black, blue, and green. My teacher explained how it is not one word used for three colors. It is one color: black, blue, and green at the same time, the shimmer on a beetle, like a scarab. The spectrum is not inevitably divided into our familiar seven colors of red, orange, yellow, green, blue, indigo, and violet. Not everyone sees orange. And where is brown in the spectrum?

When I moved into the world of plants and trees, as I did when I went on to study forestry, that insight into how we organize the world in which we live was a reminder that the frequent reclassifications and renaming that are a part of the challenge of taxonomy are not the random decisions of annoying, hairsplitting pedants but a recognition that the world is not made up of neat boxes and partitions separating like from unlike. When is a plant a shrub, and when is it a tree? Bamboos are grasses. Even within the domain of scientific botany, classifications cannot be rigid. The papaya is not very woody, but it is still considered to be a tree. Its flower would seem to place it in the same family as the passionflower; its DNA, though, tells us that it is closer to the cabbage. The boxes and partitions form one pattern when plants are organized by visual signals such as growth form. They can form another pattern when organized by the structure of their flowers, and still another when based on molecular materials.

It has been a long journey from relishing the fluidity of the compartments into which the world can be organized to tracing and narrating the transition from traditional knowledge about plants to scientific botany in China. I have been pointed in that general direction and nudged along a not very well-marked path by many people in many places over more years than I care to count.

It started with enjoying and appreciating plants of all shapes and sizes. My mother was a gardener at heart who made sure that my brothers and I knew how important plants and trees are to our lives.

Most fundamentally, I owe the very possibility of putting together and carrying out a project of this kind to those who over many years and with endless patience taught me to function and to work in Chinese, spoken and written, modern and literary, in Cambridge, Beijing, and Berkeley.

A turning point in the journey to this book came long ago, when my friends John and Gill tired of hearing me complain about the job I had and told me to "just go on and do what you want to do," which was to learn about and to work with people and plants and the environment.

Then there was a period of several years when I was working with ethnobotanists in southwestern China and Thailand, a time during which I learned about worlds that were ordered and classified in many unexpected ways but that were all perfectly systematic and logical to the people who lived and thrived in them.

A number of teachers and mentors have inspired me and pointed me in the directions I eventually followed. Kong Dejun, Richard Wang, Joseph Needham, Jeff Romm, Louise Fortmann, Fred Wakeman, Walt Coward, Pei Shengji, Xue Jiru, and Tony Cunningham all not only made me "question my answers" (thanks, Walt) but also gave invaluable guidance about how to look for, evaluate, and make use of the evidence so that it would fit together to tell a credible and coherent story.

Many colleagues, guides, and assistants have helped this project directly and indirectly. In China, thanks to Wang Zhonglang, Yang Yongping, Pei Shengji, and Xu Jianchu in Kunming; Lynn Pan and Cao Mengqin in Shanghai; Ching May-Bo in Guangzhou and Hong Kong; Hu Zonggang at the Lushan Botanical Garden; Huang Ruilan and Liao Jingping at the South China Botanical Gardens in Guangzhou; and Hu Yonghong at the Chenshan Botanical Garden in Shanghai.

Three people were instrumental in bringing me to the Huntington Library, Art Museum, and Botanical Gardens. Duncan Campbell, Jim Folsom, and June Li, thanks for making me feel welcome, encouraging me to pursue my ideas, and offering the fellowship that allowed me to research, travel, and write in such uniquely beautiful surroundings.

At the Huntington, Phillip Bloom, Michelle Bailey, and Robert Hori listened to my ideas and made helpful suggestions, while welcoming me as a colleague at the East Asian Gardens Research Center.

Librarians are the unseen facilitators who open doors and help discover the evidence that gives credibility to a project such as this one. Thank you,

Su Chen and Zou Xiuying in Los Angeles and Claremont. Thanks to Danielle Rudeen and Melanie Thorpe at the Huntington and to many others in Guangzhou, Shanghai, Nanjing, and Beijing, as well as at the British Library in London.

There are so many others to thank who may or may not know that, in one way or another, they have been a part of this book: Sir Peter Crane, Tony Willis, Jim Morris, and all the staff at the Oak Spring Garden Foundation Library; Bob Marks at Whittier College; Bin Wong at UCLA; Wang Zuoyue at Cal Poly, Pomona; and Janet Sturgeon, formerly at Simon Fraser University.

At the University of Washington Press, thanks to my editor Lorri Hagman; to Neecole Bostick, who helped me negotiate the thickets of permissions and authorizations; and to K. "Shivi" Sivaramakrishnan who, together with Lorri, saw something interesting in this idea and encouraged me to go ahead and realize it. Thanks, finally, to Joeth Zucco, Rich Isaac, and Kata Bartoloni, my guides in copy editing and proofreading this text.

I apologize in advance to anyone that I might have missed. Your absence is only a sign of my poor memory, not a comment on how valuable your contributions have been.

This year has been the unforgettable year of the COVID pandemic, which changed everything about how and where we worked. I owe particular thanks to all those at the University of Washington Press and various libraries and other institutions, who still tirelessly and enthusiastically tracked down references, scanned images, edited the manuscript, and supported turning it into a book.

Finally, thanks to Melinda, who put up with me taking over the family room table and watched it disappear under multiple mountains of possibly relevant books and papers after I had already filled all the bookshelf space in the house. Not to mention her patience in putting up with the inevitably curmudgeonly moments that could turn into days or even weeks, when I was struggling to turn my ideas into writing. I love you!

The book's epigraph from Peter Crane, *Gingko: The Tree That Time Forgot*, is cited with the permission of Yale University Press. The epigraph for chapter 6 is from *Xunzi: The Complete Text*, translated by Eric L. Hutton (Princeton: Princeton University Press, 2014), republished with permission of Princeton University Press through the Copyright Clearance Center, Inc. The epigraph for chapter 10 is republished with permission from the Missouri Botanical Garden.

TIMELINE FOR BOTANY IN CHINA

1848 — *Research on the Illustrations, Realities, and Names of Plants* (Zhiwu mingshi tukao) by Wu Qijun, the last important traditional compendium on plants, is published.

1858 — Li Shanlan, Alexander Williamson, and Joseph Edkins publish *Botany* (Zhiwuxue), their translation of John Lindley's *Elements of Botany*.

1868 — Père Pierre-Marie Heude founds the Musée de Zikawei in Shanghai.

1876 — The Shanghai Polytechnic Institution and Reading Room (Gezhi Shuyuan) opens, offering lectures on science and technology and a library. The Polytechnic publishes the *Chinese Scientific and Industrial Magazine* (Gezhi huibian) through 1892.

1893 — The first course on botany is taught at the Hubei Self-Strengthening School (Hubei Ziqiang Xuetang).

1903 — *Science World* (Kexue shijie) is launched in Shanghai. It publishes the first original paper (as opposed to a translation) in Chinese on botany, "On Plant Pollination" (Zhiwu shou jing shuo) by Yu Heqin.

1904 — The Educational Association of China publishes "Technical Terms: English and Chinese," a glossary of scientific and technical terminology.

1907 — Ye Qizhen publishes *Botany* (Zhiwuxue), the first botanical textbook written in Chinese by a Chinese author.

1909 — Huang Yiren goes to Tokyo University to study botany under Matsuda Sadihisa. He was probably the first Chinese student to go abroad to study botany.

1914	(June) Chinese students in the United States meet at Cornell to found the Science Society (Kexue She), later named the Science Society of China.
1915	(January) The Science Society of China publishes the first issue of *Science* (Kexue) in Shanghai.
1916	*Science* publishes a paper by Zou Bingwen introducing and explaining the International Code of Botanical Nomenclature.
	The first university herbarium in China is established at Lingnan University, Guangzhou.
	The Joint Scientific Terminology Commission (Kexue Mingci Shenchahui) meets in Shanghai to standardize scientific terminology.
1918	Zhong Guanguang teaches botany at the Government University of Peking (now Peking University).
	The Dictionary of Botany (Zhiwuxue da cidian) is published. The editor in chief is Du Yaquan, who leads a team of fourteen, mostly students returned from Japan.
1919	Chen Huanyong spends nine months on a collecting expedition on the island of Hainan.
	Zhong Guanguang spends four months on a collecting expedition in Guangxi and Yunnan. His field diaries are published in seven installments in *Earth Sciences Journal* (Dixue zazhi) in 1921 and 1922.
1920	Hu Xiansu carries out a collecting expedition in Jiangxi and Zhejiang.
1921	Bing Zhi, Hu Xiansu, and Zou Bingwen establish the first university biology department at National Southeast University.
1922	(August 18) The Science Society of China founds its Biological Laboratory in Nanjing.
1923	Qian Chongshu, Hu Xiansu, and Zou Bingwen publish the first textbook for college courses on botany: *Advanced Botany* (Gaodeng zhiwuxue).
1927	(April 18) The Academia Sinica (Zhongyang Kexueyuan) is established in Nanjing.
	Zhong Guanguang establishes a botanical garden for teaching purposes at Zhejiang University, Hangzhou.
1928	Chen Huanyong establishes the Botany Laboratory at Sun Yat-sen University, Guangzhou.
	The Shang Zhi Society (Shang Zhi Xuehui) founds the Fan Memorial Institute of Biology (Jingsheng Shengwu Diaochasuo)

	with funding from the China Foundation for the Promotion of Education and Culture (Zhonghua Jiaoyu Wenhua Jijinhui). Bing Zhi is appointed director, with Hu Xiansu as director of the botany department.
	The botanical section of the Natural History Museum, Academia Sinica, is founded.
1929	The Institute of Botany of the National Academy of Beiping is founded as the northern arm of the Nanjing-based Academia Sinica.
1930	Sun Yat-sen University's Botany Laboratory is upgraded to the Institute of Agriculture, Forestry, and Botany with Chen Huanyong as director.
	A Chinese delegation of six people attends the Fifth International Botanical Congress in Cambridge. Chen Huanyong presents the paper "Recent Developments in Systematic Botany in China."
1932	Hu Xiansu replaces Bing Zhi as director of the Fan Memorial Institute.
	The Botany Department of the China Western Academy (Zhongguo Xibu Kexueyuan) is founded in Chongqing.
1933	(August 20) The Botanical Society of China (Zhongguo Zhiwuxuehui) is founded in Chongqing. Half of the 105 members are taxonomists.
1934	Hu Xiansu establishes the Lushan Arboretum and Botanical Garden (Lushan Senlin Zhiwuyuan) in partnership with the Jiangxi Provincial College of Agriculture. He appoints Qin Renchang as director.
1936	The Science Society convenes a meeting with six other scientific societies in Beijing, attracting over a thousand participants.
1937	The Japanese army invades China. Most universities, colleges, and research institutes move from Beiping to unoccupied areas in the southwest.
1938	The director and staff evacuate the Lushan Arboretum and Botanical Garden. They go to Kunming and found the Yunnan Institute of Agriculture, Forestry, and Botany. Qin Renchang and staff from Lushan go to Lijiang to establish the Lushan Botanical Garden Workstation.
1948	Hu Xiansu and Zheng Wanjun publish the first botanical description of the dawn redwood (*Metasequoia glyptostroboides*) in the *Bulletin of the Fan Memorial Institute of Biology*.

1950 The newly established Chinese Academy of Sciences (CAS) consolidates existing botanical research institutes into the new CAS Institute of Taxonomy. Scientific activities continue under CAS, which restricts most international exchanges and collaboration to Communist countries until the 1980s.

1959 (May 4) At the offices of the Shanghai Science Commission, the remaining members sign the papers that dissolve the Science Society of China.

ORDERING THE MYRIAD THINGS

Introduction

On April 5, 1938, Professor Sir William Wright Smith of London's Royal Horticultural Society read a paper by Hu Xiansu, the director of the Fan Memorial Institute of Biology in Beiping.[1] Hu's paper reported on botanical exploration in China over the previous twenty to thirty years. While members of the Society were familiar with the discoveries made by Western plant collectors over the last half-century or so, they knew far less about the work of a first-generation of Chinese botanists who were establishing botany as a field of science, distinct from but rooted in the rich heritage of traditional knowledge about plants in China.

Hu acknowledged that Western botanists and plant hunters such as Père Armand David, Robert Fortune, and Ernest Wilson had already done a lot to inform the world about the diversity of China's plant life. He pointed out that, remarkable as their expeditions had been, large swaths of China had still not been studied, and they appeared to have missed or ignored many species. Chinese botanists were now working on their home ground, and they enjoyed some advantages that perhaps made their contributions more substantial than outsiders': "The collectors sent to the field are usually young men who have recently graduated from the universities. They are full of enthusiasm and energy. Being accustomed to the hardship of life in China, they can live more cheaply and work harder than their Western colleagues.... They are usually willing to endure greater hardship than their Western colleagues and succeed in penetrating into regions usually considered inaccessible" (Hu X. 1938, 389).

Hu was describing a Chinese-led exploration and their discovery of the flora of China, with a sense of pride that these young botanists were discovering more than the Western plant hunters who had been making the headlines. At the same time, Hu, who had studied in the United States, was quite condescending in his judgment of one of the few older scholars to have turned his attention to the scientific study of plants. He described Zhong Guanguang (referred to as T. K. Tsoong), a self-taught botanist, working at the time for the Chinese Academy of Sciences on the nomenclature of Chinese plants, as "an energetic old-fashioned herbalist type of botanist." Zhong was clearly not, in Hu's opinion, a fully-fledged scientist: "No notable contributions have been made by him" (Hu X. 1938, 381).

The message of Hu's paper was that botany was a new science in China. It was not the same as the body of traditional knowledge about plants. China now had a community of scientific botanists who were advancing the field and making it Chinese. Nevertheless, despite papers such as this one and others presented by Chinese botanists at venues such as the Fifth International Botanical Congress held in Cambridge in 1930 (Brooks and Chipp 1931), interest at the time in the rich plant life of China was still firmly centered on the reports reaching the West from explorers such as George Forrest (1873–1932) and Joseph Rock (1884–1962).[2]

Ordering the Myriad Things sets out to unravel the story behind Hu's words. It follows the transition from traditional knowledge about plants in China to the science of botany. In distinguishing between these, there is no intention to privilege one body of knowledge over another. My purpose is not to trace an assumed teleological progression from less knowledge to more or from irrational to enlightened. My concern is with how people choose to understand and to represent the beings that surround us. How do we organize things into categories, of which "plants" is just one of many possible groupings? What shared traits place an object together with others to constitute a category, and what traits make it unique? How do we communicate and share that knowledge?

It might be said that a traditional understanding of the world is so profoundly different from the scientific that there can only be a radical break, not a transition. This essentialist view holds that botany is part of a universal scientific system that reveals knowledge about an immutable, objective natural world. That body of knowledge is fixed and unchanging. Scientists accumulate data to uncover it piece by piece over time. The logical extension of this argument is that other epistemologies are at best outdated ("traditional") and may even fall into the category of superstition. The critique of this thesis is that it conflates nature with the way human society chooses to understand and explain the world. Science is a process of engagement, study, and analysis of observed objects and phenomena. Competing ideas explaining their workings circulate within networks of human actors and institutions, and accepted theories are the ideas that command the most support. The philosopher Bruno Latour (1999) goes so far as to propose that nature itself is produced by these networks.

This study of botany in China does not rigorously follow the premise of a nonessentialist, constructed nature but presumes that there are phenomena and objects with discernible properties, which follow predictable rules, that constitute an ontological "nature." It does, however, accept the claim that a sanctioned epistemology of science is built on those theories (or discourse) that prevail in certain conditions at certain times in certain circles. There is a distinction between an objective nature and its many possible representations, of which one might be called "traditional" and another "scientific."[3]

To identify, classify, and order things is to demarcate the boundaries and relationships between groups of things on the basis of a referential set of attributes or variables. Within Western science, the field of botanical taxonomy is just one example of fluid frames of reference. Classification in eighteenth- and early nineteenth-century natural history was marked by contestation over what constituted a true or "natural" system of classification (Stemerding 1993, 193). The eighteenth-century botanist Carl Linnaeus built his system based on the sexual parts of flowers; his near-contemporary Antoine Laurent de Jussieu and others considered morphology and the functions of plant parts as the classifying principle; more recently, botany has moved to the level of DNA and molecular analysis. The objects being classified remain the same, but the clumping or dividing is constantly changing.

The history of botany in China begins, paradoxically, with the observation that until the mid-nineteenth century, there was no body of knowledge that could be described as "botany," in which plants were the objects of observation and study in their own right. Knowledge about plants was not a named field of study, but China takes pride in a rich and ancient body of knowledge about plants documented in literature and the fine arts, in comprehensive encyclopedias and geographies, materia medicas, and horticultural works and monographs.

A word of caution is in order regarding the extent of this literature. The contemporary *Flora of China*, compiled by the Institute of Botany of the Chinese Academy of Sciences, describes 32,500 species of vascular plants in China, over 50 percent of which are endemic (Hong and Blackmore 2015, 1). *Dictionnaire Ricci des plantes de Chine*, an inventory of all the plants found in the written historical record in China, has identified some 7,000 named plants (Fèvre and Métailié 2005, 1). Only a small proportion of them were ever described, though. Li Shizhen's *Classification of*

Materia Medica (Bencao gangmu), the best known and most comprehensive pharmaceutical compendium, was published in 1596. It describes 1,096 plants. In 1848, the *Research on the Illustrations, Realities, and Names of Plants* (Zhiwu mingshi tukao) by Wu Qijun, which is often praised as an indigenous precursor to scientific botany, described fewer than 800 plants. Before 1850, scholars living and working in one of the most botanically diverse environments on earth had described fewer than 2,000 plant species. In Europe, which is far less biodiverse, the *Historia generalis plantarum* (1586–1587) by Jacques Daléchamps (1513–1588) covered 2,731 plants, nearly twice as many as Li Shizhen, his near-contemporary (Haudricourt & Métailié 1994, 409). A century later in England, John Ray devoted nearly forty years of his life to recording and classifying all the world's species known at the time. The three volumes of his *Historia plantarum* (Ray 1686–1704) had 18,600 entries, including plants from the Philippines and East Asia, North America, and the Caribbean.[4] Scholars in Europe had moved in less than a century from looking at plants in their immediate environment to a project of describing and classifying all the plants encountered everywhere on earth.

Setting aside the question of the volume or comprehensiveness of traditional knowledge about plants in China, what is of particular interest is the way in which the nature and application of this knowledge changed during the century between the publication of the *Research on the Illustrations, Realities, and Names of Plants* in 1848 and the year 1950, when scientific research institutions in China were absorbed into a new Chinese Academy of Sciences (CAS) (Liu Xiao 2013).

Traditionally trained scholars in China were concerned with the processes of change and transformation that generate the myriad things (*wan wu*), of which grasses, trees, grains, and other "things that grow" were just a few of numerous categories: "The study of the myriad things is the examination and investigation of the forms and the transformations of all things that extend across the universe" (Du 1902, 6a). Observing these natural phenomena, correctly assigning their names, and recording their uses were critical steps on the way to discerning the patterns within constant change in the natural world.

The early scientific botanists in China, by contrast, wrote about the plants they encountered as objects of study in their own right, within assemblages that together composed the natural world. They were engaged in the systematic observation of plants in order to discover the relationships between them, their spatial distribution, and their physiology. They acknowledged that earlier writers had been acute observers of the life they saw around them but criticized them for falling short of the scientific method, which advances from a single occurrence to a universal principle or hypothesis, then tests that hypothesis to establish grounds for further investigation. The scholars of the past fell short in that they devoted themselves solely to the deep study of and commentary on ancient texts (Ren 1915, 11).

The encounter with Western natural sciences demanded both a conceptual leap in the perception of the myriad things and the resolution of the problem of language and translation. Forging a consensus on the language used to describe and to articulate a new way of ordering the world was essential in paving the way to engage with the scientific paradigm.[5] The first translation into Chinese of a botanical text was published in 1858 (Li, Williamson, and Edkins 1858). The first botanical textbook by a Chinese author appeared in 1907 (Wang Zhenru, Liang, and Wang 1994, 156). The first dictionary of botany was published in Shanghai in 1918 (Du 1918). Through the 1920s and 1930s, researchers and academics met in committees and commissions to develop and standardize scientific terminology in Chinese.[6] New discoveries and advances in the sciences ensured that translation would be an ongoing task, but institutional mechanisms had been put in place to ensure the accuracy, the acceptability, and the timely dissemination of new terminology.

Botany is rooted, literally, in specific places. It is the study of plants that are associated with the soil, exposure, hydrology, climate, and other factors unique to a particular location. The connection between plants and place is easily extended to region, province, nation, or some other politically, rather than ecologically demarcated, space. The first botanical field researchers saw the flora as markers of a wider, distinctly Chinese natural world, observed and described now by Chinese scientists, not the Western plant hunters of the past. Foreign collectors had taken Chinese plants to their universities and museums, where they had published their identifications of type specimens, giving them the right according to scientific protocol to name the species. Compounding the sense of humiliation, the lingua franca of botany was Latin, a dead foreign language. Bing Zhi, a founding member of the Science Society of China, insisted that "if science is to develop in China, Chinese people will have to use Chinese words" (Bing 1926, 1349). It would be imperative to create and to standardize a new botanical vocabulary.

Translation and standardization gave botanists a language to describe their work, but the conceptual framework for the scientific study of plants still had to be actualized as praxis. When botanists visited their research sites, they were mapping barely known spaces on the margins of the familiar

Chinese world. They were documenting the plants that people used in their daily lives, and they were collecting, identifying, and naming new, unknown plants. Their field diaries and other writings give personal insights into what they were doing and what they believed they were accomplishing. They were very conscious that their commitment to seeking knowledge in the open air, outside the scholar's study, set their work decisively apart from the efforts of earlier generations of literati-scholars.

Many of the first generation of botanists had begun their education preparing for the imperial civil service exams. Some turned away from the prospect of a conventional career as a scholar-official to pursue more stimulating opportunities that were opening up, especially in Shanghai and other treaty ports. When the examination system was abolished in 1904, some ventured into unfamiliar territory in publishing or chose to pursue further education in "the new learning." They were well versed in the Chinese classics and acutely aware that they were doing something different from their predecessors' studies of the plant world. They were articulate and prolific writers, publishing in the popular press and scientific journals to explain their work to others and to reflect on what was to them not just a profession but a calling. Leaders of the Science Society of China spoke at the opening ceremony of its Biological Laboratory of a search for true principles (*zhen li*) and a mandate "to assert that science can raise the moral qualities of individuals and strengthen the determination of the people to waste no time in saving our nation from its present weakness" (*Kexue* 1922, 846).

The science of botany took root in China in three phases over a century: the encounter with botany; a period during which scientific botany became established; and a period of institutionalization. At the turn of the nineteenth and twentieth centuries, Chinese botanists navigated between alternative ways of observing and describing the world, conscious of the heritage of their past while shaping the modern. When the world learned in March 1948, shortly before the founding of the People's Republic of China, of the discovery of "the living fossil" *Metasequoia glyptostroboides* (the dawn redwood) in the mountains between Hubei and Sichuan, the Chinese botanists who had found and identified the tree rightfully took their place as recognized members of the global scientific botanical community, working as peers with their Western counterparts (see chapter 10). In less than a century, they had not only created a new botanical lexicon and learned new taxonomies of plant life but had negotiated a profound epistemological shift in their understanding of the natural world and how humans explore and question it.

CHAPTER ONE

How the Southern Mountain Tea Flower Became *Camellia reticulata*

> There is an astonishing number of varieties of camellias in southern Yunnan. Their colors exhaust all possible brilliant hues. To talk of seven extremes would not fully express their beauty; to talk of ten virtues would only hint at the truth. If one were to examine the old terms used in the past and also to refer to the more recent designations by our contemporaries, they are all admirable names, but they all fall short of reality.
>
> —FANG SHUMEI, A BRIEF ACCOUNT OF THE CAMELLIAS OF SOUTHERN YUNNAN (*DIAN NAN CHA HUA XIAO ZHI*)

THE introduction to the second section of Fang Shumei's study of the camellias of Yunnan, from which this chapter's epigraph is borrowed, captures the enduring appeal of the camellia as celebrated by classical scholars, as well as the more sober descriptions of contemporary scientists. Fang, also known by his literary name, Man of Coiled Dragon Mountain (Panlong Shan Ren), was a book collector, a scholar of Yunnan local history, and a camellia fancier. He published his *Brief Account* in 1930, drawing on the resources of his personal library and on the expertise of his friends, both scholarly and scientific, who collected, grew, and studied camellias, mostly in the provincial capital of Kunming.

Fang's *Brief Account* was something of an anachronism at a time when science was expected to be one of the agents of modernization that would save China. It resurrected a well-worn genre of monographs on a particular plant, such as the chrysanthemum, the Japanese apricot, or, as here, the camellia. The author begins with a general description of the species. An inventory of known varieties and cultivars follows, and the monograph ends with an anthology of literary works that refer to or describe the species. What sets Fang's account apart is that after a list of seventy-two varieties, he acknowledges that the accumulated wisdom of the past may not correspond with the new scientific views of the time: "From the scientific perspective, the genus *Camellia* is not limited to seventy-two varieties or cultivars. Many of those that are named in the works of Xie Zhaozhe, Deng Mei, and other scholars have disappeared and are no longer seen. Among those that I have named, there are some that existed in the past but no longer exist now, and there are some that exist now but did not exist in the past" (Fang S. 1930, juan 2, 4b/5a).

Fang acknowledged the insights of science, and at least one Western-trained Chinese botanist at the time reciprocated by treating Fang's old-fashioned compendium as a useful inventory of the genus. Yu Dejun, the director of the Yunnan Institute of Agriculture, Forestry, and Botany (Yunnan Nonglin Zhiwu Yanjiusuo), referred to it as a source for his own guide to the camellias of Yunnan, written in 1947. A report Yu submitted in 1950 to the Royal Botanical Society in London (Yu D. 1950) was the first systematic treatment of the cultivars of *Camellia* to reach botanists and horticulturists outside China, surprising even camellia fanciers in the West with the number of varieties he described.[1]

The camellia, known in Chinese as the mountain tea flower (*shan cha hua*), is a familiar and much-loved plant inside and outside of China. The large, flamboyant species in many shades of pink, known to Fang Shumei and his friends as the southern mountain tea flower (*nan shan cha hua*), is known to botanists today as *Camellia reticulata*. Tracing how the southern mountain tea flower has been described, admired, and classified from its earliest appearance in Chinese sources to its present identity as *C. reticulata* mirrors the story of the transition from traditional knowledge about plants in China to scientific botany.

DISTRIBUTION AND TAXONOMY

The genus *Camellia*, in the family Theaceae, is a shrub, sometimes a tree, with flowers that have a prominent cluster of yellow stamens in the center. Its natural distribution extends from Japan in the northeast to Indonesia in the southwest (Chang and Bartholomew 1984, 18–19). There are at least 250 species, although botanists disagree about their exact number and taxonomy. The genus includes ornamentals and camellias of economic value, the best known of which are the tea bush, now classified as *C. sinensis*, and *C. oleifera*, from the seeds of which tea oil is extracted (Macoboy 1981, 19). Current taxonomy places the tea bush and the many species of ornamental camellia in the same genus, *Camellia*. For a long time, though, the relationship between tea and the ornamental camellia was a puzzle to Western botanists.

Both tea and the camellia had first reached the West as trade with Japan and China gained momentum in the seventeenth and eighteenth centuries. Few living specimens of either the tea bush or the camellia survived the long, harsh sea journey from East Asia. European traders and scientists only had access to dried, pressed herbarium specimens; processed tea leaves; and images on imported porcelain, textiles, and wallpapers, making it difficult to determine the relationship between the two. Were there two species, one producing tea and one an ornamental species, the camellia? Were there perhaps two species of tea bush, one producing green tea and one producing black tea (known as red tea in Chinese)? Were tea and the ornamentals related in some way, and if so, how?

The question was of particular interest to the East India Company, which had grown rich on the tea trade but was entirely dependent on one source—China—for its supplies. Since 1757, the company's agents there had been confined to their warehouse, or "factory," by the Pearl River in Canton (Guangzhou), where they saw only the processed tea they purchased and whatever plants and flowers they might encounter in those parts of the city that foreigners were allowed to visit. The company did everything it could to get access to live plant material in order to establish its own plantations on territory it controlled in India (Ellis, Coulton, and Mauger 2015).

The earliest description of the camellia in Europe appears to have been in a report by the German physician Andreas Cleyer, who had visited Japan with a trade mission in 1682, then again in 1685. Cleyer's letters were published in 1688 after his return, and they include a description and an engraving of the "tzumacky," a term that is recognizable as a rendering of *tsubakki*, the Japanese name for the camellia. Cleyer mentions that the seeds of the tzumacky resemble those of tea (Cleyer 1689, observatio LXX, 133), but otherwise there is no further reference to the tea bush.

Before foreign trade had been restricted to Canton, James Cuninghame, a Scottish surgeon attached to the *Tuscan*, an independent trading ship sailing to China, had spent several months during 1702 in Chusan (today's Zhoushan, Zhejiang). At the request of his friend and patron James Petiver, a member of the Royal Society in London, Cuninghame collected and

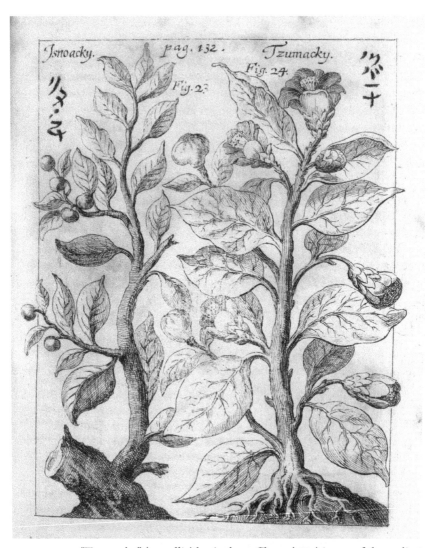

FIGURE 1.1 "Tzumacky" (camellia) by Andreas Cleyer (1689) is one of the earliest images of the camellia published in Europe. The inscriptions in hiragana script at the top right and left and the use of the Japanese term for camellia, "tzumacky" (*tsubakki* in modern Japanese), indicate that the engraving is from one of Cleyer's two visits to Japan with a trade mission. From *Miscellanea Curiosa*, Decuria II, Annus VII (Collection of W. Michel, Fukuoka). Image is in the public domain, using Creative Commons, https://commons.wikimedia.org/wiki/File:Cleyer_Tsubaki.jpg.

described the plants he came across, one of which he called *Swatea fl. rubro*. *Swatea* was Cuninghame's transliteration of the local dialect for the camellia (*shan cha* in Standard Chinese); the red flower (*flora rubro*) in the name indicates this was probably *Camellia japonica*. It was perhaps the first specimen of camellia to reach the West, where it is now in the herbarium of the Natural History Museum in London. Regarding tea, Cuninghame's account reported correctly that the three sorts of tea sold in England all came from the same plant, the difference being the season of harvest, the soil, and the processing.[2]

The relationship between tea and the ornamental camellia preoccupied botanists for some time. Engelbert Kaempfer (1651–1716) had been the physician at the Dutch East India Company's trading post on Deshima, off Nagasaki, from 1683 to 1695. His memoirs, published in 1712 as *Amoenitatum exoticarum politico-physico medicarum*, became the most authoritative source on Japan, Japanese trade with China, and the plants of Japan. Kaempfer wrote that the camellia (he used the Japanese term *tsubakki*) is very similar to the tea bush, referring to *Thea* as the genus for tea and *Camellia* or *tsubakki* for camellia. His description included an engraving of the camellia, with two pages of text describing the bush, its similarity to tea, and where it was found (Kaempfer 1712, 852).

In 1735, Linnaeus published his *Systema natura* in Leiden, establishing the system of classification of the natural world that became the foundation for modern taxonomy (Linné 1735). In his later work, *Species plantarum*, he placed tea and the camellia in two different genera. The ornamental camellia appeared as *Camellia japonica*. He then gave the name *Thea* to the tea plant now called *C. sinensis*, which he subdivided into two species, *Thea bohea* for black tea and *Thea viridis* for green tea (Linné 1762, 1:734–35; 2:982).

As late as 1859, a speaker at the Linnean Society in London began his address with the words "Great diversity of opinion exists as to whether the genera *Camellia* and *Thea* ought to be merged into one or regarded as distinct" (Seemann 1859, 337). Today, botanical taxonomy continues to evolve as techniques such as DNA sequencing lead to new groupings of species, genera, and families. The current taxonomic position of *C. sinensis* and the many other species of camellia in the family Theaceae dates from as recently as 2001 (Prince 2007).

The history of the unsettled taxonomy of camellias and tea is a reminder, if one were needed, that "scientific botany" does not constitute an unchanging, absolute truth and that "traditional" knowledge systems might organize and describe the world no less adequately than scientific systems. In Chinese, the addition of the qualifier *hua* (flower) to the single character *cha* (tea) forms the binomial *cha hua*, reliably differentiating tea and the camellia, and scholars had no need to devote their energy to unpicking the relationship between them.[3]

THE CAMELLIA AS A CULTURAL OBJECT

Books and monographs written in English for camellia fanciers often begin with a brief history of the camellia and its introduction to gardeners and horticulturists in the West. They refer to a history of "thousands of years of cultivation" in China (Feathers and Brown 1978, ch. 1; Trehane 1998, ch. 1), embellishing the mystery of an ancient history with symbolic and spiritual allusions. Some of the oldest camellia trees are, indeed, found in and around temples or the ruins of temples in Yunnan. Recent ethnographic research has shown, too, that the camellia plays a role in the rituals of some of the ethnic minority groups in southwestern China, especially in Yunnan (Xin et al. 2015). Otherwise, there is little evidence, though, that the camellia carries any particular symbolic weight in Chinese literature or popular beliefs.

The first reliable appearance of the camellia in written sources is found in *The Flowers and Trees of the King of Wei* (Wei Wang hua mu zhi), compiled by Thopa Xin between 480 CE and 535 CE. The original work has been lost, and it now exists only in fragments quoted in later encyclopedias and agricultural manuals.[4] It consists of a list of named plants, with almost no further information. The entry for camellia states: "Camellia. The camellia is like the sea pomegranate [*hai shiliu*]. It comes from Gui Prefecture." The reference to the sea pomegranate alludes to the camellia's red flower, similar in form to a pomegranate (*shiliu*) flower. The addition of the word *hai* (sea) was a common way to allude to an object coming from afar—"overseas"—which is not surprising at a time when the mountains of southwestern China, the habitat of the camellia, were still at the remote edge of Chinese territory and consciousness.

The King of Wei says little about the camellia other than to confirm that it was known at the time of writing, but the sea pomegranate or camellia was familiar enough that it was used at the time as a literary device, with its red, seasonal blooms representing the passing seasons and the transience of life. To date, the earliest known poem in which the camellia appears is by Jiang Zong (519–594)—which also refers to the camellia as the sea pomegranate:[5]

A spring day in a mountain pavilion
I have just had five days of leave
During which I perched up on a hillside.
An ancient quince crosses the nearby gully.
A precarious rock towers in front of the island.
The bank is green with fresh river willow,
The pool shines red with reflected camellias.
Wild flowers peacefully wait out the end of the month.
How can mountain insects be aware that there is an autumn?
Can humans retrace their lives at all?
A night candle cannot burn forever.

The camellia evokes a fleeting time of seasonal beauty. The quince is already dead; the fresh growth on the willow will last just a few weeks, as will the camellia blooms and the wildflowers. It is both sad and hopeful—a sign of the end of winter, bearing hope for the spring.

The poetic persona of the camellia pervades even horticultural manuals that one might expect to be practical guides rather than inspirational anthologies. An example of the genre is *The History of Flowers* (Hua shi), believed to have been printed around 1600.[6] There are two sections for each flower in the book. The first is a general description of its appearance and cultivation accompanied by a finely executed, colored woodblock print. In the case of the camellia, *The History of Flowers* continues with the names of seventeen varieties or cultivars, followed by brief instructions for the gardener. A section with literary and historical references appears later in the volume.

The History of Flowers and other manuals cover many different garden plants in one book. More detailed monographs were compiled for specific plants and flowers. In *The Compendium of Camellias* (Cha hua pu), published in 1719, Pu Jingzi described forty-three varieties of camellia he had seen in and around his home, the port of Zhangzhou, in Fujian, ten of which had come from Japan about thirty years previously (Pu 1719). After listing the names of the varieties with some historical notes, the text continues with information on cultivation and care in the garden. Pu describes propagation by layering and grafting, including grafting several varieties onto one rootstock to obtain a tree with different colored flowers that will bloom at various times over several months. The monograph ends with an anthology of literary references, with allusions to people and events of the past remembered in the more poetic names of cultivars.

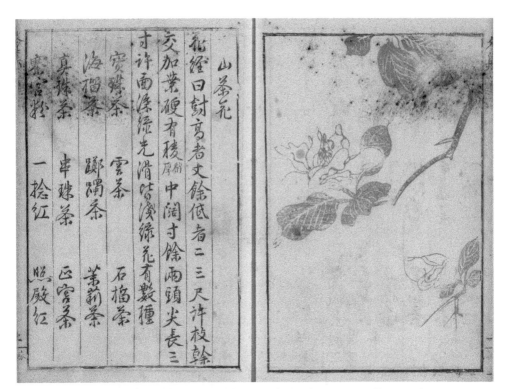

FIGURE 1.2   The first two pages of the horticultural section on the camellia in *The History of Flowers* (Hua shi), printed about 1600. The text begins with a description of the height of the camellia bush and the appearance of its branches, leaves, and blooms. A list of twelve named varieties follows, with a further five on the next page. Courtesy of the National Library of China, Beijing.

DESCRIBING THE CAMELLIA

Descriptions of the camellia in the many other sources follow a similar format. They begin with a brief portrait giving the height of the tree and the shape of the leaves, the flower, and seedpod, taking note of the visually striking bundle of yellow stamens and the flowering season. The information about varieties, the uses of the plant, and tips for cultivation differs in each kind of reference work.

The editors of comprehensive encyclopedias aimed to compile all existing literary, historical, and geographical references to a topic to preserve the complete body of knowledge about it. One of the earliest encyclopedias with a description of the camellia was *The Universal Encyclopedia* (San cai

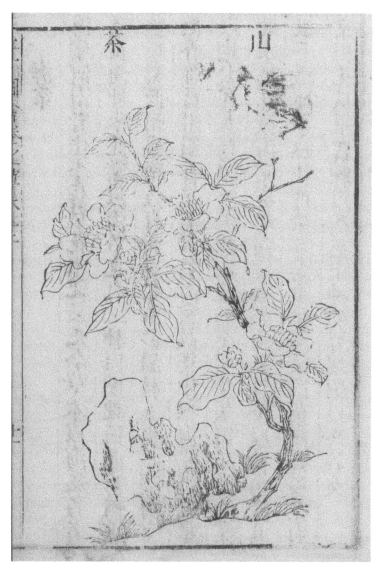

FIGURE 1.3 Camellia in *The Universal Encyclopedia* (San cai tu hui), first published in 1609. Courtesy of the Chinese Collection, Harvard-Yenching Library. Ming, Chongzhen (1609) edition. Cao Mu, juan 12, 11a.

tu hui), edited by Wang Qi and Wang Siyi and published in 1609. The entry for the camellia consists of a list of nineteen varieties, with evocative names like "Flushed Red from Pinching" or "White Precious Pearl." There is very little that would allow readers to identify the camellia in the wild or in a garden. Such information as there is appears after the first group of thirteen varieties, which "all have different foliage" and which "all bloom from the tenth lunar month and last until the second month." The entry ends with descriptions of unusual or rare varieties, including "the southern mountain camellia with flowers that are double the size of those in the central regions," almost certainly *Camellia reticulata* (Wang and Wang 1609: Cao Mu, juan 12, 11b).

The compilation that became the model of traditional encyclopedias is *The Imperially Commissioned Compendium of Literature and Illustrations, Ancient and Modern* (Gu jin tu shu ji cheng), edited by Jiang Tingxi, published in 1726. The entry in the *Compendium* begins with several synonyms for the camellia, followed by an image of a tree in bloom. There is no original description, but a compilation from well-known travel writers and other sources gives an account of its appearance, distribution, and varieties known at the time.[7] One of them, by the Song dynasty travel writer Fan Chengda, written in 1175, is of a variety in Yunnan said to have flowers more than twice the size of camellias from the central provinces, another early reference to *C. reticulata*, although it is not clear whether it was cultivated or seen in the wild. The entry concludes, again, with an anthology of poems and prose.

The *Compendium* has sections on regional and local geography, compilations of jottings, and verse about different places and the plants and produce found there. One of the jottings in the section on writings by local administrators is *An Account of the Camellias of Central Yunnan* (Dian zhong cha hua ji) by Feng Shike (1541–1617), who had served in Yunnan and Guizhou. Feng writes of seventy-two varieties of camellia, the number used in Fang Shumei's account of camellias in Yunnan nearly four centuries later.[8] He also gives a list of ten exceptional things about the camellia, which has become one of the most frequently quoted texts about camellias and still appears today in tourist brochures promoting the attractions of cities in the southwest:

- The colors are bright without being shocking.
- Even when a tree has lived two to three hundred years, it still looks as though it has only just been planted.

- The trunk can grow as high as four or five *zhang*, and it can be so thick that you can circle it with your arms.
- The veins on the surface of the leaves are dark green, smooth and dark like the patina on an ancient wine vessel.
- The branches are dark and bunch up into shapes like a deer's tail or a dragon.
- The roots coil and spiral in odd ways so that you can lean on them like on a table or you can rest on them like on a pillow.
- The luxuriant foliage forms a dense forest canopy like a tent.
- It can stand up to frost and snow and stays green through all four seasons.
- Blooms open one after another for a period of two to three months.
- They will stay fresh in a vase for over ten days with no change in the color.

Not surprisingly, the descriptions of plants in the classic pharmacological text *Classification of Materia Medica* (Bencao gangmu) by Li Shizhen, first published in 1596, are somewhat more helpful. They are detailed enough that they could serve to identify the plant, and of course each entry includes its medicinal uses. Li begins with an explanation of how the camellia gets its name *cha hua* (tea flower): like tea (*cha*), the leaves can be used to make a drink. He continues with a summary of its distribution, morphology, and flowering season, as well as a list of ten known varieties, all of which are also described in *The Universal Encyclopedia*. Among them, the "southern mountain camellia" is again identified chiefly by its flower, which is at least twice the size of other camellias. Li concludes with recommendations for the pharmaceutical uses of various parts of the camellia: the tender leaves can be boiled and eaten in times of famine; the ground flower, blended with a child's urine, ginger, and wine, is used for patients who cough blood or suffer from "wind in the intestine"; it can help to relieve pain from burns and scalds; and the seeds can be used to extract an oil for hair (Li 1596, juan 36, 71b–72a).[9]

Research on the Illustrations, Realities, and Names of Plants (Zhiwu mingshi tukao) by Wu Qijun (1789–1847) appeared in 1848 and is considered to be the last of the traditional Chinese works on plants. It only describes plants, which sets it apart from the more traditional materia medicas, which include minerals, insects, and animals used for medicinal preparations. The descriptions are very brief, concentrating on the most apparent diagnostic features of each plant, with an illustration for reference. It does not include anthologies of verse and prose from past writers, and uses only a very few quotations to link the plant to earlier texts. Its purpose, as the title implies, was to make a correct identification of known plants and to match them correctly to existing records in order to sort out the ambiguities in plant names that had accrued over time.[10]

Brevity might, in fact, indicate that Wu saw little or no ambiguity that needed to be resolved in some cases—including tea and the camellia. The entry for tea (*cha*) consists of just two lines that distinguish between tea brewed from leaves picked in the spring (*cha*) and tea brewed from leaves picked later in the season (*ming*) (Wu Q. 1848, juan 35, 13a and b). There is a separate entry of three lines for *shan cha ke* (the camellia family), with only a minimalist description of the height, the color of the bark, and the shape of the leaves, and no mention of the flowers (Wu Q. 1848, juan 34, 11a and b). *Shan cha* (the camellia) warrants two lines, which describe its medicinal uses and refer the reader to the entry in Li Shizhen's *Classification of Materia Medica* (Wu Q. 1848, juan 35, 60a and b).

Until the late nineteenth century, the purpose of texts about plants was to record what had been written about them, rather than to serve as diagnostic tools using a common terminology and methodology to identify new and unknown varieties. It is necessary to jump forward some seventy years to find the first description of the camellia in Chinese that followed what had become the vocabulary and the format of international botanical communication. The first dictionary of botany, published in 1918, places the camellia and tea in the same genus, although it uses *Thea* rather than *Camellia*. In translation, the entry for *Thea japonica* reads:

> Family Theaceae "*shan cha*." Found in temperate coastal areas. Many varieties. Evergreen bush, as tall as 20 *chi* [feet]. New growth glabrous with no pubescence. Long ovate leaves rounded with a pointed tip, thick, shiny, and alternate. Flowers during the spring. Large flowers with extremely beautiful petals. They are large and small, red and pink and striated, with single, double, and other arrangements of petals. Many stamens. Ovary is flattened and shiny. The fruit is a round capsule that splits toward the end of autumn before it is ripe, scattering two or three seeds. Seeds are blackish brown with a very hard epicarp. (Du 1918, 112)

The entry concludes with Li Shizhen's description of the camellia from 1596. The dictionary acknowledges the past history of the camellia in China, but it is quite different from its antecedents. It has been compiled as a guide for the botanist—who is expected to be familiar with a specialized scientific terminology—to identify a specimen, to classify it, and to know what distinguishes it from others. A shift had occurred as "scientific botany" had entered China, from the "traditional" interest in a plant as an object steeped in a rich cultural heritage to the plant as an organism whose significance lies in its taxonomic relationships to other plants and its broader ecological status.

THE CAMELLIA BECOMES A BOTANICAL GENUS

As late as 1848, when *Research on the Illustrations, Realities, and Names of Plants* was published, Chinese works on plants show no indication of an awareness of Western botanical concepts. There are, nevertheless, unexpected intersections between the two bodies of knowledge. Linnaeus distinguished between tea and the camellia, giving the name *Thea sinensis* to the tea plant now called *Camellia sinensis* and the name *Camellia japonica* to the ornamental shrub. Synthesizing the information in Linnaeus's *Species plantarum* and his *Genera plantarum*, the diagnostic characteristics of the camellia are as follows:

- It is a woody plant
- There is a wild variety with a simple red flower
- There is also a cultivated, ornamental variety
- There are five petals joined at the base
- The flower has numerous erect stamens
- The calyx is round, with numerous imbricate leaves
- The leaves are oval

At almost exactly the same time, in 1756, in China, the court painter Zou Yigui wrote a manual on painting, listing the key features an artist should keep in mind when painting different flowers. The entry for the camellia begins: "Camellia. A woody plant. Those with simple leaves and pink [flowers] are called the Concubine Yang [Yang Fei] camellia, and they bloom at the winter solstice. The large red variety is called the Sichuan [Shu] camellia, and it opens during the first month. They all have five petals, with a cluster of stamens the same size as the petals at the heart, a white tassel half an inch in length, and several layers of sepals; the leaves are thick, oval with ridges" (Zou Y. 1756, juan 1, 55).

It would be unwarranted to search for a direct link between Zou Yigui and Linnaeus, yet at both ends of the Eurasian land mass, completely unrelated observers took note of more or less the same six distinguishing features for the same plant and its flower. They observed the same features, but one conformed to a recently constructed framework for identifying, classifying, and describing; the other was a bridge between two cultural practices, painting and writing. The two texts reveal not just the obvious difference in languages but different concepts of the purpose of the description and what the plant represents.

The Western scientific establishment showed little interest in Chinese written sources about plants, cultivated or wild. Interest in the Chinese flora, by contrast, was high. Early accounts of China by Jesuit missionaries included descriptions of the crops, fruits, and other plants they encountered.[11] There was enough in their reports to make botanical gardens such as Kew in London and the Jardin des Plantes in Paris eager to collect specimens of Chinese species, while a rapidly expanding domestic trade in horticultural plants was on the lookout for new specimens and varieties to offer to the growing community of eager gardeners among the landed gentry and the emerging urban middle classes.

When the treaty of 1842 that ended the Opium Wars forced China to allow foreigners access to the interior, missionaries, botanists, and representatives of nurseries and seed companies all began to organize expeditions to discover new, unidentified species. The plant hunters traveled to some of the most remote but botanically rich parts of China, collecting, identifying, and classifying thousands of specimens that they sent back to their sponsors. The many species and varieties of camellia were among their prized finds, although their fame was made by more spectacular discoveries, such as "the handkerchief tree" (*Davidia involucrata*), magnificent rhododendrons, and many others.[12] They were intensely focused on the wild, the "unexplored" (by Westerners), and the "undiscovered" (by Westerners). Emil Bretschneider (1833–1901), who served as the physician at the Russian Embassy in Beijing from 1865 to 1884, was highly critical of Western botanists' motives: "Our botanists who collect plants in foreign countries do not trouble themselves generally about the indigenous names of the plants and their practical application, and they take no notice of the cultivated plants. Most of the systematic explorers endeavour only to discover new species or to create new genera in order to introduce their name into the science or to call the newly discovered plants after the name of a friend" (Bretschneider 1871, 20–21).[13]

Camellia reticulata is a case in point. The "southern mountain tea flower" is the largest of the camellia species. Individuals can grow as high as fifty feet, with a flower that can be four inches wide or more, often described in older Chinese sources as being "as big as a bowl." In the West, it was known during the nineteenth century from a single specimen brought to England in 1820 by Captain Richard Rawes of the East India Company for his brother-in-law, a member of the Royal Horticultural Society. In 1827, John Lindley identified it as belonging to a new species he named *Camellia reticulata*. Twenty-five years later, in 1851, Robert Fortune, one of the best-known plant hunters, collected another form, also red but with semidouble petals. These were the only two varieties of *C. reticulata* known outside China until 1932, when Otto Stapf, an Austrian botanist working at Kew, recognized plants that had grown from seeds collected in Tengyüeh (now Tengchong) in southwestern Yunnan by the Scottish plant collector George Forrest in 1924 as a previously unknown variety of *C. reticulata*.[14]

Since just three varieties were known, the conventional wisdom outside China held that the species was rare and limited to a few remote locations such as Tengyüeh. Yet in China, and more particularly in Yunnan and the provincial capital of Kunming, collectors and plant-loving citizens were growing many species of camellia, including *reticulata*. Not all the seventy-two varieties that Fang Shumei had listed were *reticulata*, but there were clearly many more than the three known in the West.

Eight years after Fang published his compendium, Hu Xiansu pointed out in his paper read to the Royal Horticultural Society that European botanists appeared not to have noticed or studied the more than seventy varieties of several species of camellias that could be found growing in the urban gardens of Kunming (Hu X. 1938, 387). His comments did not receive much attention. In Kunming, however, the botanist Yu Dejun had begun to identify the varieties grown in the city, trying to match them to the names found in the literature, including in Fang Shumei's work. He talked to local growers and camellia fanciers and did field research in temples and monasteries in the region. He learned that, as Fang had written, there were some old names that were no longer in use and some varieties that had been lost. He was fortunate that Liu Youtang (Y. T. Liu), the deputy manager of the Enterprise Bureau of Yunnan Province and a passionate collector of camellias, had moved recently to a villa in the village next to Yu's institute. In Liu's collection and others in the city, Yu identified eighteen varieties of *C. reticulata* and six varieties of *C. japonica*. Based on his research, he prepared a manuscript, *The Garden Camellias of Yunnan*, completed in 1947, in which he reported on his findings with descriptions and drawings of all twenty-four varieties, followed by guides to camellia cultivation, fertilization, propagation, and pest control (Yu D. 1947).[15]

FIGURE 1.4   A drawing of *Camellia reticulata* 'Takueiyeh' ('Da Gui Ye' [Large Osmanthus Leaf]) by Yu Dejun in *The Garden Camellias of Yunnan* (1947). Courtesy of Yu Dejun's family.

With the civil war between the Nationalists under Chiang Kai-shek (Jiang Jieshi) and the Communists under Mao Zedong at its height, Yu, who was studying in Edinburgh, hesitated to return to China. He gave his manuscript to Robert O. Rubel Jr. of Mobile, Alabama, a camellia collector with whom he had been in contact through an introduction from Liu Youtang before leaving Kunming. It was never published, however, until a copy of the original manuscript was reprinted in the United States in 1964 (F. Griffin

1964). Fortunately, in April 1950, shortly before returning to China, Yu had attended a conference on camellias and magnolias organized by the Royal Horticultural Society in London. His report, "*Camellia reticulata* and Its Garden Varieties," published in the conference proceedings, was based closely on his unpublished manuscript, with descriptions of the varieties of *Camellia reticulata* he had found, a key he had devised to identify them, and color photographs of four varieties (Yu D. 1950).

During the brief period between the end of the Second World War and December 9, 1949, when the People's Liberation Army (PLA) marched into Kunming, Yu Dejun and his colleague Cai Xitao were in contact with camellia collectors in the United States and in Australia. Their descriptions of the many varieties of *C. reticulata* sparked the interest of their correspondents. Dr. W. E. Lammerts of the Descanso Gardens in Pasadena, Mr. Ralph S. Peer of the Southern California Camellia Society, and Mr. Walter Hazelwood, a nurseryman from New South Wales in Australia, had each, separately, come across Hu Xiansu's 1938 presentation and were intrigued by his suggestion that there could be a number of unstudied camellias in Kunming. They were able to contact Yu and Cai and arranged for the institute to send specimens of twenty *C. reticulata* plants in March 1948 to be tested and propagated in California, as well as six to Australia.

These exchanges were only moderately successful. The plants sent to California were held up on arrival for quarantine and fumigation, and only five survived. A further shipment of sixty-five plants left Kunming in 1949 on what may have been the last flight to Hong Kong before the arrival of the PLA. These were more successful, and Peer later shipped some of them on to other parts of the world and to the Royal Horticultural Society in London (Peer 1949, 1950, 1951). It was only in the 1980s that contacts with Yu Dejun and his colleagues resumed through what had become the Kunming Institute of Botany under the Chinese Academy of Sciences.[16]

Two things stand out in the story, with all its twists and turns, of how the new varieties of *C. reticulata* found their way to the world outside China. The first is that as late as 1949, Western specialists could be rudely disparaging about Chinese botanists. One of Peer's correspondents wrote:

> In view of my knowledge of the experiences of plant collectors in China during the past 100 years and the various types of botanical "bunk" which have been disseminated from there by people desiring to sell plants to Europeans and Americans I am inclined to question Tsai's [Cai Xitao's] statement that the plants he sent you were varieties of the species *reticulata* and not

hybrids. I have checked this with such good taxonomists as William Hertrich of Huntington Botanical Garden and Dr. Frits W. Went of [the] California Institute of Technology, and they agree with me. It seems extremely unlikely to me that a species as rare as *reticulata* would have twenty or more varieties. (Robert Casamajor, letter dated July 31, 1949)

The second observation is the extent to which Western plant collectors either missed or ignored what they must have seen every day in gardens in Kunming and other cities they passed through on their expeditions. Plants that were a part of the daily lives of Chinese people were apparently invisible to them.[17] As Bretschneider had written, they appeared to have been more interested in making a mark, "discovering" new species, and assigning their own and their friends' names to their discoveries.

The story of the southern mountain tea flower's metamorphosis into *Camellia reticulata* is confirmation that by the mid-twentieth century, botany in China had become a part of the mainstream international scientific enterprise. Although camellias were being shipped to foreign countries, it was happening through contacts with Chinese botanists and a Chinese botanical institution on their terms. Knowledge about the species and its cultivated varieties was finally being built on the foundation of the rich heritage of historical records and on field research by Chinese botanists.

CHAPTER TWO

The Historical Context of an Epistemic Transition

> I am afraid that this year the planned departure date has been affected by the war in Sichuan, forcing the team to change its itinerary. Bandit activity affecting security everywhere has aggravated the situation; then there was a major disastrous earthquake in northwestern Sichuan, which blocked traffic and communications and made it difficult to transport materials. There are a number of areas where we have not been able to carry out the planned surveys. Further collecting will have to wait until it is possible to fill in the gaps at some point in the future.
>
> —YU DEJUN, A RECORD OF A BOTANICAL
> COLLECTING EXPEDITION TO SICHUAN
> ("SICHUAN ZHIWU CAIJI JI")

THE story of the camellia illustrates ways in which knowledge about one plant was organized and recorded in China, from the earliest texts compiled during the fifth century to the People's Republic of China. On the wider canvas of history, an intricate chronology framed the transition from traditional knowledge about plants to scientific botany in China.

The introduction to the report by Yu Dejun, who led a collecting expedition to the Sichuan-Tibet borderlands in 1932 and 1933, was not unusual for its time. Warlords, bandits, and natural disasters forced the team from the Science Society of China, funded by Harvard's Arnold Arboretum, to change their plans several times, until finally they ran out of funds and had to cut short their work and return to the provincial capital of Chongqing. Botany developed in China during a century of insecurity, wars and warlords, natural disasters, and more. Pressure from the incursions and demands of the Western powers, the devastation of the Taiping Rebellion (1850–64), the contested legitimacy of the Manchu imperial ruling house, and the aftermath of the 1911 Revolution made this a time of fundamental transformation in China's political structures, its economy, and the long-established moral and ethical order.

There is no specific event on a particular day or year that initiated the transition from the traditional knowledge of plants to scientific botany. The Treaty of Nanking of 1842 that ended the First Opium War, though, marked a dramatic change in the conditions governing relations between China and the rest of world. A little more than a century later, on November 1, 1949, the new government of the People's Republic of China established the Chinese Academy of Sciences (CAS). Over the next year, existing botanical research institutions became a part of the academy, making 1950 an appropriate date to mark the end of the transition (Fan H. 1999, 6 in Hu Z. 2005b, 220). From 1950 to the present, the history of botany has been a part of the history of the relations between the scientific establishment, the state, and the Chinese Communist Party, a topic that merits study in its own right.[1]

The encounter with botany took place in the decades immediately following the Treaty of Nanking, which granted Westerners access to inland regions. The specimens plant hunters sent to botanical gardens, universities, and commercial nurseries in Europe and America confirmed the extraordinary diversity of China's ecosystems, and the botanists in these institutions began the process of identifying and classifying the flora.[2] In newly opened coastal cities, missionaries translated texts and taught science, including botany, convinced that they were bringing enlightenment, the first step in turning the people of China away from superstition and toward Christianity.

China's defeat in the Sino-Japanese war of 1894–95 was a turning point. Shocked by Japan's transformation into a world power, the central and provincial governments launched a reform of education and supported sending students abroad. A first generation of scientists applied themselves to meeting the demands of the new education system: developing teaching materials and writing original texts to introduce Western knowledge to an educated and increasingly cosmopolitan audience. Chinese intellectuals drew on the deep resource of centuries of observation and writing about the natural world to generate a new Chinese lexicon for the Western sciences. For half a century, foreign plant hunters had claimed the discovery of new species and constructed biogeographies of even the most remote regions of the country. By the first decades of the twentieth century, though,

Chinese botanists organized expeditions, and they classified, documented, and stored specimens in the herbaria of new university departments and research institutions, increasingly taking the lead in establishing and practicing the discipline.

The years following the establishment in 1927 of a new Nationalist government based in Nanjing allowed for the institutionalization of botany in middle and higher educational establishments, new research centers, and professional associations, which trained a new generation of botanists. They developed collegial relations with their counterparts in Europe and the United States, and Chinese scientists attended international forums such as the International Botanical Congress. The institutions that took shape built a scientific community that had the coherence and resilience to survive the disruption and displacement precipitated by the Japanese invasion of 1937, followed by four years of civil war that ended with the founding of the People's Republic of China in 1949.

THE PREHISTORY OF THE ENCOUNTER WITH BOTANY

Before China's encounter with botany, the encyclopedias, medicinal treatises, horticultural manuals, and literary compilations in which Chinese scholars had recorded and organized known information about plants served their respective publics without defining a single systematic body of knowledge corresponding to the Western field of the natural sciences or botany. The educated elite of scholar-officials owed their first systematic exposure to Western sciences to the Jesuits who served in the Ming and then in the Qing courts, following Matteo Ricci's arrival in Macao in 1582. In their exposition of Western sciences, the Jesuits applied the Renaissance notion of natural studies (*scientia*) to accommodate the Chinese literati's established view of the natural world expressed in the term "the broad learning of things" (*bowuxue*), embedded within a wider "investigation of things and extension of knowledge" (*gewu zhizhi*) (Elman 2002) (see chapter 3). The Polish Jesuit Michał Piotr Boym (1612–1659) published an illustrated *Flora sinensis* in 1656 in Vienna, which included the names of plants and some animals in Chinese characters, with a Latin transliteration.[3] Father Pierre d'Incarville (1706–1757) was perhaps the first person to compile a herbarium of Chinese plants and to send botanical specimens back to Europe.[4] The emperors, though, were more interested in calendrical and related astronomical knowledge, which were of direct relevance in promoting agriculture and administering the empire. There is no evidence that Chinese scholars shared the interests of these botanically inclined Jesuits.

The Yongzheng emperor (1678–1735) proscribed the propagation of Catholicism in 1724, and in Rome, Pope Clement XIV dissolved the Jesuit order in 1773, bringing Jesuit influence at the Qing court to an end.[5] The trickle of botanical information still flowing from East to West now followed the trade routes dominated by the European trading monopolies,

FIGURE 2.1 Litchee painted by an unknown Chinese artist in Guangzhou probably between 1772 and 1774, in a portfolio commissioned by John Bradby Blake, of the East India Company in China. The painting follows the conventions of Western botanical drawing showing the plant in bloom, the fruit, cross-sections of the flower and of the stamens and ovary, and other details. Image #15 in vol. 1, John Bradby Blake portfolio, OSGL record # M-152. Courtesy of the Oak Spring Garden Foundation, Upperville, Virginia.

especially the East India Company operating out of London and the Dutch East India Company (Vereenigde Oostindische Compagnie) based in Amsterdam and Batavia (today's Jakarta). As the European powers expanded their empires, their trading companies collaborated with new institutions, such as the Royal Society and Kew Gardens in London, the Jardin des Plantes in Paris, and the Buitzenborg Gardens in Bogor near Batavia, to inventory the plant life in their colonies and trading entrepôts. Botany held the promise of finding plants that could become commercial crops and that might, in the case of China, offer a way to break the empire's monopoly on the production of tea.

During the eighty years or so following the Qianlong emperor's edict of 1757 restricting foreign trade to Canton, the city was the prosperous and cosmopolitan—but very confined—space in which Europeans (joined after 1788 by Americans) experienced China, interacted with Chinese merchants and officials, and gathered what knowledge they could about the empire they were not allowed to visit. Recognizing that members of the East India Company posted there were in a unique position, the horticultural and botanical establishments took steps to build rapport with them, compiling lists of plants and horticultural questions to which they should pay close attention. Their correspondents scoured the city's markets and nurseries for plants to send home and commissioned paintings by Chinese artists using emerging European standards for botanical illustration. Very little is known about the artists, but the paintings, most of which are now housed in collections at Kew Gardens, London's Natural History Museum, and the Oak Spring Garden Foundation Library in Virginia, are a rare point of intersection of things Chinese and things Western, an appropriate moment from which to turn from the prehistory of scientific botany in China to the three phases of the transition from the traditional to the scientific.[6]

IMPERIALISM AND THE ENCOUNTER WITH BOTANY, 1839–1914

For three weeks, beginning on the third of June 1839, Commissioner Lin Zexu, following the Daoguang emperor's orders, oversaw some five hundred workers as they destroyed over a thousand tons of opium confiscated from Western merchants in Canton. The merchants, who had profited for nearly a century from the illegal sale of opium produced in Bengal under the East India Company, claimed that the commissioner's action was a deliberate provocation and called for reprisals. By early 1840, Parliament in London approved an expeditionary force to demand compensation for the destroyed opium and to secure concessions that would open China to trade. British firepower and naval superiority inflicted heavy damage on the Chinese forces, leading to the signature on August 29, 1842, of the Treaty of Nanking, now referred to in China as the first of the Unequal Treaties imposed by Western powers on China during the nineteenth century.[7]

Under the terms of the treaty, China ceded the island of Hong Kong to Britain and opened four more ports (the Treaty Ports) to foreign trade. Foreigners were now permitted to venture inland, and Western botanical institutions and commercial nurseries immediately dispatched collectors to search for rarities to add to their collections or to introduce to the domestic market for garden plants. The Chinese Committee of the Royal Horticultural Society selected Robert Fortune, who set sail for China in February 1843 "for the purpose of introducing useful and ornamental plants to British gardens and of obtaining information on Chinese horticulture and agriculture."[8]

Protestant missionaries moved quickly to establish themselves in the Treaty Ports. Shanghai in particular became a vibrant cosmopolitan center for scholarship and exchange, as well as a commercial hub. In 1843, the London Missionary Society moved its press to Shanghai and renamed it Inkstone Press (Mohai Shuguan), dedicated to the translation and publication of scientific texts. In a fortuitous juxtaposition of classical and new learning, just five years after Inkstone Press had opened in Shanghai, the last "traditional" compendium of knowledge about plants in China, Wu Qijun's *Research on the Illustrations, Realities, and Names of Plants*, was published in 1848 in the inland city of Taiyuan in Shanxi.[9] Ten years later, in 1858, Inkstone Press published *Botany* (Zhiwuxue), a translation by Li Shanlan, Alexander Williamson, and the Reverend Joseph Edkins of John Lindley's *Elements of Botany*, originally published in London in 1847.[10]

The translation process involved a missionary and a Chinese scholar working together, a task facilitated by the presence in Shanghai of scholars who had fled there to escape the destruction of the Taiping Rebellion. The rebellion, which ravaged central and southern China from 1850 to 1864, was a millenarian uprising led by Hong Xiuquan, who claimed to be the younger brother of Jesus Christ. He proclaimed the Kingdom of Heavenly Peace (Taiping Tian Guo) in January 1851, launching an anti-Manchu, anti-gentry uprising. The Taiping armies controlled large swaths of central and southern China, and it ultimately took military assistance from the Western powers to defeat them. The rebellion is estimated to have caused twenty to thirty million military and civilian deaths, devastating vast areas of both urban and rural China and undermining what had been an economic and intellectual heartland in the Jiangnan (lower Yangzi) area.[11]

The Taiping Rebellion was not the only conflict to force the Qing court to make deep changes to the governance of the empire. Between 1856 and 1860, in the joint military campaign now known as the Second Opium War, British and French troops attacked the capital, forcing the imperial household to flee. Under the terms of the Treaty of Tientsin (Tianjin) of 1858, ratified in 1860, China opened more ports to foreign trade, exempted foreign goods from internal taxes and levies, removed remaining restrictions on travel, and allowed missionaries to work in the interior.

The Treaty Ports gave the Western powers (and later Japan) important footholds on the coast and enormous influence through staff seconded to Chinese agencies, such as the Customs Service. From their bases there, resident staff, missionaries, and scientists organized the collection of knowledge about natural history and plants, always depending to a degree on local guides and informants. The French Jesuit order took responsibility for the Catholic evangelization effort in China.[12] They chose to minister to the most destitute populations in the remotest parts of the country, in deliberate contrast to the Protestants, who tended to work in more urban, densely populated areas. The Jesuits continued the order's tradition of scholarship, with an emphasis now on the natural sciences. Missionaries, such as Père Armand David, Père Jean-Marie Delavay, and Père Jean-André Soulié, took advantage of their pastoral travels and lengthy residence in isolated parishes to collect and inventory the local flora and fauna, sending specimens back to Adrien Franchet, the taxonomist at the Muséum National d'Histoire Naturelle in Paris. In 1868, Père Pierre-Marie Heude founded the first museum in China, the Musée de Zikawei (Xujiahui) in Shanghai, which collected natural history specimens sent from the missions and other travelers. Their work yielded a depth of knowledge that few plant hunters were able to achieve in the course of expeditions that were briefer, though perhaps more intensive.[13]

The humiliation of the Second Opium War launched "the Self-Strengthening Movement," an effort to counteract Western aggression by establishing new schools and industrial and technical facilities, and by engaging more actively with Western learning, especially in the sciences.[14] The most urgent task was to prepare interpreters to meet the sudden demand imposed by the Treaty of Tientsin, under which the English version of all diplomatic correspondence would prevail in the event of a dispute (Li, Xu, and Zhang 1999, 43). On August 20, 1862, in Beijing, the School of Combined Learning (Tongwen Guan), a new imperial school for interpreters, began classes in English, Russian, and Japanese. A Shanghai school followed in 1863, to be incorporated the next year into the Jiangnan Arsenal (Jiangnan Zhizao Ju), recently established to introduce modern technology to build China's defense capabilities. Inkstone Press had recruited Chinese scholars to work with missionaries on translating scientific texts that they chose. A decade later, the School of Combined Learning and the Jiangnan Arsenal reversed these roles, inviting missionaries and other foreign staff to collaborate with Chinese counterparts to translate materials chosen by the Chinese institutions (Tsu and Elman 2014, 16).[15]

The late Qing era saw an explosion of print media, with new daily newspapers, magazines, and popular journals, many of which were dedicated to introducing the new learning to a Chinese audience (Vittinghoff 2004). The first periodical to publish articles about science was *Peking Magazine* (Zhong xi jian wen lu), a general-knowledge publication with occasional articles on botanical or biological topics (Wang Zhenru, Liang, and Wang 1994, 122–25). One of the most widely circulated journals was the *Chinese Scientific and Industrial Magazine* (Gezhi huibian), which appeared under the aegis of the Shanghai Polytechnic Institution from February 1876 until winter 1892. Its articles were mostly translations, covering everything from practical "how-to" advice to illustrated explanations of modern machinery and a section answering readers' questions. No lesser an authority than *Scientific American* expressed the hope that "this journal ... marks another breach in that wall of exclusiveness with which for centuries China has encompassed herself" (Fryer 1880).[16]

The hopes that the Qing court had placed in the ships and weapons produced at the new arsenals proved to be badly misplaced. Japan decisively beat the Chinese armies and navy in the Sino-Japanese War fought between July 1894 and April 1895 over the status of Korea. Not only was this another humiliating defeat but it reversed the long-standing balance of power in the region and undermined China's hitherto unquestioned cultural and military dominance in East and Northeast Asia.[17] In response, the weakened imperial authorities expanded their program to catch up with the West to include changes in government institutions, education, and the civil service examination system. Some tentative steps had already been taken in reforming education. In 1887, science and mathematics had been introduced into the examinations. In 1893, the new Hubei Self-Strengthening School (Hubei Ziqiang Xuetang) in Wuchang offered the first botany course to be taught in China (Wang Zhenru, Liang, and Wang 1994, 156). In 1896 a memorial to the emperor proposed more profound changes, including a national school system, new technical laboratories and translation bureaus, and a program to send students to study abroad (Vittinghoff 2004; Elman 2005, 379–86).

The missionary translation and education project in Shanghai continued to popularize Western science. In 1895, the Educational Association of China (Yizhi Shu Hui) published John Fryer's *Illustrated Botany* (Zhiwu tushuo), a translation of John Hutton Balfour's *Manual of Botany*, first published in London in 1851. Three years later in 1898, John Fryer's primer *Essential Knowledge about Plants* (Zhiwu xu zhi) appeared, completing what was now, together with the earlier translation of Lindley's *Elements of Botany*, a set of three basic textbooks for courses in botany.[18] These translations helped to stabilize an emerging scientific lexicon of plant morphology and taxonomy, although there were still discrepancies that would only be resolved when translations and original works by Chinese students returning from abroad began to appear over the next decade (Métailié 2001b).

The first original text on scientific botany by a Chinese author was Ye Lan's *A Verse Primer of Botany* (Zhiwuxue gelüe), published in 1898.[19] It consists of the 154 illustrations in Fryer's *Illustrated Botany*, followed by text in the form of pairs of four-character phrases expounding the essentials of botany. The rhythmic couplets suggest that the teacher would point to the relevant illustration while students recited the text in the same way that they might have learned the classics in a traditional school. It begins with an account of the formation of the world in which the first signs of life are lichens on rocks that emerge from a primordial ocean. A succession of life-forms leads eventually to plants and animals. The text proceeds to explain plant cells, respiration, nutrient uptake, pollination, and more. One page of plates (plates 98–110), for example, shows the reproductive anatomy of the flower. The related couplets explain the structures in the drawings and their functions:

> Inside there is pollen
> in a sac with an opening.
>
> When it first grows it is closed
> until the pollen is completely formed.
>
> The sac then splits open
> the pollen flies out and scatters.
>
> When it reaches the heart of the flower
> the heart begins to conceive a fruit.
> (Ye 1898, 13b)

Very few copies of the *Primer* still exist, so it may not have been widely used. Nevertheless, it was a turning point after which Chinese intellectuals and institutions took the lead not only in translating materials and standardizing new technical terminology but also in writing original papers and manuals introducing the fundamental concepts of scientific botany.

The same year that the *Primer* was published, there was a brief experiment with far-reaching political and institutional change. The Hundred Days' Reform in 1898 lasted three months before the Empress Dowager Cixi engineered a coup to put an end to the experiment. The leaders of the attempted reform were executed or exiled, many to Japan.[20] Despite the setback, some changes, particularly in the educational system, outlived the hundred days. The Imperial University of Peking (Jing Shi Daxue) (now Peking University), founded that year, took its place at the head of a new school system. The university's curriculum combined classical studies with new subjects, such as agriculture, engineering, and a general sciences course that included botany (Tsu and Elman 2014, 30).[21]

The short-lived but violent anticolonial, anti-Christian Boxer Rebellion that erupted in 1899 ended with the occupation of Beijing in 1901 by the armies of six European countries, Japan, and the United States. Among the casualties of the looting was the Imperial University, first ransacked by the rebels, then again by the allied troops. In the aftermath of the rebellion, under the terms of the Boxer Protocol of 1901, China was forced to pay a monetary indemnity, estimated to have been worth approximately twice the GDP of the empire (Wang Zuoyue 2002, 295).[22]

A decree issued in 1902 provided for a modernized educational system. In August the following year, the new Ministry of Education promulgated regulations governing the content of school instruction, modeled on the Japanese education system, with a curriculum that gave high priority to the sciences (Chen, Yang, and Gu 2011, 11–13). The abolition of the civil service examinations was announced on February 2, 1905, and implemented in September the same year (Zhang Y. 2005, 50). For centuries, success in the examinations had been the pathway to a secure livelihood open to any literate man, even of very modest means. As a consequence of the abolition, many well-educated scholars suddenly had to look for alternative careers and opportunities to which they could apply their learning. Some turned to Western learning in the new schools and colleges, some decided to study abroad, and many became political activists advocating for the downfall of the Qing and the imperial order.[23]

With scholars moving from classical studies into new fields of learning, the years immediately before and after the abolition of the civil service exams saw intensive activity in publishing, translation, and the preparation of textbooks in the sciences. The new schools and colleges needed teaching materials to meet the demands of the new curricula. In 1904, the Agricultural Association (Nongxue Hui) in Shanghai published the *Textbook of Botany* (Zhiwuxue jiaoke shu), a translation by Liu Dayou of the Japanese original written by Matsumura Jinzō, his professor in Tokyo (Liu D. 1904). In 1907, the first complete textbook on botany to be written in China appeared. *Botany* (Zhiwuxue) was a product of the times. The author, Ye Qizhen, taught science at the Beijing Teachers' College (Jing Shi Yixue Guan), and he was also the director of the Agricultural Experimental Station of the Ministry of Agriculture, Trade, and Industry, which had recently been created under the reform program.

According to one estimate, more than a hundred scientific periodicals were published between 1900 and 1919. The *Yaquan Journal* (Yaquan zazhi) was China's first indigenous scientific periodical carrying original papers and translations, with an emphasis on chemistry. Its founder Du Yaquan (1873–1933), originally from Shanyin County, Shaoxing Prefecture, in Zhejiang, had studied mathematics and chemistry and taught himself Japanese, then founded the Yaquan Academy (Yaquan Xueguan) in Shanghai in 1901. The journal was privately produced, had a limited circulation, and was superseded in 1903 by *Science World* (Kexue shijie), the first to use the term *kexue* for "science" in its title (Wang Hui 2011, 57). A year later, *Eastern Miscellany* (Dongfang zazhi) appeared, also in Shanghai, also with a focus on the sciences, and the first to be written, edited, and published entirely by a Chinese staff.[24]

October 10, 1911, is celebrated as the date of the uprising that brought about the final collapse of the Qing dynasty, with the abdication on February 12, 1912, of the Xuantong emperor Pu Yi. The 1911 Revolution marked the end of an era, but in terms of the social, cultural, and intellectual changes that China was experiencing as the old order disintegrated, it was the coup de grâce rather than a pivotal transformational moment. Military leaders, better known as warlords, hijacked the revolution, jostling for power. The northern general Yuan Shikai (1859–1916) had himself appointed president in 1912, then declared himself emperor in December 1915. With little support for his move, he gave up the title in March 1916 and died that June, unleashing a decade of disorder and political and cultural ferment.[25] It was arguably the First World War and its aftermath, though, that raised fundamental questions about unconditional acceptance of all things Western and how China could or should change to take its place in the world.

ESTABLISHING A CHINESE BOTANY, 1914–27

A dozen Chinese students studying in America met at the Cosmopolitan Club House at Cornell University in June 1914 to discuss how they could contribute to the modernization of their country. They concluded that their role was to bring science to China. Nine of the participants in the meeting founded the Science Society (Kexue She), renamed the Science Society of China (Zhongguo Kexue She) in 1915, and launched a journal, *Science* (Kexue), the first issue of which appeared in Shanghai in January 1915. *Science* innovated in using Western layout and formatting, printing text horizontally from left to right, and using Western punctuation. The introduction to the first issue said, "We are doing this in order to be able to insert mathematical equations, and formulae in physics and chemistry, not just because we just want to look novel and odd" (*Kexue* 1915b, 2).

No longer content only to be the recipients of new learning, the members of the Science Society believed that the time had come to share their professional knowledge and to promote international norms of scholarly work in their publications. They were aware that they would need to do more than publish a journal if they were to act as a force for positive change. They saw themselves as a vital part of the social movements calling for a new cultural model that would advance, not hinder, China's modernization, and organized the society as a model of what they hoped to see in a modernizing China. The society had a written charter, under which a board served for limited terms, standing for election at an annual plenary meeting, the proceedings of which were published in the journal. In 1918, the society moved its headquarters from the United States to Shanghai, where it quickly became one of the most respected scientific institutions in the country.

Despite political instability during the first years after the 1911 Revolution, it was a time of ferment and rapid development in education and research. Some of the country's most respected universities and colleges introduced courses in biology and botany and established herbaria to support instruction. The botanist Qian Chongshu (1883–1965) returned from the United States in 1915 and taught biology, with a course in botany, at the Nanjing Jiangsu Higher Agricultural College (Nanjing Jiangsu Jia Zhong Nongye Xuexiao), where he established what is believed to be the first herbarium in China (Hong, Chen, and Qiu 2008, 439).[26] Other universities that taught botany either as a subject with its own curriculum or within a broader course in biology included the Government University of Peking (now Peking University), Lingnan University in Guangzhou (formerly Canton

Christian College), Xiamen (Amoy) University, and National Southeast University (Dongnan Daxue) in Nanjing, which inaugurated the first full department of biology in 1921 (Wang Zhenru, Liang, and Wang 1994, 129).

The calamity of the First World War shocked Chinese intellectuals, who witnessed the self-inflicted carnage in the countries that had been the very source of the Western learning they admired. Shock turned to anger when they learned in 1919 that under the Treaty of Versailles, China did not regain control of the Treaty Ports and, further stoking public fury, the treaty awarded the former German concessions in Shandong to Japan. On May 4, student demonstrations in Beijing accused the government of weakness and protested against the imperialism of the Western powers. The demonstrations triggered the May 4th Movement, a wave of nationalist, antiforeign sentiment, which demanded that China reject traditional values in favor of modernization led by "Mr. Science" and "Mr. Democracy."[27]

Scientists in all fields found themselves advocating for "Mr. Science," but they were very aware, at the same time, that their disciplines existed within an epistemology that had come from the West. Botanists lived this contradiction every day as they classified plants according to a taxonomy devised in the West, used names that honored Western scientists and plant hunters, and using Latin, a dead Western language. In articles they wrote and in published accounts of their collecting expeditions, they frequently voiced disquiet at their unavoidable reliance on Western knowledge and resources, even as they welcomed opportunities to collaborate as peers with Western botanists who had been their teachers and mentors during their studies abroad. By this time, several universities had in fact invited foreign instructors and faculty to teach in China, among them Nathaniel Gist Gee at Soochow University, who took up his post in 1901 and who contributed an introduction to the first Chinese dictionary of botany, published in 1918.

Speaking to the Linnean Society of London in 1944, Li Hui-lin, who was studying at the time at the Academy of Natural Sciences in Philadelphia, said, "The beginning and the growth of Chinese botanical research, wherein Chinese workers themselves undertook a study of the vegetation of their own country on a scientific basis, belongs to the past twenty-five years" (Li Hui-lin 1944, 25).[28] It was in 1921 that the first published account of an organized botanical collecting expedition carried out by a Chinese botanist appeared in *Earth Sciences Journal* (Dixue zazhi). Following a visit to Lingnan University in Guangzhou, Zhong Guanguang, the recently appointed assistant professor of biology at what was then called National Peking University (now Peking University), traveled from Guangxi to Yunnan between June and September 1919 (Zhong, 1921–22).[29] Zhong was self-taught and did not use technical terms or the botanical names of plants in his report, but as students returned from their studies abroad, their reports began to follow the conventional scientific format more closely, with plant lists and analyses of vegetation types, written for the readers of the new professional journals in which they appeared.

The Biological Laboratory of the Science Society of China (Zhongguo Kexue She Shengwu Yanjiusuo) opened in Nanjing on August 18, 1922, housing the country's first research center dedicated to botany and publishing the internationally distributed *Contributions from the Biological Laboratory of the Science Society of China, Botanical Series* (Zhongguo Kexue She Shengwu Yanjiusuo Zhiwubu Lunwen Congkan) from 1925 to 1942. To begin building its international presence, the society sent Zhang Jingyue, a plant morphologist studying in the United States, to the Fourth International Botanical Congress in Ithaca in 1926.

INSTITUTIONALIZATION AND POPULARIZATION OF BOTANY, 1927–49

In April 1927, Chiang Kai-shek declared Nanjing to be the capital of China, unified under the Nationalist Party he led. The government implemented a vigorous program of nation-building that included an educational and scientific research structure under central control, extending to the provinces. On April 17, the party's Central Political Council called for an academy, loosely modeled on the Académie Française, to coordinate and conduct scientific research. The following day, the government announced the founding of Academia Sinica (Zhongyang Yanjiuyuan). The academy officially came into being at its first meeting in Shanghai on June 9, 1928. As the official face of research in a wide range of disciplines, Academia Sinica played a part in bringing the achievements of China's scientists to the attention of the world, especially through its journal *Sinensia* (Guoli zhongyang yanjiuyuan ziran lishi bowuguan tekan), published from 1929 until 1941 (Hong, Chen, and Qiu 2008, 439). Despite respectable academic credentials, though, the academy's official status as "the highest institution responsible for scientific research in the Republic of China" also made it the guardian and gatekeeper of science in the country, an extension of tightening government control over all areas of civil society.[30]

With the exception of Zhong Guanguang, the most active and visible figures at this time had recently returned from their studies in the United States. Qian Chongshu, Hu Xiansu (Hu Hsen-Hsu), Chen Huanyong (Chun

Woon-Young), and Zhong Xinxuan are considered in China today to be the founders of scientific botany in China, but the field grew rapidly and developed deep roots as the founders trained their successors.[31] One hundred and fifty members attended the first meeting of the Botanical Society of China (Zhongguo Zhiwuxue Hui), held in Chongqing in August 1933, most of whom were trained in China (Hong, Chen, and Qiu 2008, 439). Of the founders, Hu and Chen played an especially important role in securing funding in 1928 for two centers that were to lead botanical research until 1950. Hu became the codirector with the zoologist Bing Zhi of the Fan Memorial Institute of Biology (Jingsheng Shengwu Diaochasuo) in Beiping, and Chen became director of the Institute of Agriculture, Forestry, and Botany (Zhongshan Daxue Nong Lin Zhiwu Yanjiusuo) at Sun Yat-sen University in Guangzhou. Both institutes received funds from the China Foundation for the Promotion of Education and Culture and the Rockefeller Foundation, giving them financial security and valuable linkages with international botanical research networks, strengthened by their journals, which carried articles in English as well as in Chinese.[32]

The Japanese invasion in July 1937 put an end to the relative peace of the Nanjing decade. The government moved to Chongqing, and universities, colleges, and research centers were forced to evacuate to temporary locations in unoccupied areas of the southwest or to the areas under the control of the Communist Party in the northwest. The institute at Sun Yat-sen University continued under Japanese occupation in Guangzhou, while the Fan Memorial Institute relocated to Kunming and on July 24, 1938, in a partnership with the provincial Department of Education, founded the Yunnan Institute of Agriculture, Forestry, and Botany, with Hu Xiansu appointed director in 1940.[33]

This third phase of the transition to scientific botany drew to a close during the years of civil war that followed the Japanese surrender in 1945, leading to the founding of the People's Republic of China on October 1, 1949. The institutions that came under the Chinese Academy of Sciences in 1950 were the foundation that ensured a long-term future for the field of botany under the very different circumstances of a new order for research and learning in the sciences.

CHAPTER THREE

Nature, the Myriad Things, and Their Investigation

> Biology is the study of the lives and the appearance of animals and plants.... It includes everything that is found teeming and thriving between heaven and earth, their rise and fall, their flourishing and their decline, and everything concerning the relations between one being and another.
>
> —BING ZHI, A BRIEF ACCOUNT OF BIOLOGY ("SHENGWUXUE GAILUN")

INTRODUCING biology to readers of the first issue of *Science*, the zoologist Bing Zhi, a cofounder of the Science Society of China, had recourse to the language of the classical Chinese cosmology of change and transition, which generate the myriad things (*wan wu*) in a realm that corresponds to our present concept of nature. The modern Chinese term *ziran*, translated as "nature," encompasses the phenomena we encounter every day, be they animal, vegetable, or mineral. Among the phenomena in nature, botanists observe and study a category now classified and named as "plants" (*zhiwu*) that had formerly been known as distinct, named classes of living things, such as grasses, trees, fruits, flowers, and others. China's pioneering botanists of the early twentieth century invoked a heritage of investigation into the myriad things to signal that theirs was not an alien project imported in toto from the West, while at the same time distinguishing the practice and methods of science from the scholarly undertakings of the past.

NATURE: "THAT WHICH IS SO OF ITSELF"

In 1922, Feng Youlan, who had studied Western philosophy at Columbia University, quoted the early Daoist Zhuang Zi to explain the classical Chinese concept of "nature" to an international audience: "What is nature? What is human? That ox and horse have four feet is nature; to halter the head of a horse or to pierce the nose of an ox is human. Thus 'nature' means something natural; 'human' means something artificial. The one is made by nature, the other by man" (Fung Y. 1922, 239).

Before the encounter with Western science, the fundamental concept that shaped the Chinese understanding of the physical world was what the historian of science Nathan Sivin describes as "a dynamic harmony . . . of complementary energies" (Sivin 1973, xviii). The continuous actions and changes of the opposite forces of yin and yang set in motion the cycles of the Five Phases (*wu xing*) (the five elements: earth, fire, metal, water, and wood) through which qi, the primordial matter that is everywhere and saturates everything, is manifest. It is the transitions between the Five Phases that generate the myriad things, their waxing and waning, their emergence and disappearance. This ever-changing order of the cosmos is the *Dao*, usually translated as "the Way." There is an order or pattern (*li*) within this constant change, but order is not imposed on things; it is self-generating. The term *ziran*, literally "that which is so of itself," was used to render the concept of self-generation. *Ziran* meant "natural" in the sense of something that is not an artifact of human action—hence the modern use of the term to mean "nature." Nature, then, was not an ontological entity with which humans interact positively or negatively but a state in which the processes of change and transformation are not altered by or subjected to the actions of man. Observing natural phenomena and the myriad things was a part of the search for insights into the workings of these forces and the patterns within the constant change of the *Dao*.[1]

HUMANS AND NATURE

The botanist Cai Xitao (1911–1981), like many of his peers, wrote poetry and fiction in addition to his scientific studies and reports.[2] In 1937, he published a fable in the Shanghai periodical *Literature* (Wenxue) with the title "The Dandelion" (Pu gongying). The dandelion is as ancient as the earth, ageless but small. It hibernates in the winter and wastes no energy during spring and summer in trying to grow larger. Orchids and grasses grow under the nearby oak tree, which ultimately shades them out. Insects chew through the dead grass and turn it into compost, then begin to bore into the oak tree until finally it topples over. The dandelion enjoys the view again, until it sees a growing green horde advancing. The grasses have returned, only to be taken over in due course by a new oak tree (Cai 1937).

The fable is about cycles of aging and renewal in nature and how plants, birds, and insects coexist while continually displacing and replacing each other over generations and millennia. For Cai, the many forms of life on earth were related and mutually dependent. The transitions, cycles, and transformations that the dandelion witnesses arise spontaneously. They are not manipulated by or subjected to human actions, and it is in this sense that they are natural. They are of nature, rather than being nature itself (Liu Xiaogan 2004; López 2013, 142).[3]

Although humanity plays no role in the dandelion's world, there is no suggestion either that the world is better off without humans, or that humans might bear a particular responsibility to remain in harmony with the beings that are part of it. Nevertheless, it has always been acknowledged that human agency and activities could tamper with or modify "self-generation's pattern," and it is from this perspective that China's philosophical traditions proposed that living well would foster harmony with heaven and earth.

For Daoists, living well meant shedding desires and opinions and cultivating practices, such as meditation and "nonwillful action" (*wu wei*), that would bring the adept closer to the Dao. There is no sense, though, of an especially spiritual, harmonious relationship with nature, as some admirers of Daoism would argue. The *Dao* cannot be separated from anything in the world, including humans, "which means that it cannot be understood as a nature that may provide ethical or moral guides for how one should behave" (D'Ambrosio 2013, 407). For Confucians, living well meant that humans, as part of a universe of interconnected relationships, should develop norms and systems of ethical behavior that would maintain harmony within society and the world. Neither Daoism nor Confucianism suggested that nature offered lessons for human behavior.[4]

By the late nineteenth century, there was little evidence of harmony and order in China and the world. Nature appeared, instead, to be an arena of perpetual struggle for survival (Fan F. 2004b). In 1914, with Europe on the brink of war, Wu Jiaxu, the editor of the first issue of the *Journal of Natural History* (Bowuxue zazhi), identified four classes of things on earth: humans, animals, plants, and minerals. Of the four, only minerals appear to enjoy an existence free of conflict, and only humans are consistently able to dominate the others, hopefully for the benefit of humanity: "Humans are one of the classes of things, but only humans have been able to make use of animals, plants, and minerals. None of the others has been able to. This is entirely due to the fact that humans possess resourcefulness and knowledge. When resourcefulness and knowledge make advances from one day to the next, then they can be applied more broadly from one day to the next, and it is through study that resourcefulness and knowledge will make advances" (Wu Jiaxu 1914a, 2).[5]

PLANTS: THINGS THAT GROW

Wu Jiaxu identified plants as one of the four classes of things, using the word *zhiwu*, literally "things that grow." The plant kingdom is a category that is recognizable in most cultural contexts as distinct from other forms of life. For some, though, it may be so broad that it is not meaningful and does not correspond to the way they experience the world. In 1902, Du Yaquan wrote in an article, "The Meaning of Natural Sciences":

> People all know that plants cannot move and animals can move. . . . But there are some plants whose leaves can sense things and close up. There are also some lower-order algae that can turn and move on their own, like Dead Man's Fingers [*hai song*, *Codium fragile*, a sea algae]. Then there are "animals" such as corals and others that have movement while still remaining fixed in one place, and there are plants that live by catching insects. If we move on down to microorganisms that can only be seen through a microscope, it becomes hard to tell whether they are animals or plants.[6] (Du 1902, 6b)

Du Yaquan's reservations about what exactly constitutes a plant notwithstanding, their cycles of growth, flowering, setting seed, and withering away had historically been associated with the calendar and the passage of the seasons. "Things that grow" held an important place in the universe of things worthy of the classical scholar's attentions, an example of the natural harmony between heaven and earth in the sense discussed above of *ziran*, "that which is so of itself" (Métailié 2015, 254–393). Early writers and philosophers did not, however, identify a category of organisms collectively named "plants."

The origins of the term *zhiwu* show that even an apparently intuitive, fundamental division of physical phenomena can be a relatively modern construct. It is not uncommon in the earliest Chinese texts, appearing for the first time in the canonical *The Rites of Zhou* (Zhou li) (probably from the second century BCE), where its meaning is literally "a planted thing." While the term existed, it did not appear as a classifying division in encyclopedias and other reference works before the mid-nineteenth century, which referred instead to all or some subset of grasses (*cao*), trees or wood (*mu*), vegetables (*shu*), fruit (*guo*), grains (*gu*), ornamental flowers (*huahui*), bamboos (*zhu*), and vines and lianas (*teng*). These were categories in their own right, rather than subdivisions of a larger class known as "things that grow."[7] It was only in 1848, with the publication of Wu Qijun's *Research on the Illustrations, Realities, and Names of Plants* that the *zhiwu* of the title was used as an inclusive category under which multiple forms of plant life are identified. In the table of contents, though, plants were still subdivided into the familiar classes of grains, vegetables, grasses, and trees, with the grasses broken down into less common (though not unprecedented) subgroups such as mountain grasses (*shan cao*), grasses from rocky areas (*shicao*), and poisonous grasses (*ducao*). The historian Georges Métailié points out, too, that the editor Lu Yinggu continued to use the conventional formula "grasses and trees" (*cao mu*) in his preface (Métailié 2015, 660). Wu's innovation was limited to the title of his work and was not adopted as a new norm.

Credit for the use of the term *zhiwu* to describe plants in the modern botanical sense is due to Li Shanlan and his collaborators Alexander Williamson and Joseph Edkins, who translated John Lindley's *Elements of Botany* as *Zhiwuxue*, published in 1858.[8] In their preface, the translators did not enlarge on what they meant by *zhiwu*, although they did begin with an explanation of the main groups of plants, echoing the familiar categories of the classics: "Plants have many uses. The five grains nourish the body, the hundred fruits please the mouth, medicinal herbs cure illness, [and] wood and timber build palaces, homes, boats, carts, and all kinds of tools. Grasses and trees all have different attributes (Li, Williamson, and Edkins 1858, juan 1, 1a).

In adding the word for study (*xue*) to the binomial *zhiwu*, Li, Williamson, and Edkins coined the term for botany (*zhiwuxue*) and created a new field of science in China at the same time. Contemporary scholars have admired Li's lexical choice, noting that it corresponds to the etymology of "botany" in Western languages, while using an appropriate borrowing from the classics, an important factor in its ready adoption in China (Pan 1984, 168; Masini 1993, 59). Li and his colleagues had succeeded in introducing the new science of botany using a vocabulary and cultural references that suggested some degree of continuity rather than a total break with or a rejection of the past.

NATURAL HISTORY: THE INVESTIGATION OF THINGS AND BROAD LEARNING ABOUT THINGS

Having established plants as a category of things, the next step is to consider how the traditional scholar's knowledge about plants and natural phenomena differed from the scientific knowledge of the botanist.

Tension between the heritage of China's past and Western learning or science was not an artifact of the nineteenth century. The Jesuits had already faced the challenge of how to translate the word *scientia* in their works on mathematics, astronomy, and other branches of Western knowledge. Collaborating with their converts among the literati, they adopted an existing term, *gezhi*, literally "the extension of knowledge," which became the accepted translation for "science" until the early twentieth century (Reynolds 1991). The expression had its roots in the phrase *zhizhi zai gewu* (the extension of wisdom and the investigation of things), first found in the early Confucian classic *The Great Learning* (Da xue), which probably dates from the third century BCE. The text teaches that the sage kings of the past achieved virtue and order in their states by first cultivating sincerity in their families and in themselves through "the extension of wisdom and the investigation of things" (Fung Y. 1922, 257).

The phrase subsequently appeared in the abbreviated forms *gewu* (the investigation of things) or *gezhi* (the extension of knowledge), both referring to the endeavor to understand nature in the sense of *ziranzhi li* (self-generation's pattern). Later scholars gave an ethical and moral dimension to the accumulation of knowledge as the search for the principles of "the broad range of things" (*bowu*), a term with its own respected ancestry as a literary genre recording noteworthy events, objects, and other manifestations of the myriad things.[9] *Bowuxue* (the broad learning of things) came to refer to that part of natural studies or the investigation of things that examined the myriad things encountered in day-to-day life, and by the late nineteenth century, it had become the preferred translation of the term "natural science."[10] Natural science could therefore take its place as a legitimate and respectable concern of scholars, although it was perceived as the knowledge gained from the investigation of things rather than the process of uncovering that knowledge through observation.[11]

In the face of the Western learning of the Jesuits and the wave of scientific and technological innovations that confronted China after the Opium

Wars, the distinguished ancestry, etymology, and cultural weight of *gezhi* and *bowu* allowed scholars of the late Qing era to suggest that science was not, in fact, a foreign intrusion into Chinese culture but rather built on ancient knowledge, as recorded in the classics. Much of it, they said, was learning that had been lost (conforming to the conventional narrative of the lost golden age of the classics) and was now being recovered from the West (Zhang Fan 2009). As late as 1902, an article titled "A Study of the Differences and Similarities between Chinese and Western Natural Studies" criticized those who would pursue Western science in ignorance of the origins of "the extension of knowledge" in the Chinese classics. The author argued that the calendar, agriculture, hydraulic engineering, construction, metallurgy, medicine, finance, law, and military strategy are all to be found scattered in the Chinese classics and that as practiced now, they are all examples of the application of Western knowledge, not Western knowledge as the core (*Zhengyi tongbao* 1902, 7b).

THE INVESTIGATION OF THINGS AND SCIENCE

Following China's defeat in the Sino-Japanese War in 1895, Western knowledge and science became pillars on which reformers hoped to build a strong and prosperous China. A new expression, *kexue* (science), began to appear in the press, displacing the older *gezhi* or *bowu*. There was a sense that *kexue*—the term used today—was different and distinct. In the politicized environment of the time, the older terms became associated with a traditionalist perspective, a cypher for resistance to Western science, until ultimately they became quaint, almost forgotten expressions (Wang Hui 2011).[12]

The origin of modern scientific terminology in Chinese is a hotly contested issue, particularly where there are competing claims that a certain word or expression was first coined in either Japan or China. Scholars disagree on the origins of the term *kexue*. Some locate its origins in Chinese texts, tracing its earliest use to the late Tang era (Zhou and Ji 2009, 95). In nearly all these early cases, though, the context makes it clear that it refers to the imperial examination system (*keju*) or occasionally to the subjects the candidate had to master for the exams. The consensus is that these early references were scribal errors, given the calligraphic similarity of the characters *ju* 舉 and *xue* 學 (Jin and Liu 2005; Zhang Yaqun 2005).

The earliest documented occurrence in Japan of the binomial using the Chinese characters *kexue* 科學, pronounced *kagaku* in Japanese, is from the Edo period (1603–1868), when it was used by the writer Takano Chōei

(1804–1850), who had studied medicine at a school in the Dutch trading enclave of Deshima, off Nagasaki. Chinese scholars contend that Takano meant "branches of learning" rather than science as a broad field that incorporates multiple branches of learning (Zhou C. 2009, 185). By the late nineteenth century, though, *kagaku* (now referring to the sciences) was an important part of the curriculum in the educational system, and Chinese students in Japan would have used the term, ultimately adopting it in preference to the older *gezhi*.[13]

Looking for the early use of *kexue* in China in the modern sense of science, the historian Zhang Fan has found a reference from 1899 in *Qing Yi News* (Qing Yi bao) in an article in Chinese by a Japanese author on scientific feats, such as the opening of the Suez Canal, the Trans-Siberian Railway, and the trans-Pacific cable (Zhang Fan 2009, 105). More significant perhaps, is evidence that the political reformer Kang Youwei and his contemporary Yan Fu, best known for his translations of the works of seminal British writers and thinkers such as Adam Smith, Thomas Huxley, and Herbert Spencer, both used the term with reference to Western science in works written and published between 1897 and 1903. In a memorial to the Guangxu emperor, dated June 1898, Kang recommended the abolition of the rigid eight-legged essay format for the imperial examinations and proposed that *kexue* (science) should replace the *keju* (old examination system), indicating that he distinguished clearly between the two. As for Yan, he used the term more than once in his translations of Adam Smith's *The Wealth of Nations*, published in 1902, and of Spencer's *A Study of Sociology* (1903), in which he explained that sociology applies the laws of science (*kexue*) to shed light on what has happened and to determine the outcomes of what is being observed.[14]

Reinforcing the sense that *gezhi* was drifting away from being an all-encompassing observation and study of the myriad things, writers at the turn of the century began to describe it as one among several divisions of the sciences. Du Yaquan, in his article "On the Meaning of Natural Sciences," differentiated between *bowuxue* and *gezhixue*: "*Bowuxue* [the broad learning of things] is a part of the study of the myriad things, and it is different from *gezhixue* [the extension of knowledge].... The extension of knowledge refers to the transformations of the myriad things and their measurement. Physics and chemistry both belong in this field. The broad learning of things includes many disciplines and subjects, but it can be divided into three branches: zoology, botany, and geology" (Du 1902, 6a).

For some time, *kexue* and *gezhi* coexisted. As students returned from their studies abroad, though, there was momentum to organize scholarship into discrete, defined fields corresponding to those in which they were being trained. In the same year that Du Yaquan's article appeared, Wang Rongbao and Ye Lan, both Chinese students in Japan, published a technical dictionary with the title the *New Literary Expositor* (Xin er ya).[15] *Kexue* appears in the section on education, where it is a subheading between *guojiaxue* (the theory of the state) and *ziran kexue* (the natural sciences), described as "the study of natural phenomena such as animals, plants, physics, and chemistry." The entry says: "In research of the phenomena of the world, a systematic ordering of knowledge is called *kexue*" (Wang and Ye 1903, 59, translated in Shen Guowei 2014, 98). Later there is a section of ten pages on *gezhi* (the extension of knowledge): "The investigation of the outside appearance of material things and their transformations is called *gezhi*." With further entries in the *New Literary Expositor* on chemistry, physiology, zoology, and botany, "the extension of knowledge" appears to have evolved into a methodology of observation in contrast to "science," which referred to an analytic process seeking to understand the relations between different natural phenomena. The differences between the two were not trivial. Science stood for learning that could be subjected to the rigorous testing and proof of hypotheses, hence the preference for *kexue*, an expression that did not have the more contemplative echoes of the past associated with *gezhi* (Chen, Yang, and Gu 2011; Zhang Fan 2016).

THE SCIENTIFIC METHOD AND MODERNITY

The development of botany in China offers some parallels to, as well as significant differences with, the evolution of natural history and botany a century earlier in Europe and the Americas. In Europe, the main task of natural history, comparable to the broad learning of things in China, had traditionally been the description and recording of natural phenomena. In Europe, though, the enlightenment gave primary importance to the knowledge gained through observation, which revealed that nature operated through natural laws and structures that humans could fathom (Farber 2000, 21). Describing and classifying the components of these structures according to a scientific system distinguished naturalists from "mere collectors of curiosities and superficial trifles ... objects of ridicule rather than respect" (Ritvo 1992, 365). As natural history became botany, it moved up the hierarchy from its early status as a component of medical training to the investigation of the physiological systems of plants and how they functioned. It became a respected field of science in its own right, separating in due course into subdisciplines such as physiology or paleontology (Outram 1996, 249–65; Endersby 2008, 219).

Botany is often described as "one of the great imperial sciences, playing a key part in exploring, cataloging, and exploiting the natural wealth of the empire" (Endersby 2008, 34).[16] Although China was never formally colonized, scientific knowledge was collected and organized there during the heyday of Western colonialism by networks of foreign agents, such as missionaries and plant hunters, in an enterprise that has been called "a kind of extractive colonialism of information" (Harrell 2011, 19). The information they gathered flowed westward from China, but the practice of colonial botany was firmly located in the metropole, looking outward to identify and classify what was new and potentially useful in the vast terra incognita of the periphery. For early Chinese botanists, by contrast, botany was the scientific study of the plants that made the landscapes in which they were found unique. It represented an opportunity to expand knowledge beyond the familiar utilitarian world of materia medica and other literary genres. It was a way to rediscover and to describe flora that had become more familiar to foreigners than to the Chinese people.

At the same time, botany became a profession. In the West, at a time of growing prosperity and expansion, botanists replaced the gentlemen and women who had previously practiced natural history.[17] In China, scientists and botanists took the place of the literati of the imperial era. A profound difference was that in China, the sciences and the profession of scientist emerged at a time of internal weakness and decline. Scholars who no longer had a pathway to officialdom through the examination system and the bureaucracy became activists and visionaries searching for a response to the challenges of the future in adopting, adapting, and applying the sciences as a vital element of building a new China.

Science was an integral part of the discourse of reform and change that marked the turmoil of the first decades after the overthrow of the Qing dynasty. As one of the more self-consciously modern voices in the public sphere, the scientific community argued that Mr. Science (*Sai xiansheng*) had a critical role to play in shaping the future. While they were vehement in saying that what they were doing and the questions they were asking were not the same as in the past, ironically, they were embracing the traditional literati view that as educated scholars, they had a mission to improve the world. Furthermore, the modernizers' vision of science and technology leading the march to prosperity and freedom from foreign domination inevitably raised the question of how adopting Western science could be compatible with aspirations for a strong and independent China, free of foreign influence.[18] The modernizers answered that it was the scientific method that would lead to change rather than the specific knowledge scientists were uncovering about natural phenomena and the relations between them.

Chinese intellectuals were very aware of the legacy from the past of records in prose and verse of observations of the living world, but they contended that it was the scientific method that defined science and that it had no precedent in China. In 1915, the chemist Ren Hongjun (1886–1961) wrote in the first issue of *Science* on "the reasons that China does not have science." He argued that science is knowledge derived from observations and experiments that follow a systematic and ordered process. Scientists accumulate data from multiple discrete occurrences to construct a principle or hypothesis, which they test to develop new categories of phenomena that might warrant further study: "After the Qin and the Han, people's minds were shackled to the study of their own time. This involved the observation of things to determine their status, not to find out how they came to be that way. The methods they used pursued abstraction and avoided the concrete. It hardly needs to be said that under these circumstances there could be no science . . . the reason being that their knowledge did not involve systematic, established procedures" (Ren 1915, 8).

It is indisputable, Ren observed, that scholars of the past had limited their investigation of things to "delving deeply into ancient documents and pontificating on rationality" (Ren 1915, 11). Ren was a chemist, but his views were shared by others who were not scientists. In 1922, Feng Youlan, who had offered an affirmative interpretation of the classical concept of nature in the *International Journal of Ethics*, continued in the same paper to describe science as "the systematic knowledge of natural phenomena and of the relations between them," ending with the uncompromising conclusion that "China has no science, because according to her own standard of value, she does not need any" (Fung Y. 1922, 238).

TRADITIONAL KNOWLEDGE ABOUT PLANTS AND THE SCIENCE OF BOTANY

The camellia is also known as *mantuoluo*. It is a tree that can be as high as ten feet or as small as two or three feet. The main stem and the branches interlace and cross each other. The leaves are hard like those of the osmanthus, with ridges, thickening a little, and they can be more than an inch wide in the middle. Both ends [of the leaves] are pointed and elongated, and they can be as long as three inches. Generally, the surface is deep green and shiny, while the back is light green. They do not drop in winter. It is called *cha* because the leaves are similar to tea [*cha*] leaves, and they can be used to make a drink. There are many varieties of the flower. They flower from the tenth month to the second month. (Jiang T. 1726, Cao Mu ["Grasses and Trees], juan 296, 42a)

The description of the camellia in Jiang Tingxi's *The Imperially Commissioned Compendium of Literature and Illustrations, Ancient and Modern* (Gu jin tu shu ji cheng) is not an original account written for the encyclopedia. It is reproduced from *The Assembly of Perfumes* (Qun fang pu) by Wang Xiangjin (1630), an earlier horticultural manual, where it is part of a compilation that sets it in a cultural context, with references to other times, multiple uses of the plant, and its many varieties. It is quite different from the entry for camellia in the standard reference work on the Chinese flora today: "Camellia Linnaeus, Sp. Pl 2:698. 1753. Shrubs or small trees, rarely large trees, evergreen. Leaves petiolate or rarely sessile and amplexicaul; leaf blade leathery to thinly leathery, margin serrate, serrulate, or rarely entire. Flowers axillary or subterminal, solitary or rarely to 3 in a cluster" (Wu and Raven 2013, vol. 12, 366).[19]

In *Flora of China*, Chinese botanists have collaborated with their peers around the world to compile a reference work in a format and language that conform to a global institutional culture. Reading it, it is easy to conclude that Chinese scientists have turned decisively away from the traditional to embrace what historian of science Joseph Needham referred to as an "oecumenical science" toward which all traditions of science (Western and non-Western) converge (Needham 1967).

For many historians, traditional and scientific knowledge represent two different epistemologies. Métailié, while meticulously and respectfully documenting the record of knowledge about plants in China, refers to "an autonomous and original domain that may be called traditional Chinese botany" (Métailié 2015, 13). In their standard history of botany in China, Wang Zhenru and his collaborators assert that scientific botany is a discipline in its own right. They contend that traditional taxonomy and classification, which they use as proxies for the whole field of botany, cannot be associated with scientific botany because they always appeared as subdisciplines of other bodies of knowledge, such as materia medicas and agricultural manuals (Wang Zhenru, Liang, and Wang 1994, 149). Their reasoning echoes the arguments of the first generation of scientists and botanists as they, too, sought to draw a boundary between science and the knowledge of the past (Amelung 2014).

There may appear to be a clean and complete break from one paradigm to another, but at least some early botanists saw themselves negotiating a passage between traditional Chinese knowledge about plants and scientific botany. The construction of a class of things known as plants took place in the context of the deeply rooted idea of the myriad things (*wan wu*) as manifestations of nature in the sense of *ziran* (that which is so of itself). Plants were a category of the myriad things inseparable from and no more or less important than humans, animals, minerals, and others. In 1903, Du Yaquan wrote in the preface to the first textbook on botany for the new school curriculum that "what makes us human is our dependence on material things of which just three kinds are living beings. Plants are one of them. It is unimaginable how we could separate ourselves from them and their life cycles, and how dependent we are on them" (Du 1903, 85b–86a).[20]

In the past, literati scholars observing the natural world had been engaged in correctly describing, identifying, and classifying the myriad things, including (but not limited to) plants, the "things that grow." Since nature per se was not an entity—it was self-generation's pattern—there had been little concern with questioning its workings (Kim 2004, 102). Taking the view that the knowledge of the sages of the past had been lost or perverted, scholarship was devoted to confirming or rectifying the names of plants that appear in the classics, not in trying to understand the vast diversity of what lay outside their experience—a legacy that sheds some light on the dominance of taxonomy and nomenclature in botany in China through the first half of the twentieth century (Haas 1988a; Wang Zhenru, Liang, and Wang 1994).[21] Students of botany at the end of the nineteenth century rejected this scholarly tradition. They believed that the essential part played by plants in human existence compelled them to reject textual studies and to seek knowledge directly in the field, outside schoolrooms and the scholar's study.

It was some twenty to thirty years after the first translations of foreign works, beginning with Lindley's *Elements of Botany*, before Chinese scholars recognized botany as a field of science. The shift took place as *kexue* was becoming the preferred term for science as a methodology for observing, organizing, and classifying natural phenomena, in the course of which knowledge about natural phenomena was being disaggregated into different disciplines, one of which was botany. In "A Brief History of Botany," published in 1903, Yu Heyin, who had studied engineering in Japan, was clear that "botany is not the same as knowledge about plants." He began with a review of early writers about the natural world, from Theophrastus to those in medieval Europe, who had limited themselves to "naming plants, assembling and making a record of medicinal plants, poisonous plants, useful plants, and so on." He then turned to China and Li Shizhen's *Classification of Materia Medica*. He concluded that, valuable as that compendium might be, "its purpose was limited to investigating the medicinal characteristics of plants, their production, and their practical value, which is why I do not consider it to be the same as what we now understand as botany" (Yu H. 1903, 18).[22]

Li Shanlan's translation of Lindley had introduced fundamental concepts such as plant cells, plant anatomy, photosynthesis, seeds and germination, nutrient uptake and biogeography. In 1897, an article in the *Journal of Pedagogy* (Mengxue bao) in the form of a sequence of questions and answers was one of the first explanations of basic concepts of botany by a Chinese author (as opposed to a translation from a foreign source) for an interested but non-specialist readership (*Mengxue bao* 1897). The *New Literary Expositor* in 1902 introduced some of the disciplines in the field, including plant morphology "the investigation of the parts of plants, that is to say, the structure of their organs" and plant physiology "the study of the functions of all the plant's organs" (Wang and Ye 1903, 165). With the translation in 1904 of a Japanese textbook (Matsumura 1890) and in 1907 of the first textbook by a Chinese botanist, the core body of knowledge that constitutes the science of botany was well known by the time the first dictionary of botany appeared in 1918. The dictionary, approved by the Ministry of Education, was an affirmation of the conceptual framework of the field and an endorsement of a new language and vocabulary, which gave botanists the tools to take on the task of a systematic study of China's plant resources.[23]

CHAPTER FOUR

A New Language to Name and Describe Plants

> Botanical terminology is vast in extent, like smoke or the ocean. It shifts then comes together again. It has a finite area but ultimately it is impossible to record it in its entirety
>
> —*DONGFANG ZAZHI*, GUIDELINES FOR THE DICTIONARY OF BIOLOGY ("ZHIWU DA CIDIAN FANLI")

WHEN a critical mass of observers embraces a new way of understanding and describing familiar objects and phenomena, there is a pressing need for a language and vocabulary to communicate it to others. Ideas and terms of reference with their origins elsewhere must pass through translation before they can be shared with fellow scientists and with a wider nonspecialist audience. The translation must express new concepts precisely while anchoring them in the familiar. In today's global scientific community, the terminology must also facilitate accurate professional exchanges with international peers in the academic and research worlds.

Greek, Latin, Arabic, and French have served at different times as the lingua franca of international scientific exchange, a role that is currently played by English. Nevertheless, scholars and scientists use their own native language in day-to-day professional activities and in their interactions with others. The global lingua franca and the local vernacular coexist, calling for a continuing dynamic of translation and generation of terminology as the body of knowledge grows and evolves (Olohan, 2007; Gordin 2017). As knowledge about plants in China became botany, the camellia that had been the familiar "southern mountain tea flower," whose different varieties had been described with reference to size, color, and form, became *Camellia reticulata*, a member of the genus (*shu*) *Camellia* within the family (*ke*) Theaceae. It is one of several species (*zhong*) distinguished by differences in what are now referred to as parts of the plant's reproductive anatomy. *Shu*, *ke*, and *zhong* were all familiar classifiers in prescientific works about plants, but they have acquired a fixed place in an internationally sanctioned system of classification. Translation has done more than simply match a word in one language to a word in another language. It has created a technical vocabulary, a terminology that embodies the epistemology of global scientific botany.

THE HISTORY AND POLITICS OF LANGUAGE AND TRANSLATION

The earliest records on statecraft in China refer to intermediaries with neighboring states whose skills included translation (Behr 2004, 186–97). During the Mongol Yuan dynasty and the Manchu Qing dynasty, the official status of several languages in addition to Chinese and the vast geographical extent of the empire required translation of government documents on a large scale (Alleton 2001, 18; Schäfer 2011). The Yongle emperor of the Ming dynasty established the Office of the Four Barbarian Languages (Si Yi 夷 Guan) in 1407, charged with the translation of documents needed in dealings with neighboring people and countries, as well as training interpreters. The Qing rulers renamed it the Office for Four Translations (Si Yi 譯 Guan), replacing the term "barbarians" with a less offensive homonym meaning "translations." This office gave way in 1748 to the Combined Four Translations Office (Hui Tong Si Yi Guan). Following the Opium Wars, there was a more concerted effort to train interpreters and to carry out programs of translation, beginning with the School of Combined Learning (Tongwen Guan), founded in Beijing in 1862, followed by similar institutions in Shanghai and Guangzhou (Mu 2004; Huang and Zhu 2012).

Trade and diplomacy were a part of daily political and economic life. Crossing linguistic barriers in these spheres was a relatively straightforward matter of translating words and dialogue from one language to another. Over the two millennia of imperial history, there had been, however, three periods during which translators were confronted with materials presenting new categories of learning and vocabularies that had not existed before in Chinese.

The arrival and spread of Buddhism between the third and fifth centuries gave rise to an unprecedented program of translation of sutras and other religious texts, based in the monasteries of the capital Chang'an. A system developed in which a chain of translation began with the recitation of the original and a rough oral rendition by a foreign monk or scholar. Passing through several iterations of review and polishing by Chinese monks and scholars to produce the final text, the collaborative process was fundamental to the introduction of radically new ideas and the synthesis of Indian and Chinese thought.[1]

The model of a foreign informant working with a Chinese collaborator was followed during the latter half of the Ming and the early Qing eras by the Jesuit missionaries whose translations of European scientific works marked the second boom in the translation of ideas. Between Matteo Ricci's arrival in Macao in 1582 and the suppression of the order in 1773, sixty-six missionaries are believed to have contributed to a body of over three hundred books, of which some one hundred and twenty were on science and technology. The corpus of Jesuit scientific translations had limitations. They were sometimes a synthesis of more than one text, a reworking of the original texts, or commentary interwoven with the original text.[2] Nevertheless, the collaboration between Jesuits with expertise in the sciences they were translating and Chinese scholars well versed in the classics ensured that the terminology they developed succeeded in articulating new scientific concepts in an acceptable literary form. Some of the vocabulary they devised for mathematics, geometry, and geography is still used today.

The third wave of translation, the most relevant to the transition from traditional knowledge about plants to scientific botany, took place during the seventy-five years between the end of the Opium Wars and the relatively stable period of the republican government in Nanjing in the mid-1920s. The momentum to bring "foreign knowledge" into China during this period came from the widespread belief that an obvious cause of China's vulnerability and backwardness was the yawning gap between China and the West in knowledge of the sciences and technology that had propelled the economic development of the industrialized nations. A fundamental strategy in closing the gap was to bring foreign knowledge to China, and translation was an essential tool in that strategy.

There were sharply divergent views on whether the Chinese language was capable of expressing scientific concepts and arguments. During the late seventeenth and early eighteenth centuries, Europe had experienced a period of fascination for all things Chinese, fueled by the voluminous reports written by Jesuit missionaries. At its extravagant peak, this took the form of the fashion for chinoiserie, with its exoticized Chinese themes in home decor, porcelain, and even gardens boasting pavilions and replicas of pagodas.[3] Some fifty years earlier, philosophers and savants of the enlightenment had seen and admired what they took to be a benevolent meritocracy in China, marked by religious tolerance based on the moral and ethical system of neo-Confucianism.[4] Scholars were searching for a pure, rational language to represent directly the world that science was revealing, and many were fascinated with what they were learning about the Chinese language and script. Language reformers such as Jonathan Swift in England and Gottfried Wilhelm Leibniz in Germany toyed with the possibility that Chinese characters might be a logical, symbolic system that directly represented things unimpeded by rhetoric or ambiguity, potentially a universal language, the ideal medium with which to record and communicate the observations of nature that were the foundation of the scientific method (Porter 1996, 2001).

As China isolated itself from the rest of the world during the Qing dynasty, Western sinophilia quickly turned into disdain for what had come to be seen as a corrupt, hidebound, despotic empire. In contrast to earlier ideas that the Chinese language and its writing system were potentially an ideal vehicle for transmitting ideas across the boundaries of languages, its complexity was now seen as an obstacle to communication, an antiquated system of signs with no grammar or syntax, inflexible and incapable of expressing abstract ideas. Even Emil Bretschneider, despite his great respect for Chinese knowledge about plants, complained: "It is well known by all who have read Chinese books, how indistinctly they are written for the most part, and how confusedly separate and single ideas are thrown together. The Chinese are in complete ignorance of our system of punctuation. Few breaks are to be met with indicating the beginning of a new subject. Very often in a whole chapter, treating of several different things, no break can be found" (Bretschneider 1871, 4).

On May 26, 1886, the North China Branch of the Royal Asiatic Society held a symposium on "the advisability, or the reverse, of endeavouring to convey Western knowledge to the Chinese through the medium of their own language." Fourteen papers were read at the meeting. At one end of the spectrum, a Mr. Eric Faber ridiculed the very idea of using Chinese to write about science: "Translations into Chinese of scientific works, in the strict sense of the word—mathematics only excepted—are, for the present, either impossibilities or monstrosities. The Chinese language being yet in a state of vagueness, makes it impossible to enter into scientific details with sufficient exactness to convey definite notions" (*Journal of the China Branch of the Royal Asiatic Society* 1886, n.s., 15).

Mr. G. M. H. Playfair expressed a more nuanced view, contending that any vernacular language, be it English or Chinese or any other language, could not accurately impart the facts of science, since "looseness and inaccuracy are incompatible with science, therefore the vulgar tongue of any nation is an improper medium for scientific teaching" (*Journal of the China Branch* 1886, n.s., 16).

The participants most knowledgeable about the language were in no doubt, though, that not only was it feasible to write about science in Chinese but there were features of the language that made it amenable to generating new terminology and to expressing new concepts. The Rev. Dr. W. A. P. Martin acknowledged that the Chinese language and system of writing might be difficult to learn but:

> Are they like old bottles that cannot bear the infusion of new wine? Nothing is further from the truth; for no language, not even the German or the Greek, lends itself with more facility than Chinese to the composition of technical terms. Its elements being devoid of inflection form compounds by mere juxtaposition—each component reflecting on the other a tinge of its own color. It is not therefore an achromatic medium such as we require for some of the purposes of philosophy, but its residuary tints in most cases offer aid rather than hindrance to the apprehension and the memory. (*Journal of the China Branch* 1886, n.s., 3)

The editors of the proceedings concluded that it was possible to use Chinese as the medium for scientific instruction and communications. They emphasized that it would be essential to settle on a standardized terminology, and they recommended that translation should be carried out in collaboration with Chinese scholars to ensure a style that would be acceptable to their intended audience, the Chinese literati. Over the next ten years, this was, indeed, a focus of the missionary translation effort.[5] Successive committees reviewed different iterations of glossaries, culminating in 1904 in the publication by the Presbyterian Missionary Press in Shanghai of *Technical Terms, English and Chinese* (Educational Association of China, 1904).

While Westerners debated whether Chinese could convey scientific ideas and concepts, the issue that concerned Chinese officials and scholars was one of language reform and modernization. Language, and especially the choice of classical, literary Chinese or some written form of spoken

Chinese as the medium of formal communication, was a vital element of the ferment and change of this period (Huang K. 2008). In early 1904, the recently established Ministry of Education issued regulations for the new school system, which was to replace traditional schools after the abolition of the civil service exams the following year. They included a regulation "to prohibit schools from eliminating Chinese writing, in order to ensure that [students] can study ancient texts and the classics." The guidelines explained that since foreign languages are not capable of "the exposition of principles [*li*], of recording events, promoting virtue, and touching the emotions," foreign words should not be used, with the exception of "new words coined for new things in fields such as chemistry, engineering, and other specialized subjects." They called for vigilance against neologisms constructed from Chinese characters but which had in fact come from Japan, insisting that the Chinese language already had adequate words, as well as the grammar and syntax needed to teach the new learning. The regulations ended with a warning that any use of non-Chinese words and constructions would be "rejected and sanctioned" (*Dongfang zazhi* 1904, 64, 126–28).[6]

The Ministry of Education promulgated its regulations in 1904, the same year that the Committee of the Educational Association of China published its handbook of technical terms. The ministry's guidelines were reprinted in the third issue of *Eastern Miscellany* (Dongfang zazhi), launched that year in Shanghai shortly after the launch in 1903 of *Science World* (Kexue shijie), two journals covering the sciences, written and edited largely by Chinese scholars.[7] The tone of commentaries in *Eastern Miscellany* and *Science World* was quite different from earlier discourse on translation and the new learning. A recurring motif was anger at foreign incursions, accompanied by demands that China should regain control of her own knowledge and resources. Activists advocated that science should be a force in the overthrow of the old, spent Confucian order, clearing the way for modernization and democracy, and calls to modernize the Chinese language became embroiled in the demands for cultural and political revolution at the heart of the May 4th Movement in 1919 (Yang C. 2017, 563).[8]

Translation and the creation of a new scientific lexicon were only one element of the more fundamental question of whether the national language in China should be classical literary Chinese—or a modified version used principally in the administration known as "official language" (*guanhua*)—or the vernacular spoken language (*baihua*, "plain language," literally "white language").[9] While the language issue seemed settled in favor of vernacular language in the aftermath of May 4th, arguments about whether or not to use any form of Chinese at all in science rumbled on. As late as 1926, Bing Zhi felt compelled to make the case in an article in *Science* for why it was important to create a Chinese terminology of taxonomy: "Recently some people have suggested simply using the Western alphabet and Latin, which is the lingua franca [for scientists] around the world. [They say that] China could adopt this principle and then there would be no need to create Chinese terms. This is the kind of thing that people with no national spirit would say. There are Chinese words for phylum, class, order, family, and so on, so why should we think that they do not exist for genus and species?" (Bing 1926, 1349).

In due course, vernacular Chinese became the standard medium of communication, although some of the older generation of botanists continued to feel more at ease writing in the literary style. Published reports of botanical collecting expeditions and papers presented to professional meetings in the 1930s still often used classical and literary forms, syntax, and references (Zhong 1932a, 1932b). As prominent a figure as Hu Xiansu continued to write and publish well-regarded classical poetry about nature and natural history into the 1940s, as did many of his generation, including Mao Zedong himself (Liu W. 2012; Jiang L. 2016, 168–70).

THE LATE-QING TRANSLATION ENTERPRISE

The devastation of the Taiping Rebellion drove many scholars from the Jiangnan (lower Yangzi) region to seek safety in Shanghai, where they met and collaborated with Westerners whose interests they shared. When the weakened imperial authorities launched an intensive program in 1895 to catch up with the West after China's defeat in the Sino-Japanese war, they were able to call on experienced translators from among both the displaced scholars and missionaries to collaborate in the new offices mandated to translate technical materials.[10] In the decades that followed, government officials, missionaries, students studying abroad, and the first generation of trained botanists and other scientists were all involved in an enterprise that put in place principles of translation and standardized terminologies with which Chinese scientists could move from learning from others what was already known to conducting original research and building a body of knowledge in China about China.

In sharp contrast to the Jesuits, the actors during this period were not just trying to reach an educated elite; they were aiming to bring a comprehensive range of new knowledge to as wide an audience as possible and to produce the textbooks and other materials needed to train a new generation of scientists.[11] The scope and volume of publications produced in a matter of decades was such that annotated bibliographies maintained by scholars—valued by the educated public as a guide to reference works in different fields—struggled to create classification systems to accommodate new categories of learning that could not be placed in the traditional four divisions of knowledge: canonical classics, histories, masters, and anthologies (Jiang S. 2007; Guo J. 2012).

Missionary societies were essential players in the early years of the translation of Western science into Chinese. Under the Canton trade system, Chinese were not allowed to learn foreign languages, and foreigners were not allowed to learn Chinese. Following the Yongzheng emperor's 1734 ban on missionary activity, Macao played a vital role in training translators, and there were other opportunities to learn the language outside China, at schools in Batavia and Malacca organized by missionaries to serve the large Chinese communities there (Masini 1993, 7–12; Liang 2013, 21–23). When missionaries went into China after the Opium War, then, they had at least the foundations of the language skills they needed to use science and education in the service of their evangelizing mission.

The London Missionary Society wasted no time in moving its publishing arm, the London Missionary Press, to Hong Kong from Malacca, then again in 1843 to Shanghai. Under its new name, Inkstone Press, it became a hub for early translation and publishing, building relationships with Chinese scholars interested in Western learning and hiring some of them to collaborate on translations, mostly of religious materials but also including scientific works (Vittinghoff 2004, 89–92; Elman 2005, 283–351). One of its earliest recruits was Li Shanlan, who had escaped the ravages of the Taiping Rebellion and arrived in Shanghai in 1852. Among his early contributions were a translation of William Whewell's *Treatise on Mechanics*, William Herschels's *Astronomy*, and, in 1858, John Lindley's *Elements of Botany*.

Among the visitors to Inkstone Press were Xu Shou, his son Xu Jianyin and their friend Hua Hengfang, who had formed an informal study group in Wuxi to learn more about mathematics and the Jesuit learning. In the somewhat euphoric words of John Fryer (1839–1928), who collaborated with them for several decades, the books they discovered at the press enabled them "to leap at one bound across the two centuries that had elapsed since the Jesuit fathers commenced the task of the intellectual enlightenment of China, and bring them face to face with the results of some of the great modern discoveries" (Fryer 1880, 1).

When the Taiping rebels captured Wuxi in the spring of 1860, Xu, his son, and Hua escaped into the surrounding hills, but their books, translations,

FIGURE 4.1   A portrait of Li Shanlan, mathematician and translator. Li was the Chinese member of the group of three translators of *Botany* (Zhiwuxue), the first text on scientific botany to be published in China. From the *Chinese Scientific Magazine* (Gezhi huibian) 2, no. 2 (1877): 443.

and correspondence were all burned. Zeng Guofan, the general leading the Qing armies, recruited them during the last days of the rebellion to assemble a collection of books on Western learning in Nanjing. In 1867, they moved to Shanghai to work as engineers at the Jiangnan Arsenal, which Zeng had established in 1865 to manufacture military equipment using the most up-to-date engineering technology available. The Arsenal was a response to the weakness of the Qing armies, as demonstrated by their defeat during the Opium Wars and their inability to defeat the Taiping rebels without assistance from the Western powers. Initially, it kept rigorously to its mission of military production and had little interest in translation and publication. When Zeng did finally respond favorably to the director's suggestion to add a translation department in 1868, it could already boast a group of competent scientific translators, such as Li Shanlan and the Wuxi group, and the department quickly recruited Fryer, an accomplished technical translator, to join them. Fryer, the son of devout parents in Kent, had first taught at St Paul's College, a Church of England school in Hong Kong, then moved in 1863 to take the position of professor of English language and literature at the School of Combined Learning in Beijing. He moved to Shanghai in 1865 to teach at the Anglo-Chinese School before joining the Jiangnan Arsenal. In an inversion of the earlier model, the missionaries who had previously employed Chinese scholars now found themselves working alongside their former employees, serving the imperial government's ambitious translation enterprise.[12]

The Translation Department of the Jiangnan Arsenal operated for more than forty years (1868–1913). It was most productive between 1868 and 1880, during which it published ninety-eight works, mostly on engineering, physical sciences, and military sciences.[13] In the years that followed, there were fewer publications, and the budget for translation was reduced until the Arsenal merged with the Shanghai Military Works (Shanghai Bing Gongchang) in 1905. The Translation Department continued to exist, but after the 1911 Revolution, it was considered to be a relic of the past; it was finally eliminated in 1913 (Wang Yangzong 1988, 68–70). It is now questionable what impact the Arsenal's translations had beyond its own premises, due to their limited subject matter and small circulation, mainly through official channels. Fryer himself came to the conclusion that a more effective way to promote and spread Western scientific knowledge was through the printed media, not through state-sponsored institutional programs (Lung 2016, 49).

By this time, Shanghai had become a cosmopolitan and industrial hub at the forefront of the introduction and dissemination of science and new learning. Outside his responsibilities at the Arsenal, Fryer was one of the founders of the Shanghai Polytechnic and Reading Room (Shanghai Gezhi Shuyuan), opened in 1876 as a reading room and a space for lectures for the education of the general public (D. Wright 1996). In the same year, Fryer and Xu Shou, as coeditors, launched the *Chinese Scientific Magazine* (Gezhi huibian), later renamed the *Chinese Scientific and Industrial Magazine*, introducing a wide range of topics on science and technology in Chinese.[14] The *Chinese Scientific and Industrial Magazine*, followed in 1903 by *Science World* and in 1904 by *Eastern Miscellany*, were milestones in moving the center of gravity for the introduction of Western science and the production of reference materials in Chinese from Westerners to Chinese scholars and eventually to trained Chinese scientists. Fryer's departure from China in 1896 to take the position of first Agassiz Professor of Oriental Languages and Literature at the University of California at Berkeley in effect brought an end to both the missionary and imperial translation enterprises in China.

Chinese journals and publishers were taking the initiative in translating Western works, compiling vocabularies and glossaries, and in debating burning issues, such as the place of science in the modernization and transformation of China. State-sponsored initiatives continued, however, in the form of support for standardization and for efforts to reach a consensus on principles for coining new terminology in the future. In 1909, the Ministry of Education appointed Yan Fu to head the Office for the Standardization of Terminology (Shending Mingci Guan), with a mandate to draw up an approved vocabulary for use in textbooks and other reference materials. Yan and his staff moved slowly, preferring to construct new and sometimes obscure terms from words found in antiquity rather than to adopt the neologisms coined in Japan that were beginning to appear in China. A count in the 1930s found that just 56 of the 482 terms Yan Fu had proposed were still in use (Huang K. 2008, 4). The office was dissolved after the 1911 Revolution, but it had set a precedent, and the republican government created a similar office under the Ministry of Education. In the long term, it was the publication in 1918 of the first dictionary of botany in Chinese that marked the passage from informed wordsmithing to systematic scientific translation, an essential step in standardizing terminology and taxonomic nomenclature for the field.[15]

CRAFTING NEW WORDS FOR THE MYRIAD THINGS

In coining a new term—a neologism—the translator aims to render the foreign word in such a way that it can be understood and adopted for professional and everyday use. The translator's goal, Fryer believed, was that an ordinary person should be able to pick up the translated text and understand it as well as a Western counterpart of comparable education (Fryer 1880, 9). To be accepted, neologisms should adapt to or assume the linguistic features of the target language, in this case Chinese (Masini 1993, 128).

When the Translation Department at the Jiangnan Arsenal was first established, the staff met to formulate the basic principles they would use in developing new terminology (Fryer 1880, 9–10). The system they agreed on covered the use of existing words and established rules for coining new terms:

- Use existing nomenclature where possible, even if it might not be found in a dictionary. Look through existing works, Jesuit translations, and other sources. Inquire among those who might be using the terms, such as merchants, manufacturers, and others.
- When coining new terms, either create a new character whose sound is obvious from its phonetic portion, or invent a descriptive term using as few characters as possible, or "phoneticize the foreign term using the sounds of the Mandarin dialect."

Today these principles would be referred to as lexical substitution (using existing nomenclature), phonemic loans (using phonetic portions of characters to reproduce, as far as possible, the sounds of foreign terms), and semantic adaptation (using existing characters to create a descriptive term).

On the whole, the neologisms that have stood the test of time are those that have used semantic adaptation rather than attempts to replicate foreign sounds. The Chinese writing system is not conducive to phonetic adaptations and the incorporation of sounds other than the basic set of about 400 different syllables (or 1,300 when including tones). The phoneme or sound is represented by a graphic element—the character—which itself carries meaning, complicating the use of Chinese characters to represent non-Chinese sounds. A reader has difficulty in dissociating the sound from the meaning, limiting the effectiveness of attempts to phoneticize a foreign word or expression, which in turn reduces the likelihood that the new term will be adopted.[16]

The relatively few phonemic loans related to botany in Chinese are nearly all the names of plants, flowers, or fruits that are not native to China and were introduced in earlier historical periods. The apple, for example, is believed to have come from Central Asia or India, and the Chinese word *pingguo* consists of the character *ping* 蘋—a character that is otherwise not used on its own—abbreviating the Sanskrit *bimbara*, with the second character *guo* 菓, meaning "fruit," added to give semantic clarity to the new binomial (Chen Shu-fen 2000, 380). The lemon (*ningmeng*) has a more contested history, but it seems to be derived from a word used by maritime traders during the Song dynasty, which itself had its origins in an Arabic (*līmun*) or Persian word (*līmū*) (Laufer 1934). Overall, though, very few plant names or parts are in this form (Chao 1953).[17]

Zhong Guanguang, who dedicated much of his career to reconciling existing names for plants with those sanctioned by the International Code of Botanical Nomenclature, had two objections to phonetic transcriptions (Zhong 1932b, 27):

- Names that use phonetic transcriptions look meaningless to anyone except a botanist—who is likely to know the scientific Latin name anyway.
- The source language can be confusing. There might, for example, be an Indian fruit for which there is an English transcription from the original language, which is then used, in turn, as the phonetic basis for a Chinese rendition. It would be better to create or to use a Chinese translation for the plant or fruit and then to append the Latin botanical name for botanists.

The clear preference for adopting semantic translations is also because it is very easy in Chinese to use two or more characters to create a new word. There are no inflections or other syntactic markers in Chinese grammar, so any two or more characters can form a compound representing any grammatical form. While Chinese is often described as a monosyllabic language, this characterization really only applies—if at all—to the literary language. Studies show that 86% of modern Chinese words are polysyllabic, of which 74% are disyllabic, constructed of two characters (Masini 1993, 121). Most compounds consist of one morpheme or character somewhat general in meaning, the other (usually the first of the pair) is more specific or descriptive. Thus the camellia is *chahua*, literally "the class known as tea (*cha*) within the broader category of flowers (*hua*)."

Botany in China did not emerge from a void. The body of traditional knowledge covers nomenclature as well as plant parts and other information, such as habitat, propagation, horticulture, and breeding. Du Yaquan devoted a large part of his preface to the 1918 dictionary of botany to the complexities of navigating terminologies, pointing out that, with many different terms in use and with regional variants and changes over time, it was difficult to pick one from the many as an entry in the dictionary. The challenge was not made any easier by the rapid evolution of the field of botany in the West, opening new areas of research for which new terminologies were being coined in English, French, and German, the major Western languages in which science was conducted. It was only in the early nineteenth century that botany itself had come into existence in Europe as a science distinct from natural philosophy or natural history (Endersby 2008; Jardine,

Secord, and Spary 1996). By the late nineteenth century, it was already branching into new subdisciplines: Darwin's *Origin of Species* was published in 1859, opening the way for evolutionary botany; in 1895, Johannes Warming published *The Oecology of Plants*, launching the discipline of plant ecology; and Gregor Mendel's discoveries in plant genetics only became a part of mainstream botany at the turn of the nineteenth and twentieth centuries. Each of these had its own technical vocabulary, as did other new specialties, all of which had to be rendered into Chinese, at a time when translations and original works in Chinese were just beginning to bring the basic concepts of scientific botany to China.

In a study of early terminology relating to the sexual parts of flowers, Georges Métailié warns that we must be aware of the "new significance" of old words. Classical works on plants had words to describe flower parts, many of which have been adopted as the botanical terminology for plant anatomy. Closer study shows, though, that in premodern works, the words referred to parts that are located in more or less the right part of the flower, but they do not correspond exactly to the structures identified today in botanical plant anatomy. The example Métailié uses is the word *rui*, used today as the second character in a binomial word for the pistil when combined with the word *ci* (female), and for the stamen when combined with the word *xiong* (male)—hence *cirui* and *xiongrui*, respectively. Translators of classical texts tend to assume that *rui* means pistil or anther, depending on the context. However, tracing the use of the word *rui* in a number of texts, Métailié concludes that it referred to a yellow part near the center of the flower, which might be what we now call the anther or the pistil but in some cases might refer to something more impressionistic, like "yellow dots in the central structure of the flower." He warns against anachronism in translation, concerned that if the translator or commentator uses the modern scientific name for the term in literary Chinese, it risks giving a false appearance of early scientific knowledge (Métailié 1994, 226, 227).

Zhiwuxue, the translation of Lindley's *Elements of Botany*, is generally acknowledged to have paved the way for a new botanical terminology. Although none of the three translators was a botanist and it is wrapped in a strong element of natural theology, presenting the wonders of the plant world as a reflection of a divine creator, it introduced some basic concepts and terminology that have served as foundations for the language of botany until the present.[18]

Li Shanlan and his colleagues drew on existing terms from classical works about plants, giving them a specific botanical usage, often adding an explanation of what distinguishes the botanical term from the traditional word. An example was the use of the existing word *ke* (category or class) to refer specifically to botanical families in the Linnaean system. They drew on the past to generate the vocabulary for whole fields of study, such as plant anatomy, with terms such as *xibao* (literally "fine bladder," or "womb") for the cell, and with words for the reproductive anatomy of plants, such as *zifang* ("seed room") for the ovary and for the male and female inflorescence (*xiong hua* and *ci hua*, respectively). Many of their neologisms ultimately did not become the accepted terms, but their approach to constructing a new vocabulary became the model for much of what followed.[19]

John Fryer himself translated one of the later works on botany, John Balfour's *Manual of Botany*. Published in 1895 by the Educational Association of China as *Zhiwu tushuo* (Illustrated botany), it was a textbook with 154 illustrations, including labeled images of plant parts, and an explanatory text for each. Li Shanlan, the scholar, was more able to reach back to find existing terms from the classics and historical works than Fryer, however, making it easier for educated readers to relate the terms he constructed to familiar literary sources and to accept the new learning. Fryer's contribution was less in the creation of a scientific vocabulary than in his skill in

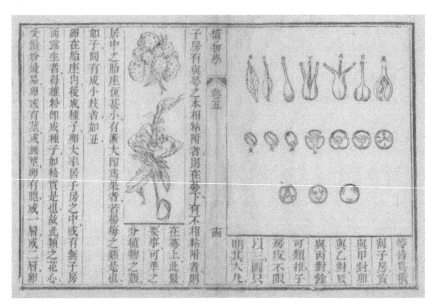

FIGURE 4.2   The opening section of the chapter on the ovary and seed formation in flowers in *Botany* (Zhiwuxue), the translation of John Lindley's *Elements of Botany*, published by Inkstone Press, Shanghai, in 1858. Courtesy of the Australian National Library, Canberra. Bib. ID 1900830.

using concise yet accurate language to define and describe both morphology and function. He also covered considerably more ground in terms of structures, anatomy, and microbiology (Luo G. 1987, 385–86).

By the turn of the century, the nature of scientific translation changed as Chinese students first in Japan and then in Europe and the United States became involved, with a growing readiness to adopt terminology from Japan. In 1903, the Literary Academy Society (Huiwen Xueshe) in Shanghai published *The Compiled and Translated Encyclopedia for General Education* (Bianyi putong jiaoyu baike quan shu), edited by Fan Diji (Doležolová-Velingerová 2014, 289–328). Part 30 was a translation of a Japanese textbook on botany covering the structure, morphology, physiology, and classification of plants (Métailié 2002, 210). The *New Literary Expositor* (Xin er ya), also published in Shanghai in 1903, contained a twelve-page section on botany, *On Plants* (Shi zhiwu) (Wang and Ye 1903, 165–76). In addition to a vocabulary that was already familiar by then, the two works included many neologisms that had been coined in Japan using Chinese characters and were therefore easily adopted in Chinese, many of which are now the accepted botanical terms.[20]

It is not surprising that Japanese neologisms were so readily adopted. In Japan, the introduction of Western science had been a deliberative process in which scholars had formed specialized committees to translate important reference works, collaborating to reach agreement on developing a technical vocabulary (Alleton 2001, 29). The more learned approach taken by Li Shanlan and Yan Fu, who were not scientists themselves, sometimes struggled under the weight of their erudition as they reached into rarefied areas of scholarship—even drawing on classical Greek or Sanskrit etymology—to formulate terminology that Japanese translators expressed in easily understood binomials using relatively common kanji or Chinese characters. There are ongoing, sometimes contentious scholarly debates about the Chinese or Japanese origins of scientific neologisms, but the current view suggests a process of crossing over several times from one language to the other, with the meaning changing subtly with each crossing.[21]

The first translations and compilations of botanical works introduced principles of botany, plant physiology, and anatomy but devoted little time to taxonomy, nomenclature, and classification. Only the final volume of *Botany* covered taxonomy, for which Li Shanlan and his colleagues coined the term *fen ke* ("distinguishing families").[22] The section began by explaining that it would only cover a selection of the 303 known families. It went on to describe thirty-seven families, nearly all of which used existing Chinese plant names, adding the suffix *ke*, to effect the transformation from a vernacular plant name to a botanical family (Li, Williamson, and Edkins 1858, juan 8). Only six of the thirty-seven families were not derived from existing names of plants. The very first family in the section is one of three semantic adaptations translating the meaning of the botanical Latin name, in this case the Umbelliferae, as *sanxingke* ("umbrella shape family"). Two use common words to create families that do not appear in older Chinese texts: *xiuqiuke* ("embroidered ball family") is used, for example, for Hydrangeaceae—an evocative description of the hydrangea flower. The term is still used today for the genus *Hydrangea*. Only one name, *danbaguke* (tobacco) is a phonemic loan. The name was ultimately not adopted, and the botanical name for *Nicotiana tabacum* is now *yancao*, the word used for tobacco in modern Standard Chinese. This methodology was their most creative and, in the long term, perhaps most significant innovation, paving the way for a comprehensive approach to botanical nomenclature, even when identifying and classifying new and unknown families, genera, and species. Métailié (1981, 68) notes that of the thirty-seven names of families in *Botany*, eighteen are still used today. Many of those that are no longer used have been lost not because they were inappropriate or rejected by botanists but because of changing nomenclature due to advances in botanical taxonomy.

It fell to the Science Society of China to develop a systematic approach to the standardization of scientific terminology. In the first issue of *Science*, the society pledged "to take the rectification of names as our basic task," using only terms that had been reviewed and approved by a specialist in the relevant discipline (*Kexue* 1915b, 2). Acting on its promise, the society organized its members into disciplinary committees to work on standardization. Articles 12–19 of its charter set out the procedure to follow:

- Assemble the technical terms that need translating. Allocate them to the relevant committee.
- Each committee then proceeds to develop translations.
- Each committee forwards its proposals to the translation committee.
- Anyone writing in the journal must use the approved translations regardless of whether they are members of the society or not.

This protocol ensured that the translations were scientifically accurate and that they would be quickly adopted for widespread use (Wang Zuoyue 2002, 300–309; Wen 2005). Before finally accepting a new term, readers were invited to comment in the Terminology Forum (Mingci Luntan) that appeared occasionally in the journal. Some comments were quite general, supporting, for example, the editors' commitment to use existing characters rather than to create new characters. Others delved in detail into issues such as the debate on the compound words to be used for stamen and pistil, which occupied the forum in issues 3 and 8 of *Science* in 1917.

This was a time when the international botanical community as a whole was struggling to agree on rules governing nomenclature. In 1905, the second International Botanical Congress, held in Vienna, adopted a code known as the International Rules of Botanical Nomenclature, modified five years later in Brussels.[23] The Science Society of China took the lead in ensuring that nomenclature in China would converge with international norms. In 1916 *Science* introduced the principles of botanical nomenclature with a translation of key sections of the rules as modified in Brussels. In the article, Zou Bingwen (1893–1985), who had returned to China in 1915 with a degree in agricultural engineering from Cornell, explained the binomial system and provided translations for the terms to be used for all levels of the taxonomic hierarchy, from the kingdom to the individual. He also discussed the value of Latin as a means of communication across languages and the rationale for rules such as attribution to the first published author, the need for a drawing, and a description in Latin (Zou B. 1916). To illustrate the principles of the code as applied to Chinese flora, the December issue followed with a list of eighty-one botanical plant names using the Latin names with their Chinese equivalent (Wang Yanzu 1916).[24]

The example of *cha hua* (camellia)—a compound name with a long history—is a reminder that historically, there had been procedures in China for constructing plant names and placing them in some form of hierarchy that established relationships between categories of plants. Chapter 6 will look at how botanists went about ordering and organizing the plant life around them, referring to or consciously rejecting existing classifications and groupings. In the past, scholars had been concerned with correctly naming and ordering known plants. In the transition from traditional to botanical nomenclature of plants, a more difficult issue for Chinese botanists seems to have been the question of how to map existing plant names onto the Linnaean system.

A vital tool facilitating the consolidation of botanical terminology was the *Dictionary of Botany* (Zhiwuxue da cidian), which appeared in 1918. It was the outcome of over a decade of work by a committee of thirteen, including students who had returned from Japan, working under Du Yaquan, the editor-in-chief (Du 1918). The dictionary had 1,590 pages, with 1,002 illustrations and descriptions of 1,700 species. Each entry gave the Chinese, Latin, and Japanese names, and in some cases, a German name, too. Entries listed family, genus, and species following the International Code of Nomenclature, with some physiological information, notes about uses of the plant, and historical references to works such as Li Shizhen's *Classification of Materia Medica*. The dictionary was well received, going through several reprints over the following twenty years or so.[25]

The year before its publication, an article in *Eastern Miscellany*, probably penned by Du, set out the principles the committee had followed in compiling the dictionary. The first was to recognize that there were two components in the new botanical lexicon: words and language from China and their equivalent in Eastern (Japanese) and Western languages. Other guidelines explained that botanical names in Chinese, English, German, Japanese, and Latin were all included, where known, for each plant, together with an illustration, so as to ensure that the user of the dictionary had all the information needed to correctly identify a plant they were looking at. The committee stated that their definitions were based on current scholarship but took care to remind readers of the legacy of the past: "The body of the entries uses material from the most recent and proven body of knowledge from botanists. It is acceptable to supplement the entry with a modest number of classical references" (*Dongfang zazhi* 1917, 16).

The publication of the *Dictionary of Botany* was a major step in the process of developing terminology and nomenclature and linking them formally to the International Rules of Botanical Nomenclature. Twelve years later, in 1930, an official delegation of five Chinese botanists took part in the Fifth International Botanical Congress in Cambridge. Chen Huanyong (referred to in the proceedings as Chun Woon-Young) of Sun Yat-sen University was the first Chinese botanist to deliver a paper to a panel at the congress, and he was elected to the Permanent Bureau of Nomenclature. For the first time, there was a Chinese voice in the deliberations that named newly published species (Brooks and Chipp 1931; Qin R. 1931). The momentum had moved from creating new vocabulary for the science of botany to identifying China's flora, classifying it, and correctly assigning Chinese names that conformed to international nomenclature.

CHAPTER FIVE

Observing Nature, Practicing Science

> Over the last hundred years, Yunnan has been the playground of botanists and horticulturists from many countries in Europe and America.... But I am surely not alone in feeling ashamed that while Westerners have left their footprints in the most remote parts of this region, you hardly hear anything about Chinese collecting expeditions.
>
> —HU XIANSU, EDITORIAL ON THE FIRST ISSUE OF THIS JOURNAL ("FAKAN CI")

HU XIANSU wrote these words in 1941. The war against Japan was at its most intense, with a large part of the academic and scientific community in exile in Chongqing or Kunming. He wanted to remind his readers that even under these dire circumstances, Chinese botanists were active, conducting field research where previously the world had heard only about the adventures of Western plant hunters. Over the last ten years in Yunnan alone, his assistants and students had undertaken numerous collecting expeditions during which they had "hunted down and collected more specimens of plants and seeds than anywhere else in the world" (Hu X. 1941, 1).

In the field sciences, the collector travels through the landscape to observe and to describe. Observation and description are at the heart of botany and geology, two field sciences that have played an especially important role in the formation of modern states through their mapping of the nation's spaces and their inventories of its natural resources (Scott 1998; Fan F. 2007, 529).[1] In a classic study, Suzanne Zeller argues that during the nineteenth century, Canadians used botany and the natural sciences to construct a distinctive national identity (Zeller 1987, 181–268). In her study of Ding Wenjiang, one of China's first geologists, Charlotte Furth (1970) shows how geology was deployed to map the country's resources as China struggled to recreate itself as a modern nation. In contrast to the Canadian project to create a nation, though, mapping China's resources might be described as an effort to recover them after decades of incursions by foreign powers.

Introducing botany to the readers of *Chinese Students' World* (Zhonghua xuesheng jie) in 1915, Wu Jiaxu wrote four articles about collecting plants, preparing and mounting voucher specimens, and recording botanical data (Wu Jiaxu 1915a, 1915b, 1915c, 1915d), activities that best exemplify what field botanists do. Debates on how scientific knowledge is generated and how it enters into public discourse distinguish between the sciences that find their facts and the sciences that do not find their facts but create them de novo. Field naturalists find facts as they observe objects while traveling, often through new and unfamiliar territory. The laboratory scientist, on the other hand, selects objects to observe them at length in the closed and controlled environment of the laboratory (Outram 1996, 259–62; Kohler 2002, 192–95 and 2017, 450).

The distinction between field and laboratory sciences may be a constructed binary, while in practice there is a spectrum from the open spaces of one to the enclosed confines of the other. The difference between the two ways of observing has resonance, though, in China, where scholars of the past started with selected objects (texts), which they studied in depth in their libraries in order to understand the world and the myriad things. To derive knowledge from firsthand observation of the myriad things in the open air was a new and distinctly different practice (Shen G. Y. 2009). Wu Jiaxu made sure that his readers understood that it was collecting in the field that made botany different from earlier scholars' studies of plants: "Collecting is the first step in mastering how to stick to the facts. Collecting is something that can have a profound impact, and it is the methodology of collecting that ensures that the knowledge obtained from research will always be trustworthy, not misleading.... To transmit knowledge while collecting will definitely be far superior to any amount of lecturing in a classroom" (Wu Jiaxu 1915b, 1).

During the nineteenth century, the professionalization of science in Europe and the Americas was marked by a parting of the ways between the field naturalist and the laboratory scientist. Formerly, the same person had observed, collected, and classified. Increasingly these activities diverged, with one person observing and collecting in the field and another person in a study or laboratory examining, classifying, and documenting the collected specimens. Building a collection involved networks of actors, from the collectors in remote field sites to the scientists or academicians in central institutions in the colonial metropole, London, Madrid, or Paris.[2] Sir Joseph Banks, president of the Royal Society from 1778 to 1820, and Sir Joseph Hooker, the director of Kew Gardens from 1865 to 1885, both enlisted ships' captains, travelers, colonial officers, and settlers to collect specimens, which they sent back to London for identification and safekeeping in herbaria. Their correspondence and institutional records document the sometimes-tense relations with their collectors, who not unreasonably often felt that they were more knowledgeable about the local flora and deserved more credit for their work in isolated and difficult settings.[3]

The divergence in China between the literati scholars' experience of nature and the early botanists' did not follow the same trajectory as the evolution that had taken place in Europe and the Americas. While some of the first Chinese botanists had been candidates in the imperial examination system, when they turned to the new learning, their practices of observation and information gathering to build knowledge about plants were quite different from those of scholars before them. For the first part of the twentieth century, at least, the field collector was the same person as the laboratory scientist, and there was no lineage of practitioners leading from the literatus of the past with an interest in the plant world to the modern scientist finding, identifying, and classifying specimens.

DESCRIBING THE FLORA OF IMPERIAL CHINA

A table in *Plants of China*, the companion to the referential *Flora of China*, names 111 botanists, 41 of whom were born before 1920, who are said to have been the founders of botany in China (Hong and Blackmore 2015, 213–14). These botanists may not have seen themselves as the latest in a direct line from the past, but there had been a long premodern history of studying the natural world. Several Chinese scholars have been celebrated as precursors of today's natural scientists. They include writers better known as essayists, as well as the more obvious candidates for the title: Li Shizhen, compiler of *Classification of Materia Medica*; and Wu Qijun, author of *Research on the Illustrations, Realities, and Names of Plants*.

Shen Gua (1031–1095) was a scholar-official of the Song dynasty, known for his *Dream Pool Essays* (Mengxi bitan) of 1088. The collection consists of short essays on phenomena he encountered in daily life and on his travels. They cover archeology, botany, geography, hydraulics, meteorology, paleontology, and more. He has been admired for his method of inquiry, which was to record an observation, then to collect and sift through "facts" about it from a diversity of sources, including hearsay, texts, and interviews, to find explanations for what he had seen. To some historians, Shen's dedication to empirical inquiry and his careful evaluation of evidence have earned him a place in the pantheon of early "protoscientists." Others, while acknowledging his distinct voice and insights, argue that he did not organize his essays around a conceptual core that would build a coherent body of knowledge from his many, often unrelated observations (Brenier et al. 1989, 345; Zuo Y. 2018).

Several centuries later, in 1608, Xie Zhaozhe (1567–1624) published *The Fivefold Miscellany* (Wu za zu) with a similar mix of anecdotes and commentary on observations of the world about him. It has also been cited as evidence of early scientific analysis and writing, again following the rationale that the essays demonstrate patterns of inquiry and reasoning comparable to the scientific method (Xu Guangtai 2011). Mark Elvin has compared Xie's work to that of his English near-contemporary, the Reverend Gilbert White, author of *The Natural History of Selborne*. Elvin identifies some commonalities in the two men's interests and investigative methods. He concludes, though, that "Xie's lively but volatile curiosity in a certain sense lacked investigative stamina because of his deep attraction to the idea that the ultimate intellectual purpose of life was less to *understand* the cosmos than to attain a state of mind in which one *merged* into it" (Elvin 2014, 33).

Georges Métailié builds a strong case that the Qing scholar Cheng Yaotian (1725–1814) was developing an incipient botany based on textual and empirical research (Métailié 2012). In 1804, Cheng published *Notes on All the Arts* (Tong yi lu), which includes two treatises on plants, one on grains, and one on grasses, in which he hoped to resolve questions regarding the names of plants found in the classics. He extended his investigations to inquire into the nomenclature of crop varieties as well as the names of plant parts, and he relied on direct observation and experimentation when faced with conflicting accounts of plant phenology or appearance. These treatises were still in the mainstream of scholarship that sought to rectify errors in understanding the past, but his methods brought his work closer to what would now be called botanical observation. Comparing Cheng's work to Li Shizhen's, Métailié concludes that while both scholars made critical use of existing texts, the fundamental difference between them is that "the contemporary evidence Li Shizhen sometimes reported was generally hearsay or borrowed from other books, whereas Cheng Yaotian always adopted a concrete and empirical attitude and brought firsthand remarks [into his work]" (Métailié 2012, 254).

In the West, there were some published sources on the flora of China based on direct observations in the field that predated the forced opening of the interior after the Opium Wars. During his nine years in China (1643–52), the Jesuit father Michał Piotr Boym traveled to Hainan, where he sketched plants and took notes on their pharmacological uses. The plants in his *Flora sinensis* are mostly tropical fruit he would have come across on the island, such as the papaya, cinnamon, and pepper, although it also includes species from other regions, such as the lychee, the loquat, and ginger (Boym 1656; Szczesniak 1955) (see chapter 2).

Before the decree confining foreign maritime trade with China to Canton, the few European trading vessels to visit the southern Chinese coast had sent plants that they found to gardens and botanists in their home country. These specimens sparked interest in the flora of China, but they were fortuitous discoveries during brief stops in ports of call, not the result of any concerted collecting effort. James Cuninghame was almost certainly the first to carry out systematic fieldwork, albeit over very limited areas in the immediate vicinity of the harbors where the ships on which he sailed during two voyages between 1697 and 1709 had stopped to trade. The specimens and paintings of plants he sent were of such interest to the scientific community that he was elected to the Royal Society in 1699 while in London between his two journeys to China.[4]

The first annotated collection of plants to be sent to Europe came from the Jesuit mission in Beijing. Father Pierre d'Incarville (1706–1757), a student of the French botanist Bernard de Jussieu, collected in and around the capital, and sent a total of 149 specimens back to the Muséum National d'Histoire Naturelle in Paris. D'Incarville prepared each plant, attaching a label with its name in French or Latin where known, as well as in Chinese; recording its provenance; and distinguishing those that had been found in the city from those found in the surrounding hills (Franchet 1882).[5]

Field collecting was not possible under the Canton trading system. If anything, though, the mystery of the forbidden only heightened interest in the flora of China. The East India Company agreed to house residents at its Canton factory, who sent any interesting specimens they found in the city back to Kew Gardens in London (Le Rougetel 1982; Fan F. 2003a). In Uppsala, Linnaeus himself received duplicates of some of these, as well as specimens sent by his students, who became known as his "apostles," attached to the Swedish East India Trading Company, also in Canton (Nyberg 2009; Skott 2014).[6]

This was a time of increasing commercialization of plant material in the West, driven by the search for economic crops that might be introduced to the colonies and a boom in gardening that made seeds and cuttings of new and showy ornamentals valuable commodities.[7] Seed companies and nurseries were quick to take advantage of the articles of the Treaty of Nanking that opened China to foreign trade and to others, such as plant collectors and botanists. In February 1843, the Royal Horticultural Society sent Robert Fortune to search for garden plants. His later efforts on behalf of the East India Company to bring tea bushes and seeds to British plantations in India became the stuff of legend (Fortune 1847, 1852, 1857, 1863; Ellis, Coulton, and Mauger 2015, 93–114). Fortune and the plant hunters after him were lionized in the press as heroic adventurers on a quest to discover new treasures in the wilds of China.[8]

Hong and Blackmore (2015, 215–23) have identified fifty-one foreign collectors who were significant players in the botanical exploration of China between 1842 and 1949.[9] In addition to these professionals, Kilpatrick (2014) also writes about fourteen French Jesuits, four Franciscans, and one German Protestant missionary who collected and sent specimens back to Europe from their remote parishes and their pastoral travels. Chinese scholars estimate that over the two centuries from 1750 to 1950, foreign plant hunters and botanists collected approximately one million plant specimens and over one thousand varieties of seedlings and seeds, describing and naming over 10,000 species and 158 new genera (Wang Zhenru, Liang, and Wang 1994, 158).

The one-way flow of knowledge affected Chinese researchers directly. Early Chinese field botanists struggled to determine whether the plants they had collected were newly identified species or whether they were, in fact, examples of known species whose type specimen might be located at Harvard or in Vienna, London, or Paris. In a study of vouchers collected in China and housed at Kew Gardens, the historian Lu Di found a total of 11,653 specimens, the vast majority of which were collected between 1840 and 1940 and were labeled with the name of a Western collector. The labels on a small number of vouchers have no name, reading "Chinese Collector" or "Native Collector" or simply "Native" (Lu D. 2014).

In 1934, during a collecting expedition in Yunnan, Wu Zhonglun arrived in Tengyüe (now Tengchong), where the British plant hunter George Forrest had died of a cardiac arrest in 1932. He hired Wang Hanchen, who had worked for Forrest collecting rhododendron. Wu appreciated his good fortune in finding a trained collector, but he noted in his diary on August 26 that all the specimens Wang had so carefully collected had been sent directly to Kew Gardens through the British consulate in Tengyüe. "Most rhododendron come from China, especially from Yunnan, where they are especially abundant. But if anyone wants to study the plants from the place where they are most prolific, they have to go to England to see the type specimens" (Jin T. 1999, 186).

There appears to have been little or no direct interaction between Chinese botanists and the Western plant hunters. Sometimes Chinese and Western collectors may have been working in the same place at almost the same time, but there is no evidence that they met or were even aware of their parallel expeditions. In 1934, Sun Yat-sen University sent a young botanist, Zuo Jinglie, to collect on Hainan Island. On his return, Zuo published a report on his experiences rich with botanical and ethnographic information (Zuo J. 1934). A year later, Linley Gressitt of the University of California at Berkeley, who was associated with Lingnan University, also in Guangzhou, spent three months collecting on the island. In his report, published in 1936, Gressitt never mentions having any contacts with or even hearing of a Chinese group before or after his visit, even though he passed through several of the same places as Zuo (Gressitt 1936).

Despite the frustration of not being able to study specimens that were kept in distant herbaria, Chinese botanists admired the plant hunters' willingness to face difficult and dangerous conditions and recognized the value of the work that some of them had done. An obituary of Augustine Henry and Ernest Henry Wilson written in 1931 concluded: "Wilson was a friend of China. Other plant hunters were dismissive about China and the Chinese and said bad things about our customs. Wilson was different. He insisted that everyone has different customs, that some are good and some are bad, and so it is with China. He was very different from those who constantly use terms like 'ignorance' and 'naive'" (Qiao 1931, 171).

WHO WERE THE BOTANISTS?

The first Chinese field collector was probably Huang Yiren, who is believed to have been the first Chinese student to study botany abroad, with Matsuda Sadahisa at Tokyo University. Huang collected in Jiangsu in 1905, sending specimens of some 270 species back to Japan, where Matsuda published them, thus getting the credit for the first modern flora of China.[10] Huang himself was one of the editors of the 1918 *Dictionary of Botany*, before being appointed professor of agronomy at National Peking University (now Peking University) (Wang Zhenru, Liang, and Wang 1994, 157).

Chinese and Western collectors were venturing into rarely visited parts of the country. The guides, porters, and the assistants they hired shared knowledge about the local plants and wildlife that was critical to their efforts. The botanists learned from healers, elders, and others about where to find plants; their seasons of growth, range, flowering, and fruiting; and their uses as medicine or food (Dasgupta 2014). The absence in the plant hunters' and botanists' diaries and reports of any recognition of their assistants' part in the work is so striking that it has taken heroic efforts of archival research and reconstruction to bring even one of them, Zhao Chengzhang, to life and give him the credit that is his due (Mueggler 2005, 2011). There are very occasional references to these informally trained collectors, as when Wu Zhonglun hired Wang Hanchen in Tengyüe, but they mostly remain invisible—the "Native Collector" on the label on the voucher in the herbarium at Kew Gardens.

The biographical sketches that follow are of four botanists whose careers represent the transition from traditional to scientific over the first half of the twentieth century. Zhong Guanguang began studying for the imperial civil service exams, then opted out and taught himself Western learning and botany. Hu Xiansu was classically trained, then studied abroad. He retained a deep respect for his early education, writing in classical, literary Chinese to the end of his life. Chen Huanyong was born in Hong Kong and educated in Shanghai, then the United States. He used his international connections and fluency in several languages to bring China into the worldwide botanical community. Cai Xitao had no experience of the imperial order. He was born and lived in the twentieth century, learned botany on the job, and loved the adventure of fieldwork in Yunnan. Fittingly, surviving pictures of them often show Zhong and Hu wearing the traditional scholar's gown. I have yet to find one of either Chen or Cai in traditional wear.

Zhong Guanguang, 1868–1940

Zhong Guanguang was born in September 1868 in Yao Jiang An village, Zhejiang.[11] He took the civil service examination in 1887, earning the *xiucai* degree. He did not pursue an official career though, and studied Western science at a local school, then moved to Shanghai. In 1901, he was one of the founders of the Shanghai Scientific Instruments Factory (Shanghai Kexue Yiqi Guan), setting up a small plant producing sulfur. It closed after just six months, and Zhong went to Japan for a brief study tour (Yu and Chen 1985, 3; Wang X. 2013, 100). On returning to Shanghai, he established a science study group, which attracted a wide range of people, including Cai Yuanpei

(1868–1940), later to be the first minister of education after the 1911 Revolution, then president of the Government University of Peking (now Peking University) and the first head of Academia Sinica. In 1903, Zhong founded the journal *Science World* (Kexue shijie) together with Du Yaquan and Yu Heqin, both from the Scientific Instruments Factory (Jiang W. 1941, 58; Wang X. 2013, 100).[12]

The turning point in Zhong's career came in 1908. While in Hangzhou convalescing from an illness, he read Li Shanlan's translation of *Elements of Botany*, which encouraged him to start collecting specimens, beginning a lifetime commitment to botany (Jiang W. 1941, 57; Chen Jinzheng and Zhong 2008).

In 1911, Cai appointed Zhong to the post of chargé d'affaires in the Ministry of Education in Nanjing. On weekends and holidays, they would go botanizing together in the hills around the city, a pleasure they continued to pursue when the ministry moved to Beijing. In 1913, with the rise of the warlords, Cai resigned and Zhong returned to the south to accept an appointment in 1915 as assistant professor of sciences at Hunan Normal College (Jiang W. 1941, 58–59).

FIGURE 5.1   Zhong Guanguang (center) and his son Zhong Buqian (right) with an unknown assistant during fieldwork in Hunan, 1915. Courtesy of Ningbo Botanical Garden, China.

When Cai was appointed president of the Government University of Peking in 1918, he offered Zhong a position as assistant professor of biology. Zhong set up a herbarium on the campus, then wasted no time in organizing a collecting expedition to Fujian, Guangdong, and Guangxi. In 1919, he continued by steamer to Vietnam, then by train to Yunnan. While in Guangzhou, he was invited to give a lecture by George Weidman Groff, an American agronomist who was the head of the Department of Agriculture at Lingnan University. Groff introduced Zhong to E. D. Merrill, then director of the Manila Botanical Garden in the Philippines, with whom he continued to correspond for many years.[13] In 1932, when Merrill identified an unknown species Zhong had sent him, he named it *Tsoongia axillariflora* Merrill, giving them both credit in naming a new genus and species (Zhu and Liang 2005, 826; Yu and Chen 1985, 5).

Zhong carried out several more collecting expeditions between 1920 and 1930 (Li Hui-lin 1944, 26). In 1927, while teaching at the College of Agriculture of the Third National Sun Yat-sen University (Guoli Di San Zhongshan Daxue) (now Zhejiang University) in Hangzhou, he established a garden for teaching purposes, considered to be the first botanical garden in China.[14] In 1930, Zhong was appointed to the scientific nomenclature committee of Academia Sinica, where he combined his early literary studies with his botanical expertise in a project to identify the plants mentioned in the classics and to incorporate new information about growth, habitat, and other botanical data.[15] After the Japanese invasion in 1937, he returned to Zhejiang, where he continued working on this project until his death on September 30, 1940.

Hong and Blackmore consider Zhong Guanguang to have been the first Chinese botanist to engage in systematic plant collecting, and the collections he made in Fujian and Guangdong in 1918 are said to have constituted the beginnings of what has become *Flora sinica* (Hong and Blackmore 2015, 223; Lu D. 2014, 57)

Hu Xiansu, 1894–1968

Hu Xiansu (also known as Hu Buzeng and Hu Hsen-Hsu in English-language publications) was born in Xinjian, Jiangxi, on May 24, 1894. He grew up in an educated household, with a tutor who taught him the Confucian classics and poetry.[16] After the abolition of the civil service exam, he enrolled in the Imperial University of Peking in 1909. The university's mission was to introduce Western learning while retaining the core of Chinese values through the study of the classics, a philosophy that Hu followed through his whole life (Jiang L. 2016, 163).

In 1912, Hu left to study at the University of California at Berkeley, graduating from the College of Agriculture in 1916 with a BA in botany. While there, he became one of the founding members of the Science Society of China and even contributed some of his own money to help found the society's journal, *Science*.

After returning to China, Hu taught botany from 1918 to 1922 at the Nanjing Teachers' College. With the support of Ernest Henry Wilson at Harvard's Arnold Arboretum, he organized a collecting expedition in Zhejiang and Anhui in 1920. An account of his travels, written in a classical style and including several poems, appeared in the *Critical Review* (Xueheng), sometimes known as "the flagship journal for traditionalists" (Hu X. 1922; Jiang L. 2016, 172–73). Hu went back to the United States in 1923 to study dendrology at the Arnold Arboretum, graduating in 1925, the first Chinese student to receive a doctorate of science degree from Harvard. While there, he devoted time to transcribing and cataloging the herbarium's collection of Chinese plants, giving Chinese botanists access to otherwise inaccessible reference materials collected by Westerners (Yu D. 1985, 77).

Hu is revered as the father of modern botany in China not only because of his collecting and his contributions to taxonomy but also in recognition of his role in building the institutions that trained and supported young scientists. Hu and the zoologist Bing Zhi were founding members of Academia Sinica in 1927, and in 1928, they established the Fan Memorial Institute of Biology in Beiping. In 1934, Hu was one of three founders of the Lushan Arboretum and Botanical Garden (Lushan Senlin Zhiwuyuan), China's first botanical garden outside a university campus.[17]

Following the Japanese occupation of Beiping in 1937, Hu entered into discussions with the Yunnan provincial authorities to facilitate the transfer there of the Fan Memorial Institute, if necessary (Hu Z. 2001, 239). They responded positively, and so the Yunnan Institute of Agriculture, Forestry, and Botany (Yunnan Nonglin Zhiwu Yanjiusuo) opened in 1938. Hu was appointed director in 1940 but left later that year to become president of the National Chung Cheng University in Jiangxi (Hu Z. 2001, 244).

After 1949, Hu was among those who supported bringing the Fan Memorial Institute into the Chinese Academy of Sciences, a fateful move (Hu Z. 2013b, 41). In the 1950s, the Ministry of Education accused him of questioning Soviet influence in scientific research, leading to public criticism by the Chinese Academy of Sciences and in the *People's Daily*. He was partially rehabilitated in 1962 but attacked again during the Cultural Revolution from 1966 until his death from a heart attack on June 15, 1968. Hu

FIGURE 5.2   Hu Xiansu (seated second from right) with leaders of the Fan Memorial Institute of Biology in front of the institute in Beiping, 1928. Courtesy of Hu Zonggang, Lushan Botanical Garden, Jiangxi.

was cleared posthumously of all charges in 1979, and in 1983, his ashes were buried at the Lushan Botanical Garden (Luo Z. 1981).

Chen Huanyong, 1890–1971

Chen Huanyong was born in Hong Kong.[18] His father was a diplomat who met his Cuban wife while serving there as the Chinese consul-general (Hu Z. 2013a, 6). When he was seven, the family moved to Shanghai, and when his father died in 1905, an American friend of the family took him and his older sister to America. On graduating from high school in Seattle in 1909, he enrolled at the Massachusetts Agricultural College, then transferred to the New York State School of Forestry in 1912, going on in 1915 to study dendrology at the Arnold Arboretum (Haas 1988b, 10; Wu D. 2008, 1).

In 1919, Chen received a fellowship to organize a collecting expedition in southern China. Charles Sprague Sargent, the director of the Arboretum, gave him an introduction to J. Tutcher, the superintendent of the Hong Kong Botanical Gardens, who helped him prepare to spend a year on the island of Hainan. Chen set sail in October 1919, but in July 1920, he was bitten by a venomous insect, shortly after which he also came down with

FIGURE 5.3   Dr. Chun Woon-Young (Chen Huanyong) in Nada, Hainan, during his collecting expedition in 1919. Courtesy of the South China Botanical Garden, Guangzhou.

malaria and had to cut short his time on the island. While he was convalescing in Shanghai, the specimens he had packed for shipment back to Boston burned in a fire on the docks.[19]

Chen spent the next seven years (1920–27) teaching in Nanjing, first at the University of Nanjing, and then at National Southeastern University (Guoli Dongnan Daxue). In 1922, *Chinese Economic Trees*, Chen's first major publication in English, appeared, and he organized a collecting expedition in Hubei with recent Harvard graduates Qian Chongshu and Qin Renchang, together with Huang Zong, a colleague from the university; and "Old Yao," a guide who had collected with Augustine Henry several decades earlier. They sent some of their specimens back to Harvard, but in a repetition of what had happened in Shanghai, most of what they had kept in China burned in a fire at the end of 1923 (Haas 1988b, 13–17).

In 1927, Chen left to study the Chinese plants at the Hong Kong Botanical Gardens. He moved to Guangzhou in 1928 to serve as director of a new Institute of Botany (Haas 1988b, 17). During the first ten years of his tenure there, Chen mobilized his network of international contacts to help build the library and herbarium, initiating exchanges with some sixty countries and obtaining more than 30,000 foreign specimens (Wu D. 2008, 3; Huang R. 2016, 774). In 1930, he launched *Sunyatsenia*, an English-language journal to promote exchanges with the international botanical community.[20] In 1930, he led the Chinese delegation to the International Botanical Congress in Cambridge, where his paper, "Recent Developments in Systematic Botany in China," was the first by a Chinese botanist to be presented to the congress (Brooks and Chipp 1931, 524; Qin R. 1931) (see chapter 4).

When Guangzhou fell to the Japanese in 1938, Chen escaped to Hong Kong with the institute's herbarium and library. When Hong Kong fell on January 1, 1941, he returned to Guangzhou, where he accepted an appointment at the puppet régime's Kwangtung University. After the Japanese surrender, he was accused of collaboration and related charges. Although he was ultimately acquitted, the episode returned to haunt him during the Cultural Revolution.[21]

Chen's institute was able to remain independent until 1954, when it became the South China Institute of Botany under the Chinese Academy of Sciences (Chen F., Li, and Huang 1984; Wu D. 2008, 4). Although he was able to keep his distance from many of the political movements of the first decades of the People's Republic, he came under attack during the Cultural Revolution. His health suffered, and he died in January 1971.

Cai Xitao, 1911–1981

Cai Xitao became a botanist almost by accident and set off on his longest field trip because he loved the idea of adventure and "the tigers, the wild animals, and the small Yunnan horses and horse caravans that only existed in Yunnan" (Cai 1980, 2). He extended botanical exploration to the tropical fringes of southwest China, and he is remembered as a mentor who passed on his passion for discovery to his students.

Cai was born in Cai Zhai Zhen, Zhejiang on March 12, 1911. He wanted to study literature, but his family fell on hard times, so after one year at Guanghua University in Shanghai, he dropped out and went to Beiping to look for a job (Cai 1980, 1). A chance meeting with Hu Xiansu led to a position at the Fan Memorial Institute. Hu then sent him to learn his job on what proved to be a successful collecting expedition in Hebei, from which he brought back over six hundred specimens for the institute's herbarium (Xu Z. 2011, 3; Hu Z. 2018, 31).

Cai had read some of the plant hunters' memoirs, and in 1932, he offered to go to Yunnan to collect in areas that even Westerners had not visited. Hu approved, so Cai set out in March 1932 with two volunteers from the institute. In the end, he spent nearly three years in Yunnan, returning to Beiping at the end of 1934 (Hu Z. 2018, 31–44). He was not only interested in plants but also intrigued by the lives of the ethnic minority communities where he worked and felt inspired to write about what he saw. He submitted short stories to the Shanghai journal *Literature* (Wenxue), which published several of them between 1933 and 1937, including the fable *The Dandelion*.[22] The editors thought so highly of the story's creative blending of science and literature that they made it the lead piece in that issue (Chen Siqing 1987).

While in Yunnan, Cai took time to speak with officials responsible for education and economic planning, explaining what he was doing and discussing possible future collaboration with the Fan Memorial Institute (Cai 1980, 2). It was largely thanks to his carefully nurtured contacts that the province approved Hu Xiansu's proposal to establish a Fan Memorial Institute in exile after the Japanese occupation of Beiping. It opened on July 24, 1938, at Heilong Tan in the suburbs of Kunming under the new name, the Yunnan Institute of Agriculture, Forestry, and Botany (Xu Z. 2011, 4–5).[23]

After the Japanese surrender in 1945, the colleges and institutions that had taken refuge in Yunnan and Sichuan drifted back to Beiping. When civil war broke out, the Fan Memorial Institute was no longer able to offer financial support, forcing Cai and the remaining staff to find creative ways

FIGURE 5.4 Professor Cai Xitao with a local guide in the Nujiang (Salween River) Gorge during his collecting expedition in Yunnan from 1932 to 1934. Courtesy of the family of Cai Xitao.

of generating income, which included growing flowers and vegetables for sale, along with birds and other pets in a shop they ran in the city (Xu and Pei 1985, 442; Pei Shengji, pers. comm., January 2018).

In 1950, the institute was one of the first to be brought under the Chinese Academy of Sciences, with Cai as director. Cai went on to enjoy a long and distinguished career, remaining active until just months before his death in 1981. He is remembered for his dedication to developing the field of tropical botany in China and founding the Xishuangbanna Tropical Botanical Garden, now recognized internationally as one of the leading research centers in its field.

DOING BOTANY: OBSERVING, IDENTIFYING, AND RECORDING IN THE FIELD

The first published accounts of plant collecting sometimes read like reports of school outings to historic scenic spots with the added pleasure of some gentle botanizing.[24] A report of a collecting trip in the mountains near Nanjing begins with an introduction reminiscent of many classical poems: "It was a holiday, and we agreed to Chen Zizong's suggestion that we should go to a scenic spot high on a hill.... Niu Shan is a famous place celebrated by poets since Eastern Jin times, where monks learned in the Buddhist scriptures had sojourned during the Qi, Liang, and up to the Tang and Song dynasties" (Wu Y. 1914, 99).

It was not long before expeditions ventured into remoter parts of the country, braving the dangers of travel in these insecure times. In 1924, pirates threatened students from the Jiangsu Provincial First Normal College who were on their way by boat to collect in Shandong. On their return, many were pressed into the warlord armies and were not able to graduate (Wu Zixiu 1924). As late as 1934, Yu Dejun's report of the botanical survey he led in Sichuan reads like a litany of disasters. The threat of an attack by bandits forced the expedition to take a detour, and ultimately the party had to turn back. The party then needed an armed escort to pass through more bandit-controlled territory to reach Chongqing. After leaving Chongqing on May 13, 1933, for the second leg of the survey, an outbreak of hostilities between warlords in northwestern Sichuan forced them to change their route. At the beginning of August, in the area around Mao Song, military operations by the 24th and 28th Armies forced another change in their route. Though safe now from warring factions, a huge earthquake at Diexi on August 25 blocked roads and caused flooding. "When traveling and collecting in border areas," Yu wrote wearily, "every time there is a delay because of the security situation, it becomes difficult to carry out your plans" (Yu D. 1934, 451). Nevertheless, he believed the expedition had made a good start on mapping the major vegetation types in the province, and he was optimistic that future surveys would fill the gaps.

Many expeditions followed in the footsteps of foreign plant hunters in remote regions such as Yunnan, the Tibetan borderlands, and the tropical south. There were areas of the interior, though, that were hardly better known than the periphery. Hu Xiansu chose to collect in Zhejiang and Anhui, two long-settled areas where few Westerners, who tended to be more interested in the "discovery" of new species, had worked.[25] He traveled between July and November 1920 from Hangzhou through rarely traveled mountainous parts of Zhejiang, then traced a loop back around Tiantai Shan and Yandang Shan, both of which are now protected scenic areas. His daily reports in the *Critical Review* (Xueheng), not a science-oriented journal but one with a well-informed readership, were comparable in form and content to what his peers in the West might have written at this time, although they were written in classically inflected literary language. He reported on the topography and vegetation through which he passed, and he noted his botanical finds of the day, sometimes including the Latin botanical name of plants. He recorded new discoveries and collected specimens that found their way in due course into the documented flora of the region:

> September 24th, Friday. Woke up at 6 am. . . . The trail was made up of steep stone steps with dense vegetation. A dark forested landscape pressed in on us. Bushes reached up all the way into the sky. There were large numbers of *qing li* [*Quercus glauca*] and *ku zhu* [*Castanopsis sclerophylla*]. There was one tree in particular, a large chestnut, *Castanopsis tibetana* Hance, whose tall trunk towered into the clouds.[26] The leaves are large, almost a foot in size. From a distance it looks like *nanmu* [*Phoebe nanmu*]. The shell and the nut itself are both the same as the common chestnut. It is some fifty to sixty feet high and one to two feet in diameter. It must be excellent timber. (Hu X. 1922, no. 6, 1)

In 1932, distilling his experience from several collecting expeditions, Zhong Guanguang advised younger, less experienced field botanists that "members of collecting expeditions should always record the names of plants. They should talk to herbalists, woodcutters, guides, and other informants. They should be sure to identify all the plants mentioned in the 'local products' section of the local gazetteer. Members of the expedition should have consulted the gazetteers before they leave. This means of course that the expedition should include people who are able to read the gazetteers" (Zhong 1932a, 6–7).

Zhong's accounts of his collecting in Guangxi and Yunnan between June and September 1919 reflected both his classical education and his deep knowledge of the plant world. He read the relevant sections of the Qing gazetteer of Yunnan and planned his itinerary from the provincial capital to Dali and Erhai Lake so as to retrace the route of Xu Xiake (1587–1641), the late Ming travel writer, whose final journey had been to Yunnan. He did not only search for the rare and unknown; he was also interested in the everyday. On August 14, 1919, on arriving in one of the small towns between Kunming and Dali, he drew up a list of fruits and vegetables found in the local market: peaches, pumpkins, potatoes, eggplants, chili peppers, several mushrooms, indigos, pomegranates, beans, and more. He deliberately looked for known plants in villagers' gardens and sent his assistants to look for new local varieties (Zhong 1921–22, vol. 11, no. 12, 25).

Hu Xiansu and Zhong Guanguang were traveling through remote places, describing familiar things—plants by the road and in gardens—as well as things that their readers had never seen. They wrote about their pleasure in experiencing landscapes in which, as botanists, they could identify the plants and understand the world around them. On August 9, on Taihua Shan near Kunming, Zhong described the ecosystems through which he climbed:

> The trail now climbs higher, coming out onto a slope planted
> with pine and cypress. Occasionally there are beautiful plants of
> the orchid family under them. . . . The top of the mountain is
> covered in rocks, sharp and fearsome, with green creepers over
> them and a lot of coarse grasses. I rested for a while on a rock
> and looked at the swell of the lake water, with its surface as
> smooth as a mirror. Villages and houses were scattered on
> the islands. With the color of the trees and the shadows of the
> clouds, the interplay of light and dark was like a painting. (Zhong
> 1921–22, vol. 11, no. 11, 64)

Zhong and Hu and their contemporaries were botanists, but they were reporters of a kind, too, who saw themselves as participants in China's march to modernization. Looking over the city of Kunming from Taihua Shan, Zhong reflected on the hydroelectric potential of the mountains and lakes: "Looking down over the lake . . . where its shores meet the city, where shops and homes are clustered close together, with electric lights just beginning to glimmer under the glow of starlight, I felt that this place in Yunnan really is a mountain city. It is surrounded by mountains like Jin Ma and Bi Ji and the large lake that supplies the water for Kunming. Where this water flows out, there must be a tremendous reserve of natural energy that could be put to work. Unfortunately, I do not know how to implement [this idea]" (Zhong 1921–22, vol. 11, no. 11, 60).

THREE WAYS OF LOOKING AT A MOUNTAIN

With some exceptions, Western plant hunters who wrote about their travels adopted a swashbuckling, adventurer persona. Chinese collectors, on the other hand, portrayed themselves as visitors learning and writing about unfamiliar places not so far from home. Where a Chinese and a foreign plant collector visited the same area within a few years of each other, their accounts of the landscapes reveal as much about divergent responses to the natural world and the environment as they do about the places they described.

Zhong Guanguang spent the week of the September 18–26, 1919, on Jizu Shan, which translates as Chickenfoot Mountain, in northwestern Yunnan. Four years earlier, in May 1915, the Austrian botanist Heinrich von Handel-Mazzetti (1882–1940) had also spent several days there, writing in his memoirs about what he had seen (Handel-Mazzetti 1927). Almost four centuries before them, Xu Xiake, whose itinerary Zhong Guanguang was consciously retracing, visited Jizu Shan, first in 1638 and again in 1639.[27] Reading Xu's account of his visit together with the two twentieth-century descriptions affords some insight into what was distinctive about Zhong Guanguang's reporting with respect to both the Chinese tradition of travel writing and to the practices of his Western contemporaries.

Jizu Shan lies to the east of the Dali Bai Autonomous Prefecture in northwestern Yunnan, at an elevation of 3,240 meters. It is an important pilgrimage destination for Buddhists and Daoists, with numerous temples and nunneries on its slopes. The complex was at its most flourishing during the late Ming and early Qing periods, when there were more than 360 shrines and around five thousand monks and nuns on the mountain.[28] The description in the gazetteer of Dali Prefecture, dated Jiajing 42 (1563 CE) reads: "Bingchuan County. Jizu Shan. Also known as Nine Curve Crags. It is a remarkable and strange range of hills. There are three curved ridges spreading out like a chicken's foot. There is a stone gate at the summit,

FIGURE 5.5 The central peak of Jizu Shan (Chickenfoot Mountain) from the *Gazetteer of Jizu Shan* (Jizu Shan zhi) (1692), showing the main peak with the Golden Hall (Jin Dian) at the summit and monasteries and nunneries on the pilgrim trail up the mountain. Used with permission from the Chuan Si Nian Library of Academia Sinica, Taibei (pp. 0053–0059)

majestic like a watchtower on a city wall. It is said that the Buddha's disciple Jiayepo [Kāśyapa in Sanskrit] kept watch on the Buddha's robes at this mountain, waiting for Mile [Maitreya]" (Li Y. 1563, 65)

Xu Xiake, Heinrich von Handel-Mazzetti, and Zhong Guanguang came to Jizu Shan from three different worlds.

Xu was born in 1587 at Jiangyin, near the mouth of the Yangtze River, and died there shortly after returning from southwestern China. *The Travels of Xu Xiake* (Xu Xiake youji) is an edited transcript of his diaries assembled after his death by close friends. Xu set out in 1636 accompanied by Jingwen, a Buddhist monk who was on a pilgrimage to the mountain, carrying a copy of the Lotus Sutra written in his own blood. He died on the way, and Xu buried his ashes on the mountain. The diary ends with the outlines of a never-completed geography of the mountain with ten poems about scenic spots he had visited (Ward 2001, ch. 3 and 168–175).

Handel-Mazzetti earned his doctorate in botany at the University of Vienna in 1907. In 1914, the Austro-Hungarian Dendrological Society of Sciences sent him to China to carry out botanical research. A planned trip of some months became an enforced stay of nearly five years, after the outbreak of the First World War prevented him from returning. During this time, he traveled widely in the southwest, collecting specimens that are now housed in the herbarium at the University of Vienna. He returned to Austria in 1919, serving from 1925 as curator at the Natural History Museum. In addition to publications on the flora of China, he published the memoirs of his travels, which have been translated into English as *A Botanical Pioneer in South West China*.[29]

All three visited Yunnan in dangerous times. Xu was traveling during the last years of the Ming dynasty, a time of unrest following a period of profound economic and social change, when central control over the fringes of the empire was tenuous at best. Handel-Mazzetti and Zhong were both traveling during the early years of the Republic, at the time of China's humiliation at the Versailles conference and the turbulence it unleashed.

In China, there have long been powerful associations between mountains, enlightenment, and spiritual awareness. Xu experienced the mountain as a different world, a living entity with movement in its rivers and waterfalls: "I could see the summit floating in mist, suspended in the highest heavens, while on the sheer cliff in falling snow below were inlaid the deepest earths while the combination of the brightness of the clearing blue sky and the floating gleam of the blossom led me to feel I was not in the realm of mortals" (Ward 2001, 178; XXKYJ, 841).

Handel-Mazzetti was the confident outsider celebrating his own achievements in filling in the blank spaces in (European) maps and assembling collections of rare plants. A mountain represented a challenge for him, both scientifically and personally: "On 28th May I arrived in Lijiang, with a splendid collection of plants, many of them, because the season was so much more advanced, quite different from those I had gathered in 1914, and gazed once again on the giant peak—a sight to quicken the pulses of any mountaineer" (Handel-Mazzetti 1927, 132; Winstanley 1996, 61).

Zhong was as much an explorer in the new science of botany as a traveler in a remote part of his native land (Meng, Liu, and Yu 2018). Each mountain was a unique topography and environment. As a botanist, he saw the landscape as well as the temples and structures on it as habitats harboring the diversity of plant life: "The streambed went over a deep ravine then northwards toward a high gorge, plunging into its base. Since there was a lot of spray, all kinds of plants grew in profusion. There was a wide road next to it, which climbed to the Hall of the Bronze Buddha and then on to the Jin Ding temple.... It really was a wonderful place for collecting—and it was right next to my lodging, making it convenient to come and go" (Zhong 1921–22, vol. 12, no. 5, 59).

Scholars of classical Chinese travel writing have observed that the journey to a mountain was a journey to put the cares of the world behind you. As Xu wrote in his poem "Reflections on a Wine Cup":

> The moon's reflection divides in the distance over the Boddhisatva's offerings,
> And darkness and light come and go over the mountain scenery,
> Soon I will have pulled free of mortal dreams,
> Now my old chilled bones face the wine glass.
> (Ward 2001, 180; XXKYJ, 1149)

In a radically different historic and cultural context, Handel-Mazzetti also found escape in the mountains, in his case, an escape from the unhappy news about the war in Europe. As for his awareness or interest in the extraordinary phase of history that China was going through, it was a part of the backdrop of the landscape in which he was working. He saw gruesome sights, but his interest would move immediately, without further comment, back to the landscape and vegetation: "Furthermore, but for this detour I might have encountered a band of army deserters, allegedly some thirty strong, who had plundered a caravan there four days previously. Two of them had been captured, and their heads, with a suitable notice, had been hung up in cages above the road; the others had fled into the mountains. Above Midian the valley broadened once more into the familiar landscape: groups of green bowls between low rounded hills clothed with pine forest" (Handel-Mazzetti 1927, 127; Winstanley 1996, 59).

Zhong, by contrast, was one of the generation of scientists in China committed to using science to save China. Far from being an escape from the world, the very remoteness of the mountains only highlighted how current events touched every corner of the country. On the road, he asked about how people lived and how current affairs affected them. On September 20, when staying at a "small and primitive inn" in Shangguan, he observed: "The women running the inn were extremely welcoming and polite. I was quite surprised. I later learned that the family is called Duan, and they have shown an aptitude for scholarship. The son is very bookish and is now studying at the middle school in Dali. His mother and sister manage the inn and cultivate the fields at the same time in order to earn enough for his school fees. This is where the idea of enlightenment comes from!" (Zhong 1921–22, vol. 12, no. 5, 55).

All three travelers had an interest in the flora of Jizu Shan. Xu's diary has a "Note about the Plants of Central Yunnan" (Dian zhong hua mu ji) (XXKYJ, juan 9, 4–5) It consists of just four or five lines, depending on the edition, about camellias and azaleas, noting that camellias are popular in temples and private gardens in Kunming and that azaleas are more commonly seen growing wild in the mountains. Elsewhere, he describes the trees and vegetation on Jizu Shan as elements in the scene before him that prompt a reflection on the passage of time or on the illusory nature of form and substance: "A film of green moss climbed up the eastern side of the cliff like stunning colored silken velvet. It was blue-green like a drop of water, which then transformed itself into dabs of color, neither rock nor haze, instead forming something illusory. . . . The rocks here were all the color of grass, and the trees were all shaped like rocks. In their appearance, none of them was true to its substance" (XXKYJ, 840; my translation).

Not surprisingly, Handel-Mazzetti and Zhong, the two botanists, share an interest in and pay close attention to the plants and vegetation on the mountain. Handel-Mazzetti uses the botanical names of species to inventory the plant communities that make each place unique: "Scattered pines and yews crowned the crest; lower down there was *Lithocarpus* forest and holly-leaved oak scrub, and below them aspens and willows spread their delicate foliage. The summit ridge ran east and west and was the only place where I found *Vaccinium delavayi*, a dwarf shrub resembling bilberry, and *Gaultheria cardiosepala*, which looks like *Erica*, both of them grew on humus soils among cushions of moss. (Handel-Mazzetti 1927, 129; Winstanley 1996, 60).

Handel-Mazzetti's collecting may have been in the service of science, but his science was firmly situated in the world of imperial botany and plant hunting. His writing exudes a sense of competition and claims to possession, using terms such as "booty," "spoils," "a rich haul," and so on. Zhong, on the other hand, found all the plants he collected, however apparently unremarkable, to be interesting: "When Mr. Zhang got back to the Bronze Buddha Hall, he said that he and the assistants and Huang the hired laborer had reached the summit and had been collecting around the Jin Ding Temple. They said that there was not much there and that they had not done what they had hoped to do. But when I looked at the plants, it turned out that they were all quite special. Although there were not many of them, they were all very precious" (Zhong 1921–22, vol. 12, no. 5, 60).

In China today, Xu Xiake is heralded as a great geographer, a "scientist" before the term had been invented. His interests, though, were less in discovering and describing new things than in setting the record straight about the flow of rivers and the location of temples and important landmarks. In the early twentieth century, Zhong Guanguang and his contemporaries embraced fieldwork as an antidote to the old habits of book learning, which they believed had held back the country's modernization. Zhong was not rejecting the past. He was very aware of the history of the places he visited. But as a field botanist, his mission was to identify and classify both known and unknown plants to discern the patterns and principles that help in understanding the world.

By 1919, Latin botanical names were known among Chinese botanists, if not yet very familiar. Zhong was both knowledgeable and dedicated to identifying and correctly naming plants, but in his account of this collecting expedition, he did not give detailed descriptions of individual plants nor did he give their scientific names. It may be that his purpose was to encourage his readers to think about plants as part of the wider landscape rather than as pharmaceuticals, foods, or ornamentals in gardens, which was how they had been inventoried and studied in the past. Now, at a time of intense questioning of the established order, scientists were traveling through the landscape, representing it as a natural world unique to China, filled with plants and features with particular meaning to its people. In their fieldwork, their identification and classification of the plants they collected, and their writing, this first generation of botanists used a modern framework to document a distinctly Chinese environment, whose diversity placed China on par with the rest of the world, both in terms of the richness of its natural endowment and with respect to achievements in the world of scientific endeavor.

CHAPTER SIX

The Inventory of Nature

> Thus, even though the myriad things are very numerous, sometimes one desires to refer to them all together, and so one calls them "things." "Things" is a case of large-scale group naming. By drawing analogies, one groups things together, grouping and grouping, until there is nothing more to group, and then one stops. Sometimes one wishes to refer to them partially, and so one calls them "birds" and "beasts." "Birds" and "beasts" are instances of large-scale differentiated naming. By drawing analogies, one differentiates things, differentiating and differentiating, until there is nothing more to differentiate, and then one stops.
>
> —XUNZI, *CORRECT NAMING* ("ZHENG MING PIAN")

EARLY in the third century BCE, Xun Kuang, the Confucian thinker better known as Xunzi, taught that groups of things and their names are not predetermined; they are constructed by agreed convention and usage. The naming of things is a social activity, and ordering and naming things correctly are therefore fundamental to establishing or maintaining a harmonious society (Djamouri 1993, 66–69; Goldin 2018).[1] The importance of correctly naming the myriad things shaped the study of the natural world in the past and gives context to the importance to Chinese botanists during the first half of the twentieth century of the classification, taxonomy, and nomenclature of the plant life they were collecting.

The *Yunnan Scholarly and Critical Weekly* (Yunnan xueshu piping chu zhoukan) ran a series of articles between 1916 and 1917 asking, "What is science?"[2] Shen Huanzhang was one of several readers who defined science as a process of recording and bringing order to things: "To call something science, one has to record absolutely everything about a kind of object, and the recording must follow a methodology that orders things systematically. The categories of this systematic order must fit into an overall concept that is able to accommodate all the categories. This initiates a process that will, when completed, be scientific" (Shen H. 1917, 7)

Observing and recording everything about the plants they collected in the field were the practices by which members of the newly fledged profession of botanist defined themselves. The task that followed, making an inventory of their specimens and ordering and naming them was not, in itself, a departure from the heritage of the extension of knowledge (*gezhi*). What made their work scientific (*kexue*) was the methodology they used and the categories they constructed in identifying and classifying.

LUMPING AND SPLITTING, ORDERING AND NAMING

An inventory of named categories and the things in them turns a vast, unwieldy collection of observations into accessible, retrievable, and predictable information. To order and to classify involve making judgments about what differentiates one object from another, as well as distinguishing one or more characteristics that suggest an affinity between objects. The principles that determine when to split objects into separate categories and when to lump objects into one single category reflect the classifier's understanding of the organization of the world in which she encounters those objects or phenomena (Alfonso-Goldfarb, Waisse, and Ferraz 2013, 551). The historian Harriet Ritvo observes that in the Judeo-Christian tradition, the seven days of the creation can be seen as "a series of founding discriminations," beginning with the division of light from the darkness and naming them "day" and "night," until the sixth day, when God made "the beast of the earth after his kind, and cattle after their kind, and everything that creepeth upon the earth after his kind" (Ritvo 1992, 364). In the Chinese context, the world is not formed from a series of divisions into different categories of things, but the association of the myriad things with yin and yang and the Five Phases situates them all within an orderly pattern: self-generation's pattern (*ziranzhi li*).[3] Locating and naming things in that pattern are the foundations of a peaceful and well-ordered society.

If grouping, ordering, and classification are ultimately a subjective evaluation of affinities and differences, it is relevant to question how objective the identification and naming of species can be. In practice, taxonomists agree that while the process cannot be completely detached from subjective judgment, neither is it completely random. Theophrastus, who is considered the founder of botany in the West, used observed physical features to organize plant life into four taxa, or groups, on the basis of growth form: trees, shrubs, undershrubs, and herbs (Theophrastus 1916, book 1, 23–25; Pavord 2005, 21–30). Certain visual and sensory cues can be agreed on as indicators in differentiating between things so that classification becomes a process not unlike the diagnosis a doctor performs on a patient by evaluating a checklist of recognized symptoms (Stevens 1984; Strasser 2012, 333–34).

Given a taxonomy based on a consensus on the diagnostic characteristics of a plant, the question then becomes what the characteristics should be on which to cluster plants into taxa.[4] For a long time in both China and the West, the classification of plants was closely associated with pharmacology. In the West, medieval taxonomies barely deviated from the classical models of ancient Greece and Rome that described plants from the Mediterranean basin, listing them alphabetically or according to their uses, mostly medicinal. By the late sixteenth century, the naturalists of the Renaissance had to deal with a growing number of plants from beyond the Mediterranean that were being collected and described by scholars and teachers all over Europe, only some of which had pharmaceutical properties. Andrea Caesalpino (1519–1603), the director of the botanical garden at the University of Pisa, set out to devise a classification system that was based on what he described as the natural order of things, not on qualities such as color, flavor, fragrance, or root structure, which were contingent on cultivation, soil, or location. In *De plantis libri XVI* (1583), he broke with tradition and ordered plants according to their morphology. He used the reproductive organs as the primary interspecies markers, anticipating Linnaeus's hierarchical, binomial system with its classes based on the number of stamens, subdivided in turn by the number of pistils. While later botanists emphasized their significance as mechanisms of reproduction, Caesalpino made his choice based on his determination that these were the anatomical features with the least variability from one individual to the next.[5]

In Europe, naturalists at this time maintained that their calling was to pursue philosophical questions that Plato and Aristotle had posed about the true forms of beings, the classification of living things, and the nature of phenomena and change in the world (Endersby 2008, 156–57). They were beginning to construct "scientific" taxonomies that were distinct from comprehensive systems of ordering the world, and they rejected the criticism that they were mere catalogers. They believed that in collecting and recording specimens, studying them, and placing them in a logical system, they were building a body of knowledge that could be applied to advance human progress (Ritvo 1992, 365).[6] In China, classification systems originally consisted of a small number of broad categories, which grew over time to more numerous, more fine-grained categories. Until the encounter with Western botany in the late nineteenth century, though, they remained embedded in the fundamental cosmological scheme that had been established in the earliest written records and that accounted for the formation of the myriad things.

Two ostensibly contradictory analytic principles are fundamental to identifying the generic and the specific, the two basic taxa for purposes of organizing and classifying. They are: grouping things that have properties in common, and distinguishing between things on the basis of unique characteristics. The dyad of generic and specific is deeply rooted in the human experience of nature. Ethnobotanists maintain that folk taxonomies always identify plants using a term for a life-form (tree, vine, grain, etc.) and two signifiers. The first situates the plant as a member of a genus, a class based on relationships to others, both similar and different. The second designates the species, a distinct kind of being within the genus, identified by a limited set of unique attributes or features (Berlin, Breedlove, and Raven 1973; Berlin 1992). Folk and scientific taxonomies are logically comparable, but they both leave unresolved the fundamental problem of how to locate the boundaries between genus and species.

The issue of whether to lump or to split is at the heart of debates about the difference between a kind and an individual. In the light of recent insights into constant evolutionary change and the stability or otherwise of molecular-level structures such as DNA, contemporary taxonomists continue to grapple with the problems inherent in distinguishing between the generic and the specific, between the intrinsic properties of individual organisms and the extrinsic relations between organisms (B. Briggs 1991; Ereshefsky 2007, 296). Despite skepticism over the existence in nature of the taxa into which biologists have organized living beings, defenders of the concept argue for pragmatism: "'Species' is well entrenched in biology and elsewhere. Students are taught the term from their earliest encounters in biology. Field guides and taxonomic guides use the word 'species.' And the term is even found in our governments' laws. From a practical standpoint, it would be hard to eliminate the term 'species'" (Ereshefsky 2010, 420).

Nomenclature is the linguistic expression of the act of ordering and classifying things. Problems inherent in one are inseparable from the other. Xunzi believed that naming a group of the myriad things is what procures agreement on the ordering of objects: "Names are the means by which one arranges and accumulates objects" (Hutton 2014, 274; SKQS, juan 16, 153). As natural history in Europe became the science of botany, taxonomic nomenclature was fraught with struggles over who had the right to name genera and species. Botany's association with imperial expansion added a new dimension to the power of classification and naming, since bringing order to tropical nature and colonized societies was at the heart of the concerns of the natural historians, horticulturists, and agronomists mobilized to carry out the "civilizing mission" of colonialism.[7] During the nineteenth century, the issue became so contentious that the keepers of natural history collections, the self-appointed arbiters of identification and naming of species, blocked all discussion on the subject (McOuat 2001; Witteven 2016). The inevitable discord and arguments led to the convening of the International Botanical Congress in Paris in 1867, which set itself the task of resolving the problems dividing botanists and formulating rules on naming newly published genera and species, how to deal with revisions of genera and species within the overall taxonomic system, and how to deal with specimens known only from small samples or drawings (Nicolson 1999; Ereshefsky 2010). The Paris Congress did not settle all the arguments, but it did lead to the formulation of successive codes of botanical nomenclature approved at the International Botanical Congress, now held every six years.[8]

Li Shanlan's translation of Lindley's *Elements of Botany* introduced China to the binomial classification systems pioneered by Linnaeus, although the preface still referred to traditional classifications of plants from the Chinese classics. The section on taxonomy gives little explanation of the principles according to which plants have been organized, simply stating that there are a total of 303 families (*ke*), avoiding potentially contentious questions about the boundaries between genera and species. The text describes thirty-seven families, to thirty of which Li and his colleagues assigned existing Chinese plant names (Li, Williamson, and Edkins, 1858, juan 8).[9] Over time, as Chinese botanists turned to the task of recording the flora of China, standardizing the botanical names of plants in Chinese proved to be problematic. The international system was built on the work of European natural historians and botanists, mostly using known European genera and species, making for a relatively easy match between common name and botanical name. With a rich and ancient existing body of named plants in China, it was important but not easy to match literary, common, and botanical names. Gong Lixian, one of the authors of a set of botanical textbooks, articulated a hope shared by many when he asked,

"Wouldn't it be better to take what is beneficial and keep things convenient, with one consistent way of expressing many names?" and suggested that "when the scientific names of families and genera in botany are translated into Chinese, where there are examples that accurately [match] the reality, we should record it as a mark of respect to our ancestors" (Gong 1934, 134).

ORDERING THE MYRIAD THINGS: TRADITIONAL TAXONOMIES

The eighty volumes of the definitive *Flora Reipublicae Popularis Sinicae* record 31,228 species of vascular plants in China (Wu Zhengyi and Chen, 1959–2004; Hong and Blackmore 2015, 1). Each entry has a Latin botanical name together with a Chinese botanical name sanctioned by the Institute of Botany of the Chinese Academy of Sciences.

Flora has been in preparation for over eighty years, realizing the dreams of early botanists such as Zhong Guanguang, Hu Xiansu, and their contemporaries, who had planned and worked on it ever since the founding of the Botanical Society of China in 1933 (Ma J. and Clemants 2006, 451). A first step in this ambitious project was to align the traditional and scientific taxonomies and nomenclature for plants and to integrate them into the fast-growing body of botanical material from ongoing collecting and from herbaria in China and other countries. Their starting point was the seven thousand named plants found in the written record in Chinese before 1850 (Fèvre and Métailié 2005, 1).

One of the enduring tropes of Orientalism in the West has been of a bureaucratic China compiling encyclopedias and constructing libraries organized in baffling categories that embody a mysterious, unfathomable other. Setting aside these misrepresentations of Chinese scholarship, it is fair to observe that the ordering of knowledge in China has a history almost as ancient and respected as the texts venerated as the classics that formed one or more of the divisions of learning in their own right.[10] The words of Confucius about *The Book of Poems* (Shijing) are often quoted as the starting point for a review of the classification and naming of plants: "The Master said: 'Little ones, why don't you study the Poems? The Poems can provide you with stimulation and with observation, with a capacity for communion, and with a vehicle for grief. At home, they enable you to serve your father, and abroad, to serve your lord. Also, you will learn there the names of many birds, animals, plants, and trees'" (Leys 1997, 87)

The Book of Poems, also known in English as *The Book of Odes*, is a collection of 305 poems or songs dating to the Shang and Zhou periods (seventeenth to third centuries BCE). It names 130 different plants, most of which are from the Yellow River basin, the cultural heartland of China at the time (Wang Zhenru, Liang, and Wang 1994, 6–7).[11] The songs are organized into six groups corresponding to their origins and the ritual functions ascribed to them, but there is no suggestion of any ordering or classification of the plants or animals that appear in the anthology. Confucius's injunction to his students was to impress on them the importance of knowing the names of the birds, animals, plants, and trees, not to encourage them to devote themselves to studying the pattern of nature (Mittag 2010, 318).

The earliest appearance of some form of classification of plants is found in two nearly contemporaneous texts, *The Rites of Zhou* (Zhou li) and *The Literary Expositor* (Er ya). They divided life-forms following principles very different from those followed in later texts. Ritual works such as *The Rites of Zhou* organized and classified living things in the context of the cosmology of the Five Phases. Reference works such as *The Literary Expositor* functioned as guides to the correct naming of the myriad things.

The Rites of Zhou, from the third century BCE, delineates the responsibilities of various officials during the Zhou dynasty (1046–256 BCE). The Great Director of the Multitudes (Da Situ) surveyed the kingdom to determine taxes. Mountains, forests, and waters were separate categories of land, and the Great Director divided the things in each habitat into plants and animals, each of which was divided, in turn, into five classes. The five kinds of plants were distinguished by what we would now call their reproductive morphology (Carr 1979, 51):

zaowu	plants with acorns
gaowu	plants such as the poplar and willow, with a "quiver" or protective covering over the buds
hewu	plants such as the plum or peach, with a kernel in the fruit
jiawu	plants with a seedpod
congwu	plants that grow in clumps, such as reeds

The Literary Expositor, described as "the archetypal nucleus of early Chinese lexicography" (Carr 1979, 1), probably dates from the late third century BCE. Its longest sections are devoted to seven categories (not five as in *The Rites of Zhou*) of living things: plants (*cao*),[12] trees (*mu*), insects (*chong*), fish and other scaly animals (*yu*), birds (*niao*), beasts (*shou*), and livestock (*chu*). Only the chapters on plants and trees (chs. 13 and 14, respectively) are botanical (Carr 1979, 48–53). The original content of *The*

Literary Expositor consists of very brief entries with the name of a plant, one or two alternative names, and occasionally a note on the time of flowering and fruiting or a clarification of the name. This format, well suited to the unending task of the rectification of names, persisted in different compendia and encyclopedic works for centuries, with the contents expanded in successive commentaries (Carr 1979, 39–40; Nappi 2009b, 22).

The best-known and perhaps best studied taxonomies are pharmaceutical, found in the body of materia medicas. One of the most comprehensive catalogs of this genre lists a total of 278 named, extant works from the fifth century BCE to the nineteenth century CE (Long B. 1957, in Needham, Lu, and Huang 1986, 222; also Hong, Chen, and Qiu 2008). Over such a long period, both the classifications and the names of the plants varied, and one of the rationales for compiling and publishing later works, such as Li Shizhen's *Classification of Materia Medica*, was to identify plants correctly, to rectify errors, and to resolve confusions that were undermining the knowledge of the past. *The Classical Pharmacopoeia of the Heavenly Husbandman* (Shennong bencao jing) (second to first centuries BCE) had three grades (*pin*) of plants: those in the highest grade were nonpoisonous with some medicinal benefits; those in the middle grade were nonpoisonous or mildly poisonous and of medicinal benefit; those in the lowest grade were highly toxic. As the genre developed, the classification system became based on plant forms.[13] By the fifth century CE, *Collected Commentaries on the Materia Medica* (Shennong bencao jing jizhu) by Tao Hongjing (456–536 CE) listed grasses, trees, fruit, grains, and vegetables (Chen Jiarui 1978, 108; Métailié 2015, 456). The mid-Ming *Sources for Materia Medica* (Bencao yuanshi) by Li Zhongli (1612) had more than five hundred entries organized in ten categories: grasses (*cao*), trees (*mu*), grains (*gu*), vegetables (*cai*), fruit (*guo*), rocks or minerals (*shi*), wild animals (*shou*), birds (*qin*), insects (*chong*), and fish (*yu*).[14] Each entry includes the place of origin, basic information about the material, a description, information about the material's taste and medicinal properties, and prescriptions for using the material (Zhang and Zhang, 2010).[15]

Li Shizhen's *Classification of Materia Medica* was far from the last of the genre, but it is the best known and most frequently quoted, still used and most often translated. In it, Li developed a sophisticated, three-tiered classification of sections (*bu*) (or sometimes tribe (*zu*)), categories (*lei*), and kinds (*zhong*) (Métailié 2015, 90). The features used to assign plants to the sections are quite varied. They include the nature of the stem (herbaceous or ligneous), whether the plant is wild or cultivated, whether it is used for food or only for medicinal purposes. The categories are also classified by features that are not necessarily related to appearance, including habitat (mountains, aquatic, growing on rocks), the part that is consumed, unusual properties (toxicity, fragrance), its growth form (liana, upright, shrubby), and the type of plant (moss or lichen, mushroom, epiphyte) (Needham, Lu, and Huang 1986, 313–17; Métailié 1988).[16] Li's system, based on observable physical characteristics of plants and their habitat, indicates that he intended his work to be a guide to easy, accurate identification of plants (and other materials) for pharmaceutical and medicinal use. Although it is often praised as a precursor of modern botanical classification, *Classification* was not and could not be used as a guide for the purpose of identifying an unknown plant in its natural habitat.

Handbooks on agriculture, horticulture, and fungi used taxonomies designed to serve their users. Grains, vegetables, fruit, plants for fiber, timber, and the extraction of oils were among the categories used in manuals for farmers or the managers of agricultural estates. In addition to the frequently used categories of plants, trees, vegetables, and fruit, gardeners reading *The Assembly of Perfumes* (Qun fang pu) (1621) could look up plants grown primarily for their flowers (*hua*) or plants that were grown for the ornamental value of the whole plant (*hui*), an interesting distinction (Chen Jiarui 1978, 109). One historian has even reported a taxonomy of animals and plants found in the codes of some Buddhist monasteries in northern China during the early Tang era (seventh century CE) specific to the plants and livestock of importance to the monastic community (Chen Huaiyu 2009). As living beings, they were first divided into four kinds of births (following a religious text, *The Four-Part Vinaya*), then into thirteen categories defined by their growth form (root, seed form, cane, and so on). Used in managing the monastic estate, the four kinds of births and the thirteen categories allowed the community to bring the plants and animals that were the source of their livelihoods into the spiritual domain of their striving toward enlightenment.

Familiarity with the existence of different systems of classification used for different purposes was no doubt a factor in the openness of the first generation of botanists to the new scientific taxonomy when it reached China from the West. Even the concept of a binomial nomenclature would not have been especially perplexing, since traditional systems such as Li Shizhen's sections tended to follow the pattern of general and specific lexemes, as well as higher-order categories, as proposed by Berlin, Breedlove, and Raven (1973) in their studies of folk biology.

ORDERING AND CLASSIFICATION BY BOTANISTS

Li Shanlan, Alexander Williamson, and Joseph Edkins introduced the Linnaean system of botanical taxonomy in *Botany*, but there is little evidence that it made much of an impression on the nascent scientific community in China. Occasionally a story appeared in a journal or periodical with accounts of unusual trees, plants, or other botanically related curiosities, but they did not offer any pedagogical materials for their readers. It was nearly twenty years later, with the launch of the *Chinese Scientific and Industrial Magazine* in February 1876 that readers could find encyclopedia-like articles introducing the different branches of the sciences. In the tenth issue of the first volume, an article on botany began by covering the basic physiology of plants and photosynthesis. It went on to introduce botanical taxonomy, beginning with the Linnaean system based on reproductive parts, including an explanatory illustration of a flower labeled in Chinese. The article noted that there were "problems" with the system and that it had now been superseded by de Jussieu's natural classification. It briefly explained that this was based on physical characteristics or "ways in which flowers open" using Cruciferae ("in the shape of a cross") and Papillionaceae ("in the shape of a butterfly") as examples, concluding that "most naturalists use this system." From taxonomy, the article went on to discuss pollination, plant nutrition, and responses to light, ending with a section on plants that are useful to humans in which the author fell back on a classification reminiscent of the traditional divisions of plant life: grains, plants with edible leaves, plants with edible roots, flowering plants, fruit, and trees for timber (*Gezhi huibian* 1876, section 174, 201).[17]

The post-1905 education system generated demand for reference works, manuals, and textbooks on the natural sciences. In his preface to the first textbook on botany, Du Yaquan marveled at the richness of China's plant life and expressed awe at just how little had been described, identified, and named. He reflected on the enormity of the task that lay ahead: "The varieties of plant life can be counted in the millions. Of them all, fewer than one in a thousand have been named either in the ancient records or in modern descriptions. I cannot be certain that all of [the plants] recorded by the ancients are real things. Nor am I certain of all the scientific names of those described by modern writers" (Du 1903, 85).

An important reference work for the new curriculum was *Zhiwuxue jiaoke shu*, a translation of Matsumura Jinzō's *A Textbook of Botany* (Shokubutsugaku kyōkasho), published in 1904. The second volume outlined the principles of botanical taxonomy, the purpose of which, it explained is "to shed light on the relationships between plants, to distinguish between them, and to classify them on the basis of their similarities and differences . . . in order to reveal the categories they belong to" (Liu D. 1904, juan 2, 1a). The text introduced six classes (*bu*) of plants: fungi and algae (*jun zao*), mosses (*xiantai*), brackens and ferns (*yangchi*), gymnosperms (*luo zi*), monocotyledons (*dan ziye*), and dicotyledons (*shuang ziye*). It continued down the taxonomic ranks, with descriptions of some families (*ke*), tribes (*zu*), genera (*shu*), and species (*zhong*) within the six classes and illustrations of key diagnostic features. The presentation was not comprehensive, but it followed the internationally recognized botanical ranking. It was a clear break with the past, although it did not present a complete review of the system, nor did it include the Latin botanical names, limiting its value to a collector looking for help to identify specimens in the field or in a herbarium.

It took some years for a Chinese author to offer a practical guide on how to classify freshly collected plant specimens. In 1914, in the first issue of the *Journal of Natural History* (Bowuxue zazhi), Wu Jiaxu reported on his recent visit to Jiangsu Province, possibly the first published account in Chinese of a collecting expedition by a Chinese botanist. Eight of the ten pages of his report consist of a table with an explanation of how to classify and record specimens, which he said is the most urgent task on returning from collecting in the field: "The layout of this table has been designed for practical use while doing research. Each species has been placed in its taxonomic position, from the lowest to the highest rank. I have included uses of the plant below its name. Where there are some that will probably not be familiar to most readers, I have included an explanation" (Wu Jiaxu 1914b, 135)

The second row in the table describes the taxonomic relationships between related plants ordered from the phylum (although he does not use the term *phylum*) to the family, omitting genus and species. The entry in the third row, which he labels "Name," is often the recognized botanical name for the species. Table 6.1 shows the format of Wu's table using his classification of the Chinese chive *Allium tuberosum* (*jiu*) on the second page.

The importance of Wu's table was acknowledged by his peers, although with some misgivings. In 1915, in the fifth issue of *Science*, Qian Chongshu, who was studying taxonomy at Harvard, applauded Wu for paving the way in developing teaching materials for students in China. He worried though that Wu had only included the Chinese terms for taxonomic divisions and had used the common Chinese names for plants, which could cause

TABLE 6.1 A taxonomic table devised for readers of the *Journal of Natural History* by Wu Jiaxu in 1914

EDIBLE PLANTS	CATEGORY OF USE
[Wu enters the name of the phylum directly, without the general term for phylum, which today would be *men*] Angiosperms (*beizi zhiwu*)	Taxonomic rank
Class (*wang*)—Monocotyledons (*danzi ye*)	
Order (*mu*)—Liliflorae (*bai he hua*)	
Family (*ke*)—Liliaceae (*bai he*)	
Jiu 韭 (*Allium tuberosum*)	Name
	Use
Found in all sixty counties	Distribution

Wu's table used only the Chinese terms for the taxonomic divisions. The table reads from right to left, with the user filling in the details of the plants they have collected in the left-hand column. The example here is the entry for the Chinese chive (*Allium tuberosum*) (*jiu*). (Wu Jiaxu 1914b, 138)

confusion at a time when there was still disagreement on the correct translations. It would be preferable, he suggested, to use both the Chinese and the Western (Latin) terms: "When a name includes the scientific name in Latin, there is absolutely no doubt as to its meaning. Every country has different names for its plants and animals, but the scientific Latin names are always the same in all countries. So once the name has been determined, it is appropriate to include the Latin name. Science belongs to the whole world, not to any one nation" (Qian 1915, 606).

NAMING THE MYRIAD THINGS

It was perhaps a little arrogant of a student in the United States to expect that his counterparts in China, with little access to the most up-to-date publications, should be familiar with the Latin botanical nomenclature. It was only the following year, in 1916, that *Science* published its introduction to the International Rules of Botanical Nomenclature and a list of eighty-one botanical plant names in Chinese next to the equivalent Latin binomials. In an effort, perhaps, to bridge the conceptual gap between traditional classifications and botanical taxonomy, the list was in two sections. The first consisted of forty-two species of "flowers and trees." The second group of thirty-nine species had the title "The Five Grains, Gourds, Fruit, Vegetables, and Materia Medica," an arrangement that closely echoes the familiar groupings found in earlier works on plants (Wang Yanzu 1916).

Understanding the taxonomic system itself, with the legitimacy and the written rules of the International Rules of Botanical Nomenclature behind it, was not a major hurdle for botanists. They faced a more difficult task, though, in deciding where to place Chinese plants within the scheme and how to match vernacular names with the Latin binomials. First there was the vexed issue of whether a Chinese plant name referred to a genus or a species or another, possibly higher rank. Then there was the question of identifying a specimen correctly, selecting one name as the agreed botanical term in Chinese (where there might in fact be several regional or local names for the same plant), then finally assigning a correct Latin binomial to the Chinese name. The task was complicated by the fact that if a plant had already been published and assigned a Latin name, the voucher with the type specimen was most probably in the herbarium of one of the European or American collections, inaccessible to Chinese botanists.

By this time, serious work was already underway to address the problem Du Yaquan had articulated in 1903: how to chart a course between the apparently innumerable plants to be found in nature, only a small number of which had been named in ancient Chinese texts, and the many modern names proposed by botanists for plants both known and unknown in the past. Hu Xiansu published papers in 1915 and in 1916 in which he identified plants described in the early-second-century dictionary, *The Analytical Dictionary of Characters* (Shuo wen jie zi) and matched them to their modern scientific names, usually to the family level, in some cases to the genus or species (Hu X. 1915 and 1916).

Having stated the problem, Du used his authority as editor-in-chief of the 1918 *Dictionary of Botany* to draw up guidelines regulating botanical nomenclature. The editorial committee adopted the principle that, as far as possible, the common Chinese name or a name found in earlier works on plants should be used. Where a plant with a vernacular name had been assigned a correct botanical name, this would be recognized as the standard. Where a plant with a Japanese name using Chinese characters had been assigned a botanical name, it would be acceptable to use the Japanese name, incorporating any vernacular Chinese names as part of the explanatory description for that entry. In cases where one plant had several Chinese names, they would appear in the dictionary, but only as alternatives to the botanically correct names for the genus or the species. The committee determined that the Latin botanical name should be included in all entries, as well as the English and German names. Different fonts and parentheses would help distinguish between the different sources of the nomenclature, the original authors in the botanical literature, and, where appropriate, references to earlier Chinese sources (*Dongfang zazhi* 1917, 16).

The principles and rules for ordering and naming plants were in place, but it was difficult to monitor their implementation. By 1919, the proliferation of names was "making scholars ill" (An and Zhang 1920, preface). Articles in the scientific press worried about the problem of different plants sharing the same name, which was likely to get worse as more Chinese botanists organized collecting expeditions, sending specimens back to their home institutions labeled incorrectly with local names, or careless, idiosyncratic identifications (An and Zhang 1920, preface; Liu S. 1919, 1920, and 1922). To bring some order to the growing confusion, Liu Shoudan, a professor of biology at Wuchang Normal College, drafted a corrected list of plant names, published in the college journal in three installments between 1919 and 1922. His method was to give the common name or names, followed by the correct botanical Latin name, a detailed botanical description, and a commentary explaining how he had matched one common name to one botanical name, often referring to Li Shizhen or to Wu Qijun and his *Research on the Illustrations, Realities, and Names of Plants* as authorities for his identification (Liu S. 1919, 1920, and 1922).

An alternative approach, also acceptable under the guidelines for the 1918 *Dictionary*, was to adopt the nomenclature devised by Japanese botanists using Chinese characters. An Ji and Zhang Zongxu, who may have been students in or just returned from Japan, compiled a directory using a list they had obtained in Tokyo from Matsumura Jinzō of 3,000 Chinese plants in his herbarium (An and Zhang 1920). The directory began with a list of the Latin binomials, followed by the Chinese name as far as it could be ascertained, cross-checking with names found in earlier works—again, most frequently Li Shizhen and Wu Qijun, but also referring to *The Literary Expositor* and more recent local gazetteers, encyclopedias, and literary essays. To facilitate searches, there was an index of the Chinese names (arranged by the number of strokes in the first character) and an alphabetical list of the romanized Chinese names.

In 1926, Bing Zhi wrote a review in *Science* of the difficulties zoologists and botanists shared in rendering binomial taxonomy into Chinese, especially in cases of animals and plants that were familiar to people and for which there could be many different names in literary works and in common speech. He pointed to practical difficulties in adapting Chinese to the Linnaean system. In Chinese, an adjectival or descriptive qualifier usually precedes the noun or generic word. The opposite is the case for Latin scientific names, where the genus comes first. Bing rejected a proposal that the Chinese order should be reversed in scientific writing in order to conform to the Latin order. He pointed out that this would only cause confusion, even among specialists, let alone among students just beginning to study taxonomy. He reminded readers that "This can be likened to Chinese names, where the family name is the equivalent of the genus and the given name . . . could be said to be the equivalent of the species" (Bing 1926, 1347). If it was acceptable to continue to use the traditional order for personal names (family name then given name), the same should hold true for scientific names, with misunderstandings among scientists being avoided by including the Latin binomial after the Chinese name (Bing 1926, 1349).

Bing felt strongly that scientific practice should not become exclusive, limited to a small caste of insiders. The essay recommended that in naming animals or plants, it was preferable to use existing characters rather than to invent new characters. While he recommended including the Latin binomial after the Chinese scientific name, he also rejected the more radical suggestion that it would be better for scientists simply to use the Western alphabet and Latin binomials and not concern themselves with devising new Chinese terminology: "This is not right, because it excludes ordinary people from learning about science. Science is an important part of the country's development, so it is wrong to limit it to a few specialists who can read the Western alphabet" (Bing 1926, 1348–49).

In his review, Bing acknowledged the efforts of the Joint Scientific Terminology Commission, a multinational consortium of medical missionaries, publishers, and professors of medicine under the auspices of the Jiangsu Provincial Educational Association (Jiangsu Sheng Jiaoyu Hui) that first met in 1916 (Peake 1934, 198; Luesink 2012, 56–83).[18] In 1924, the commission held a meeting on botanical terminology at which participants reported errors in the translations of many plant families, genera, and species. Political instability limited how research facilities and institutions might respond, though, and it was only after 1927, with the establishment of the Nationalist-led central government in Nanjing, that the conditions were in place to consider implementing any of the commission's recommendations. Regulations issued on July 4 that year governing the organization of a new University Council (Daxue Yuan) announced that it had established the Academia Sinica to organize and fund research in China. Biological research was to prioritize collecting and studying China's own indigenous flora and fauna as symbols of the nation, reinforcing an increasingly strident nationalist discourse of reclaiming sovereignty from foreign powers.[19]

At its inaugural meeting in 1928, the Academia Sinica established its own group to review botanical nomenclature based on the Joint Committee's earlier report. In 1932, it invited Zhong Guanguang to take charge of bringing order into botanical taxonomy and nomenclature (Jiang W. 1941, 59). The academy explained that "not only is Mr. Zhong very familiar with current knowledge about botany, he is also well versed in the historical materials. He has always said that botanists should take the old learning and match it to the new, fusing the two together like a tally, which would be a great achievement with broad applications. If this does not happen, the guardians of the national spirit will not understand any of the terminology, and it will be left to those who are steeped in Western learning to take care of it as though it were an imported good" (Zhong 1932a, 1 [footnote]).

Zhong accepted the invitation. In the introduction to his first report, he criticized scientists who had turned their backs on Chinese knowledge about plants and were both unwilling and unable to name plants in the field or to recognize anything that was not mounted on specimen vouchers housed in a herbarium (Zhong 1932a, 4). He identified three issues on which there had barely been any progress since 1924:

- There had been little or no effort to align scientific names with existing common names. Not only did this cause confusion, it made it difficult for botanists in the field to gather local information about the plants they were collecting.
- The use of Japanese names led to incorrect identification as well as confusion between Japanese and Chinese meanings and pronunciations of the same character.
- Many plants were known by several different names: classical, common, and Japanese. Since few botanists were knowledgeable about classical Chinese texts or understood Japanese, confusion had spread into textbooks all over the country, "becoming completely uncontrollable, like a mob of ox ghosts and snake spirits" (Zhong 1932a, 2–3).

Zhong believed that the names of a country's products bear the imprint of history and geography. He echoed the time-honored refrain that the first task of a ruler is to rectify names, and those who love their country will treasure those names, showing their esteem for customs and propriety. His goal was to promote the use of indigenous Chinese names, but the names should be correct, following the rules of nomenclature to show the relationships between families, genera, and species (Zhong 1932b, 12). They could then be used in education, in research and in professional botanical communications, and they would also be meaningful to nonbotanists.

The body of his report took the form of tables correcting errors the 1924 committee had made in classifying and naming plants. One table addressed the most basic error, the simple misidentification of families and species. Another table gave alternatives for many names that had been created using phonetic loans, on the grounds that they were meaningless to anyone other than a botanist, who would probably know the correct scientific Latin binomial anyway. Yet another table recommended standardized terms for plant anatomy to be used in constructing names by translating common Latin terms: *chang ye* (long leaf), for example, to render *magnifolia* in a binomial such as *Populus magnifolia*.

As he set about rectifying names, Zhong acknowledged that there were cases where he was not sufficiently familiar with indigenous names and had used a provisional term until such a time as a correct name could be determined (Zhong 1932b, 38). He was confident that if scientists were to make informed choices based on what they observed and if they were consistent in their use of what they determined to be correct, then common usage would catch up with the scientific terminology, and the scientific name would in due course become the common name. In conclusion, Zhong accepted that it would be difficult to achieve the philosophers' ideal state in which all objects are correctly named. His intent, though, was to refine the system of nomenclature so that it would be "more elegant than in the past, more concrete than would be the case outside the world of the learned, and more meaningful than the European rendition" (Zhong 1932b, 39).

Zhong Guanguang's report formalized the rationale and a methodology that transposed not just the terminology but also the conceptual framework of the internationally accepted system of botanical taxonomic nomenclature to China, assimilating, not rejecting, the deep reservoir of indigenous knowledge about plants. It was the outcome of several decades of work in which Chinese botanists trained abroad had been particularly prominent. Ironically, it was Zhong, who had not studied abroad, who had been given the responsibility of codifying an authoritative set of guidelines and steering them through what was becoming an institutionalized process of standardization and communication.

CHAPTER SEVEN

Botanical Illustration

> An illustration emphasizes drawing from life as a record of what has been observed. It is a complete description of the form [of the plant] and is especially fine if it is executed in color. It is ten times more useful than a specimen. Most specimens fade and lose their color. They are dull and they have lost their original appearance. This is why an accomplished botanist will ensure that each specimen is matched with a colored illustration drawn from life.
>
> —WU JIAXU, THE METHODOLOGY OF RECORDING PLANTS—CONTINUED ("ZHIWU JIZAI FA—XU")

IN the last of a series of four articles on plant collecting, Wu Jiaxu explained that a written description, while essential, is unable to capture all the information that is needed to identify a plant. Nor is a specimen in a herbarium adequate to convey the appearance of the live plant. An illustration complements the preserved specimen and the written record with a visual rendering that can reconstitute structures and colors lost in the process of preparing the voucher. Artistic techniques such as composition and framing can pick out plant structure, filtering can reduce the image to its essential elements, and juxtaposition makes it possible to compress time by including parts drawn at different seasons in one image.

Flowers and plants are favorite subjects of drawings and paintings in all cultures. The genre of flower and bird painting has a distinguished history in China, with several styles recognized by art historians. Some, such as the ink-wash style (*shuimo hua*) convey mood and spirit rather than accurate representation. Others such as the fine-brush style (*gongbi hua*) are prized for meticulous realism in depicting their subjects.[1] In China, as elsewhere, botanical illustration is considered to be a category separate from artistic expression, however aesthetically appealing the image might be. Illustrations of flowers and plants were essential to the materia medicas, agricultural manuals, and horticultural guides, but these representations generally fell short of what is expected of botanical illustration today (Zou X. 1998, 31–32; Mu and Wang 2017, 319).

A well-executed botanical illustration shows what the plant has in common with members of its genus or family, while identifying what distinguishes it as a species or variety. The scientifically complete portrait of a plant should show its roots, stems, inflorescence, and seeds. It should depict the morphology and structure of leaves and indicate the scale of the whole specimen and the relative size of its parts. Botanical illustrations work with the accompanying text as a primary reference. A successful illustration calls on the skills of an accomplished artist together with the specialized knowledge of the scientist. Learning to draw is often considered to be an integral part of botanical training, and the act of drawing itself can constitute an element of the botanist's close observation of nature (Saunders 1995, 85–100, 141–48; Endersby 2008, 112–36).[2]

Images are especially effective tools in marshaling the evidence to convince others of scientific facts (Latour 1986, 14–16). A botanical drawing validates a discovery and gives a plant formal standing in the body of botanical knowledge (Cholet 2019, 7). Botanical illustrations can be a record of things in a particular place that may not have been seen by the viewer. They are mobile, and they have been a means of "transporting" travelers' observations from distant places in the periphery back to the metropolitan (Lamy 2008; Bleichmar, 2011, 2016).

The nineteenth century was a golden age of botanical illustration in Europe. The nurseries to which plant hunters such as Robert Fortune sent specimens and field sketches from China took advantage of the appeal of vivid pictures of previously unseen new varieties in popular catalogs advertising their wares (Desmond 1977). The accurate portrayal of newly identified species overlapped with a more aesthetic sensibility in opulent portfolios, sold to subscribers, that blurred the distinction between scientific representation and fine art. There is, nevertheless, a boundary between pictures of plants as art and pictures of plants as a scientific record. For some, the fear is that the objectivity of science could be compromised if an image shows any trace of individual interpretation of natural objects, any suggestion of human hopes or aesthetics. During the late nineteenth century, the compilers of flora and botanical reference works took every precaution to exclude any hint of subjectivity, using mechanical devices such as the camera obscura, projection, direct printing, and tracing to guarantee the accuracy and objectivity of their work (Daston and Galison 1992, 81–87; Galison 1998). In the 1950s, Feng Chengru, now honored as the father of botanical illustration in China, dedicated a chapter of his manual on biological illustration to the use of these aids, with diagrams and photographs showing how to set them up and with formulae for scaling and sizing. The student is constantly reminded that this is scientific work (Feng C. 1959, ch. 4).

For all the emphasis on objective science, the illustrator is creating a picture. A botanically accurate picture that is also aesthetically appealing is more than a helpful complement to the text. It fixes facts about its subjects in a visual language that can be more legible to the viewer than words (Pang 1997, 142–45). The botanical illustration is confirmation that art and science are not incompatible. They enjoy a symbiotic relationship in which intent rather than ability separates the two genres.[3]

THE VISUAL REPRESENTATION OF PLANTS IN CLASSICAL CHINA

In a frequently quoted essay, *A Brief Account of the Graphic Arts* (Tupu lüe), the southern Song writer Zheng Qiao (1103–1162) wrote that "an image is the warp (*jing*), while the text is the weft (*wei*). The warp and the weft intersect to make a pattern in the cloth. The image is a being that grows, and the text is a being that moves. The being that moves and the being that grows depend on each other to generate change" (Zheng Q. 1150, 1b) As a practical matter, he believed that "the shape of insects and fish, the form of plants and trees cannot be identified without images" (Zheng Q. 1150, 8a).[4] Zheng's essay makes it clear that image and text are both essential to a proper understanding of the beings that are generated in the course of the constant changes of yin and yang, of qi and the Five Phases (Bray 2007, 39).

Illustrations have enriched many of the works dedicated to plants. An image may be no more than a simply drawn reference to information embedded in the text. In other cases, it might be a finely rendered picture of a plant that accompanies a celebratory anthology of verse and prose. The text and the image work together, but the weight given to either medium has not always been the same.[5]

It can be quite problematic to imagine what the original illustrations in the oldest texts might have looked like. Are the images in later, extant editions accurate copies and a good indication of the state of botanical illustration at the time? Or are they more recent reprints from a later edition, bearing little or no relation to the original? By the Ming dynasty, a flourishing publishing sector ensured that books were produced in sufficient quantities such that many have survived to the present. Movable type existed, but woodblock printing was cheap and could easily accommodate the inclusion of illustrations. Carving woodblocks is a technique that favors line drawing, eliminating distracting details such as shading and depth. The subject appears, almost without exception, isolated against a plain white background, a property that would be described today as decontextualization. These qualities are intrinsic to the medium itself, and they also happen to be conventions in modern botanical illustration (Saunders 1995, 15).[6]

The best known of the classical works about plants are the materia medicas, of which forty are illustrated, with a total of over ten thousand images (Zheng J. 2018, 151). The oldest extant illustrated compendium is believed to be *The Illustrated Pharmacopeia* (Bencao tu jing), compiled during the Song dynasty by Su Song (1020–1101) and published in 1061. It was an imperially commissioned survey of medicinal plants from the whole country and is said to have included over nine hundred illustrations of more than three hundred plants. The original images have been lost. Some of them are said to have been copied and used in later compendia, although it is not easy to know which were copied and to what extent the copies differed from the originals.

The earliest known work with colored illustrations is *The Cliff Walker's Herbal* (Lü chanyan bencao) by Wang Jie, compiled in 1220. Wang was an artist rather than a specialist in medicinal herbs, but he had an interest in the plants that grew wild around his mountain retreat outside the Southern Song capital, present-day Hangzhou. Two hundred and two of the original 206 illustrations still exist, each one filling a page, with a brief account of the plant's medicinal uses on the opposite page but with no description of its morphology. The images are exquisitely drawn, almost certainly from life, carefully proportioned and highlighting the subjects' distinguishing features, making textual description unnecessary.[7] The original was probably hand-colored. Zheng Jinsheng (1989) has described the painstaking research involving bibliographic studies, textual comparisons, and even studies of the late Song Hangzhou dialect that was needed to be able to conclude that the one remaining copy of *The Cliff Walker's Herbal* is from the Ming dynasty, well over a century after Wang Jie first published it, but that the illustrations are likely to be faithful reproductions of his work, and that the text, at least, is very close to the original. There were few copies in circulation even at the time it was published, so it did not have much impact on later herbal compendia. Only one copy, a facsimile from the Ming dynasty, has survived to the present.[8]

Another category of the materia medica consisted of guides to edible plants for use in times of famine. They were prepared for administrators who were unlikely to have much knowledge about plants, so images were important aids to ensure that the victims of famine found the correct foods without the risk of poisoning (Needham, Lu, and Huang 1986, 328–55). The first of these guides, *Treatise on Plants for Use in Emergency* (Jiuhuang bencao), published in 1406, has 414 illustrations, 276 of which are known to have been commissioned from professional artists by the author, Zhu Xiao (1360–1425), the fifth son of Zhu Yuanzhang, the founder of the Ming dynasty. As with *The Cliff Walker's Herbal*, the text is brief, but the images, known to be the originals, are sufficiently detailed that it would not be difficult to recognize the plants in the wild.[9] As might be expected, the emphasis in both the text and the illustrations is on the parts of the plant that can be eaten safely, such as the bark and the leaves. The entry for the camellia, for example, states where it is found, points out the distinguishing features of the bark and leaves, then gives instructions on how to prepare the leaves for food or as an infusion in boiled water to drink (Zhu X. 1406, vol. 2, 372–73).

Despite its iconic status as the most comprehensive, carefully researched of the genre, it is widely acknowledged that the eight hundred illustrations of plants in Li Shizhen's *Classification of Materia Medica* are of markedly inferior quality than the text. They have been described as "crude and ungainly" (Zheng J. 2018, 156) and "somewhat astonishing" (Haudricourt and Métailié 1994, 388). The images are not integrated with the text but are collected at the beginning of the work. In some cases the illustration does not match what is said about the plant in the text, showing, for example, seedpods or root structures of a different shape from Li's description (Li and Chen 2015, 166). The body of exquisite woodblock prints from this period is evidence that the poor quality of illustrations in *Classification* cannot be attributed to a lack of drafting and printing skills at the time.[10] The images were probably not intended to be used to identify the plants but more as a reminder to jog the memory of users as they stocked their pharmacies or assembled the ingredients for prescriptions.[11]

A striking contrast to the disassociation between text and image in *Classification* is the way that Li Zhongli, a near-contemporary of Li Shizhen, integrated them in his *Sources for Materia Medica* (Bencao yuanshi), printed in 1612. He is known to have made the 442 drawings himself from examples he found in apothecaries. His illustrations accurately record the appearance and morphology of each plant or substance. They include cross-sections and grouped images showing the organism in several different states or at different seasons. Li's written descriptions include the place of origin, notes about taste, medicinal properties, and the most common applications of the substance, sometimes with prescriptions in which it is the active ingredient. Text wraps around the image to form one integrated page, unlike the more conventional layout of text and illustration on opposite pages or organized in separate sections of the work. Most importantly, the illustrations include annotations within the space occupied by the picture, with notes attached to the drawing to indicate what to look for, a presentation that helps the user ensure they are getting the desired product.[12]

The entry for tea (figure 7.1) shows both the layout and the content of the description. The heading, at the top of the right-hand page, is the single character *ming*, highlighted as a white character against a black background. *Ming* refers to tea that is picked later in the season. The more familiar *cha*, used for leaves picked earlier in the season as well as for the beverage, appears in small characters below the heading. The first section of text describes in large characters where it is grown and the season that new leaves grow, and gives several alternative characters found in literary sources. The next three lines, in smaller characters, describe the taste of the leaves, the medicinal properties of tea, and some of the ways it can be prepared with other ingredients, such as ginger and onions. The drawing of the tea bush is simple but correctly shows the narrow, pointed leaves and the pairs of leaves at the end of each branch that are picked for processing. Directly above the illustration are notes about the time and season tea should be picked, the preferred size of leaves, and a note about cultivation. The final section is an anthology of anecdotes about tea in other herbals and sources, such as monographs on tea. *Sources for Materia Medica* was very successful, was reprinted several times, and is said to have influenced the composition of later Ming and Qing materia medicas (Zheng J. 2018, 155).

Materia medicas represent just one category of illustrated texts pertinent to botany. As befits a culture known for its gardens, horticultural manuals and monographs are another source of information about plants, especially ornamentals.[13] Some manuals were almanacs, practical guides with reminders of what to plant each month and which tasks, such as propagation, pruning, grafting, or fertilizing, should be done, with instructions for how to carry them out. They were often illustrated with woodcuts of varying quality. *The Mirror of Flowers* (Hua jing), for example, by Chen Haozi, published in 1688, has ninety-four illustrations of plants, many copied

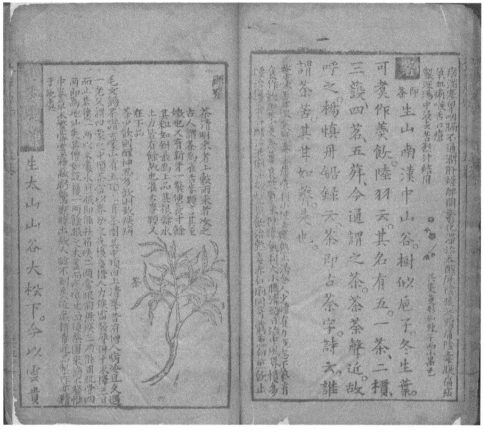

FIGURE 7.1 Tea (*ming*) from *Sources for Materia Medica* (Bencao yuanshi) published in 1612. The right-hand page follows the conventional form, describing where the plant grows, its origins, and its medicinal uses. The second (left) page of entries, in a format unique to this work, wraps text about parts of the plant around a drawing that is a recognizable rendition of the plant. The text above the drawing describes the difference in appearance and the taste when brewed of leaves of different sizes when picked. Courtesy of the National Library of China, Beijing.

directly from Li Shizhen's *Classification of Materia Medica*. The placement and ordering of the illustrations make it clear that they were included to evoke the pleasures of the garden rather than to identify its plants (Halphen and Métailié 2006, 158–83). Other manuals, such as *The History of Flowers* (Hua shi) from the Ming dynasty, are sumptuously beautiful, with accurately drawn, colored illustrations; fine calligraphy in the titles; and a curated literary anthology for each plant (figure 1.2). *The History of Flowers* is organized by the season of flowering, but otherwise there is little that would make it of practical value to a gardener. The intent seems to be to provide an elegant companion to scholarly reflection in the garden rather than a guide to identifying flora.[14]

Monographs on popular plants, such as the bamboo, plum, chrysanthemum, or camellia, might be expected to make good use of illustrations. Their content covers a spectrum, from solid horticultural advice to albums that are almost entirely visual, created for aesthetic appreciation. Very few of them integrate text and images. *Gao Song's Compendium on the Chrysanthemum* (Gao Song ju pu), published in 1550, has 112 illustrations by Gao Song (1522–1566), a scholar known for his poetry and paintings. The monograph begins with seven pairs of rhyming couplets, which capture the flower's most distinctive features in just seventy characters, showing off Gao's skill as a writer. Ten pages of drawings follow, illustrating the couplets, then two more, each spread over two pages, showing the chrysanthemum's most prized virtues for the gardener: it blooms in the autumn and it readily produces sports (mutations), making it easy to propagate new cultivars from cuttings. The body of the work consists of drawings of a hundred varieties organized by color: yellow, white, red, and purple, each accompanied by a poem (although the book itself is printed in black and white). The first two pages illustrating the first four couplets are reminiscent, at first sight, of the layout of a modern botanical drawing, with different views of the inflorescence and tags identifying different parts. On closer examination, though, it turns out that it is a guide for the chrysanthemum fancier to the desirable features to look for when selecting or breeding new varieties. They are not intended to be nor do they work as botanical drawings to identify varieties and cultivars of the genus.[15]

Wu Qijun's *Research on the Illustrations, Realities, and Names of Plants* does not fit any of the conventional categories of works on flora and vegetation. Wu set out to identify plants correctly rather than to record their medicinal or other properties, so his point of reference was the written corpus of classical sources going as far back as *The Literary Expositor* and earlier. The 1,800 illustrations were drawn from life in the course of his field observations, and they are more accurate than anything that came before them, precisely highlighting the most distinctive features of the plant, flowers, leaves, and fruits, as appropriate.[16]

Wu served in an official capacity in Guangdong and Fujian, where the presence of foreign traders had generated a market for paintings by local artists of Chinese scenes, gardens, and plants. He also spent time at court in Beijing, where he might have seen work by the Jesuit court painter

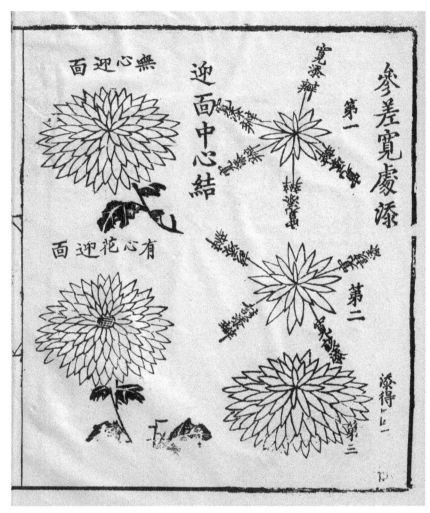

FIGURE 7.2   Introductory page of *Gao Song's Compendium on the Chrysanthemum* (Gao Song ju pu) (1550), with tags showing the desirable features the chrysanthemum fancier should look for in new varieties. From a facsimile of the original published in 1959.

Giuseppe Castiglione (1688–1766) and others.[17] It is at least possible that his drawings might have been influenced by these encounters (Li and Chen 2015, 168). Although there is no evidence to show this might have been anything other than a fortuitous convergence of time and opportunity, this was a time, though, when Chinese artists began to draw and paint Chinese

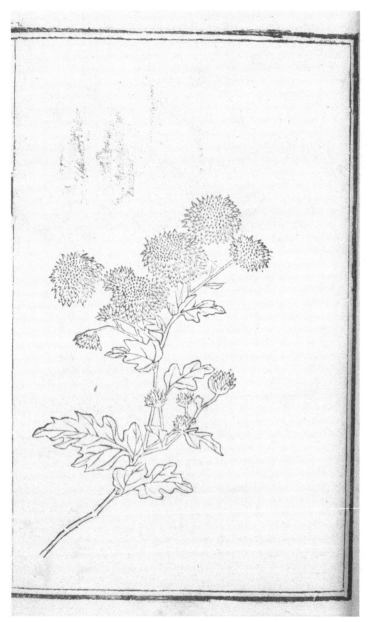

FIGURE 7.3 Wu Qijun, *Research on the Illustrations, Realities, and Names of Plants*. Starfilled Sky (*Man tian xing*) *Gypsophilis paniculata* L. (commonly known as baby's breath). Guangxu 6 (1880) reprint of Daoguang 28 (1848), Taiyuan Fu Shu edition, juan 27, 4a. Courtesy of the Asian Library, Special Collections and Libraries, Claremont Colleges Library.

plants more or less according to botanical conventions for audiences in London, Paris, St. Petersburg, or Boston (Mu and Wang 2017, 302–4).

BOTANICAL ILLUSTRATION: INTERNATIONAL PRACTICES COME TO CHINA

The herbarium vouchers James Cuninghame collected during his two visits to China (1697–99 and 1699–1709) are some of the earliest physical specimens of Chinese plants, but they are now dried and have lost their color. Associated with that collection, there is also a portfolio of several hundred paintings now in the British Library, commissioned from Chinese artists in Amoy (today's Xiamen), which Cuninghame brought back to James Petiver in London for study. Among these paintings are some with a note that they were sent by Christopher Brewster, who had traveled to China on the same ship as Cuninghame. A collection of 376 paintings commissioned by Brewster in Amoy, now at the Oak Spring Garden Foundation Library, has an additional annotation that they were "Done at Emuy in China by Doctor Bun-Ko, bro[ught] thence by Mr Chr[istopher] Brewster 1701." Most of the paintings include the Chinese name of the plant and a romanized rendering of the pronunciation. Some of them are very much in the traditional school of flower and bird painting. Some are comparatively crude, and some are quite finely executed. A very few show the flower, leaves with both surfaces visible, and the seeds or fruit all in the same composition. Whatever their shortcomings as botanical illustrations, these are almost certainly among the first images by Chinese artists for the purpose of botanical identification, and Dr. Bun-Ko would be the first Chinese botanical artist known by name.[18]

Under the very different conditions of the Canton trade system, with Westerners confined to their factories, the horticultural and botanical establishments reached out to employees of the different countries' trading companies with requests for plant materials and any information they could glean about the flora of China (Fan F. 2003a, 67–74). Very few of the plants sent to Europe survived the hazardous sea journey, so there was great interest in paintings commissioned from Chinese artists. John Bradby Blake was one of these correspondents. He spent seven years with the East India Company in Canton, from 1766 until his death in Macao at the early age of twenty-nine in 1773, long enough to assemble a carefully selected collection of botanical paintings. The paintings were executed by Chinese artists following the emerging standards in Europe for botanical illustration. Blake is said to have spent eight or nine hours a day "laying out the

FIGURE 7.4 Pear (*lizi*) from a portfolio painted for Christopher Brewster in 1701 in Amoy (Xiamen), with the attribution "Done at Emuy in China by Dr. Bunko." Brewster portfolio 1, no. 50. Courtesy of the Oak Spring Garden Foundation, Upperville, Virginia.

natural specimens as they were from time to time gathered, [and] dissecting the parts of fructification." Those who saw the paintings in London were fulsome in their praise, describing them as "elegantly and scientifically disposed" and so accurate that the "ingenious and learned botanist, Dr. Solander . . . has classed and arranged the plants they represent according to the great Linnaeus's system from their parts of fructification" (*Gentlemen's Magazine and Historical Review* 1776, 349). Only one of the artists is named: Mai Xiu, transcribed at the head of one of the pages in the archive, using the Cantonese pronunciation, as "Māāk or Mauk Sow-u. The Name of Mr. Blake's Painter who went every Day to Mr. Blake's Rooms—about 33 Years of Age. Middle Size. Lives at Canton." Mai was almost certainly not the only artist working with Blake, but he was the first Chinese artist to have received some training in scientific botanical illustration whose name is known today.[19]

Some twenty years later, in 1817, the Horticultural Society of London engaged another employee of the East India Company to send plants and drawings from China. John Reeves, who was in Canton from 1812 to 1831, is said to have taught the artists he engaged there, supervising their work and supplying them with suitable paper from Europe, watercolors, and pencils. Four artists are named on the drawings: Aku, Akam, Akew, and Asung—unfortunately only in a transcription from the Cantonese pronunciation, not in the Chinese characters. Reeves had at least four copies made of each painting and sent them to the Horticultural Society and other institutions, such as Kew Gardens and the Royal Botanic Garden in Edinburgh, where they were added to collections accessible to botanists and to interested members of the public.[20]

By this time, there was a flourishing export market for Chinese paintings, with some studios in Canton specializing in natural history and flowers. Much of their work was derived from traditional bird-and-flower–style painting, but they produced some botanical art for a clientele in Europe interested in natural history.[21] Chinese artists had become proficient in the genre and worked to satisfy the demand from abroad. A scholar such as Wu Qijun might have seen their work while in Canton, but it does not appear to have had a direct influence on him or his peers outside the foreign enclave involved in the broad learning of things (*bowuxue*) or natural studies (*gezhixue*).

The botanical texts published in Shanghai by Inkstone Press and the Educational Association of China were well illustrated. The translation of Lindley's *Elements of Botany* had some two hundred drawings. John Fryer's *Plants Illustrated and Explained* was built around a selection of 154 plates—also from Lindley's *Elements of Botany*. Not only did they introduce a new vocabulary but they also brought the most up-to-date imagery and printing technology to their readers. The same set of illustrations reappeared in journals such as the *Chinese Scientific and Industrial Magazine*, familiarizing readers with what was by now a systematic visual language to represent not just the appearance and morphology of plants but also their anatomy and structure, using dissection, cross-sections, and microscopy. It was some time, though, before Chinese artists began to explore this new territory (Sun, Ma, and Qin 2008, 776–77).

CAI SHOU, 1879–1941

In March 1907, the monthly *Journal of National Essence* (Guocui xuebao) launched two new columns, "Natural History" (Bowu) and "Art" (Meishu). In June, they added "Natural History Drawings" (Bowu tuhua), sometimes just a single plate, sometimes two to a page, sometimes two on facing pages.

Over the next four years, ending with the September 1911 issue, the journal published a total of 128 illustrations of animals, fungi, and crustaceans, but mostly of plants and fruit. The captions and seals in the pictures tell us that they were all by Cai Shou from Shunde in Guangdong. The plates are as varied in style as their subject matter. Some are close to traditional brush-and-ink paintings with a calligraphic inscription. Some are simple line drawings with no wash or shading. Some show just the subject against a plain background, while others include elements of a landscape or habitat. Sometimes an explanatory caption refers to the Chinese classics or gives the Latin botanical name with notes from Western scientists on subjects ranging from plant anatomy to evolution. The text often refers to a place where Cai saw the subject, confirming (or rejecting) from his own observations what he believed to be problematic statements about it in the classical literature (Guo Y. 2013, 60–62).

One of the very few biographies of Cai says that "he took part in all the movements and joined the scholarly associations of the time without ever attracting anyone's attention" (Guo Y. 2013, 59). He was born in Longjiang township, Shunde, Guangdong. His name was originally Cai Xun, but he changed his given name to Shou and used numerous other names and styles at different times. In English, he signed his work as Y. S. C. He enjoyed painting and poetry when he was young and left home at seventeen to study in Shanghai. He probably learned English at Aurora University (Zhendan Daxue) between 1903 and 1905.[22] A political activist, he had to leave Shanghai in 1905, moving around southern China, finding himself in 1907 in Beihai (present-day Guangxi) as an interpreter at the British consulate, where he had befriended William Fletcher, a sinophile diplomat with whom he was in touch for many years. He spent a year in Hong Kong then returned to Shanghai in 1909.

In 1905 Cai joined the Society for Preserving National Learning (Guoxue Baocun Hui) and worked on its *Journal of National Essence*, founded that year to rescue and to protect China's artistic and cultural heritage. The society was virulently anti-Manchu, but it directed its ideology toward a rediscovery and revitalization of the essence of traditional values, arguing that "just as the Renaissance revived Western classics, what China needed was a revival of pre-Qin learning which flourished in the late Zhou" (Fan F. 2004b, 415). The embrace of the old learning did not preclude accepting Western scientific concepts such as evolution, for example, which, in the society's eyes, allowed for the satisfactory possibility that Manchu and other non-Han people had fallen behind in the march of evolutionary progress. The society also affirmed a link between a land, a people, and their culture, which stimulated an interest in local geography, with its history inscribed on the landscape as temples, monuments, and memorials, each a particular confluence of qi and the land nurturing its own unique assemblage of the myriad things—hence the society's interest in publishing a series of natural history illustrations.[23]

In his drawings, Cai wanted to show that China had a wealth of traditional knowledge about the natural sciences and that there was evidence in the classics to show it. The quality that differentiated his natural history drawings from art was that they were the result of direct observation. Captions tell us that he drew from specimens collected while he walked or hiked, or from things he found in the markets. The text associated with the image of the "hanging bell flower" (*diao zhong hua*) (figure 7.5) says that there are usually nine to fifteen or sixteen flowers on each inflorescence. "People who say that the only place that you can find them with twelve flowers is Dinghu Shan are wrong. I have hiked around Dinghu Shan. There are only a few plants growing there and it is forbidden to cut them, so those that are being sold in the city are certainly not from this mountain." Cai may have been rather optimistic about the effectiveness of a harvesting ban, but his note confirms that his comments in these illustrations are based on personal observation, not just on hearsay.[24]

Cai himself formulated standards for his natural history drawings. The image usually fills the frame—even when it is in the traditional ink-and-wash style in which empty space would be a part of the composition, accentuating the calligraphy that complements the image. The drawings are precise and accurate, although most of his illustrations did not include dissections. He usually followed botanical practice in laying the specimen flat, spreading it out to show the upper and lower surfaces of leaves and flowers. Cai sometimes included the Latin botanical names in the captions. He referred to classical Chinese sources such as *The Literary Expositor*, different materia medicas, and local gazetteers, but when he doubted what he found there, he searched for answers in the Western literature, too. Cheng Meibao has identified some images that are close copies of English-language reference works that Cai had probably consulted at the North China Branch of the Royal Asiatic Society in Shanghai. He was clearly aware that natural history drawings needed both scientific knowledge and artistic skills.[25]

Cai was interested in curiosities, things that were not found in the Chinese literature. His drawings included exotic fruits such as the jackfruit (*Artocarpus heterophyllus*) (*boluomi*) from Southeast Asia and the pitcher plant (*Nepenthes* sp.) (*ping cao*) from Borneo. He believed that his task was to identify objects correctly and to verify whether they had been known in

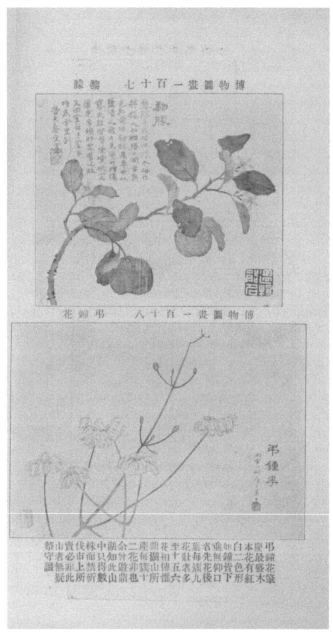

FIGURE 7.5 One plate with two natural history illustrations by Cai Shou published in 1911 in the *Journal of National Essence* (Guocui xuebao). Top: lemon (*li meng*). Bottom: "hanging bell flower" (*diao zhong hua*) (*Enkianthus quinqueflorus* Lour.). Courtesy of the Shanghai Library.

the past. For the scholars who had preceded him, the classics were the point of reference. If the object did not match what they read in the texts, it was not relevant, it was nameless (*wuming*), as Wu Qijun noted in such cases. For Cai, this only meant that he had to look more widely for information. He asked local people about what he saw and recorded what he learned until he was confident that he knew what he was drawing.

In September 1911, the *Journal of National Essence* published Natural History Drawing number 128, the last in the series. In the same year, the journal was renamed the *Journal of Ancient Learning* (Guxue huibao). The priority for the Society for Preserving National Learning was no longer to develop a new Chinese learning based on the old, judiciously enriched with some content from the West. It was now urgent to recapture the ancient learning before it was too late. It seems that at this point, Cai Shou lost interest in the new directions in illustration he had been exploring and moved into a more conventional world of traditional art, antiquarianism, and archaeology, which he pursued with his wife, Tan Yuese (Cheng 2007, 48). In the 1930s, he was working at the Palace Museum in Beiping. In 1931, Tan was appointed one of the founding directors of the Guangzhou History Museum, and he moved there with her. Cai died of heart failure in 1941 at the age of 62 (Guo Y. 2013, 60–62). He is now remembered as a painter and calligrapher, a poet, an antiquarian, and an archaeologist. His early botanical illustrations are all but forgotten.

FENG CHENGRU, 1896–1968

Cai Shou was a scholar steeped in the classics who had innovated in creating natural history drawings combining elements of scientific illustration and traditional Chinese painting styles. His subjects were drawn from the natural world, but his concern was in how they were emblematic of what made China's environment, history, and culture unique. He followed a short-lived trend of returning to the past that gave way after the 1911 Revolution to the modernizing spirit of the May 4th Movement (Guo Y. 2013, 63). It was that modernizing spirit, by contrast, that drove Feng Chengru, an artist who had deep knowledge about and respect for the history of illustrations of plants, animals, and insects in China but saw a clear difference between traditional representation of flora and fauna and what he identified as "biological illustration" (*shengwu huitu*). Over some thirty years, he dedicated himself to perfecting his craft and to teaching a cohort of botanical artists who later became the core of the *Flora of China* project that was finally completed in 1999.

Feng Chengru was born in Yixing, Jiangsu. He studied arts and crafts at the Third Normal College of Wuxi (Wuxi Di San Shifan Xuexiao), graduating in 1916. In 1920, the Nanjing Higher Education Normal School (Nanjing Gaodeng Shifan Xuexiao) invited him to prepare wall posters and other visual teaching aids for classes in Chinese, history, and geography. While there, he enjoyed hiking in the hills around the city and sketching what he saw. He became skilled in drawing from life and developed an interest in Western oil painting from his visits to art galleries in the city. By 1922, Bing Zhi and Hu Xiansu of the Biological Laboratory of the Science Society of China knew of his work and asked him to draw large colored posters for the classes they taught at National Southeastern University, and illustrations for some of the laboratory's publications. One of Feng's first assignments was to illustrate papers on the genetics and morphology of goldfish by Chen Zhen (1894–1957).[26] At the same time, he was collaborating with Chen Huanyong on *The Economic Trees of China* (Zhongguo jingji shumu) and an *Illustrated Manual of Trees* (Shumu tushuo), both published in 1925. These are acknowledged to be the first illustrated botanical atlases in China, with Chen's description of each species accompanied by a lithographed plate drawn by Feng.[27]

When Bing and Hu left Nanjing for Beiping to establish the Fan Memorial Institute of Biology, they invited Feng to join them as the institute's resident artist and director of its printing and publishing arm. Over the next ten years, until the Japanese invasion, he worked at the institute, producing 250 plates for *Illustrated Plants of China* (Zhongguo zhiwu tupu), a five-volume flora of China published in 1927, and 200 plates for a four-volume study, *Icones filicum Sinicarum* (Zhongguo juelei zhiwu tupu) (Illustrated manual of the cryptogams of China) compiled by Qin Renchang and Hu Xiansu (Qin and Hu 1930–37).

It was during his years at the Fan Memorial Institute that Feng refined and perfected his techniques of botanical illustration. He is credited with developing an approach that combined the heritage of Chinese traditional artistic techniques with Western scientific botanical illustration. In practice, this meant that he used fine calligraphic strokes to create line drawings of his subjects. He rarely used shading or perspective, although he believed that, used sparingly, they could clarify some of the details of a drawing that were needed as diagnostic features: "Think about how to depict a spider's legs. In traditional Chinese painting, one brushstroke would be enough. In scientific illustration, the artist must show each joint, the sensory pads that you would see using a lens and the drawing must show the right number and the correct arrangement of the pads because

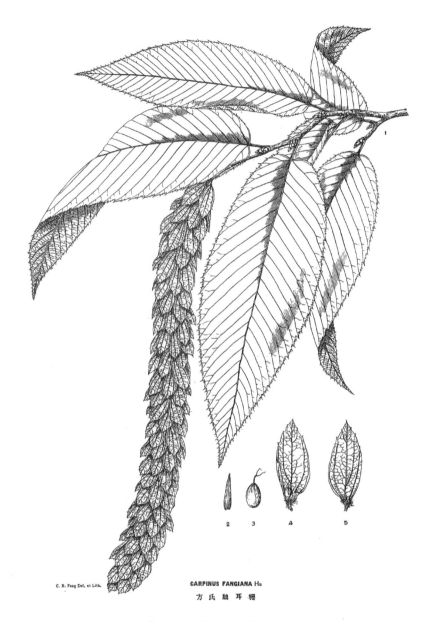

FIG. 102.—CARPINUS FANGIANA.

FIGURE 7.6 Drawing by Feng Chengru of *Carpinus fangiana* (hornbeam). This is perhaps the first botanical drawing by a Chinese artist to be published outside China. From the *Journal of the Royal Horticultural Society of London* 63, no. 8 (1938): 389. Courtesy of the Botanical Library at the Huntington Library, Art Museum and Botanical Gardens, San Marino, California.

these are among the features that the taxonomist will use to identify the spider. So it will not be enough to render it with just one stroke of the pen or brush" (Feng C. 1959, 3).

In the introduction to the manual on scientific drawing that Feng compiled from lecture notes assembled over some twenty-five years of teaching, he articulated his philosophy of scientific drawing, saying that it is essential for the artist also to be a specialist in the relevant sciences: "Even if there are flaws from the artistic point of view, it is important to get the scientific aspect correct. It must be based on observation. It should be as though there were a fine screen over the viewer's eye so that there is an unblemished view of the object. If the image is missing any identifying characteristic at all, it might look fine as an artwork, but it will be of little scientific value because it is not completely accurate" (Feng C. 1959, 2).

Admirers of Feng's work refer to his innovations in the tools of his art. In the search for objective accuracy, he recommended the use of ink rubbings, tracing, drawing from projected images, and even carbon copies as ways to capture intricate details, such as serrated leaf edges, pubescence, or complex root systems (Feng C. 1959, ch. 4). He invented a small calligraphic brush with a retractable tip that could draw lines of different weight:

> I found a brush used in Japan for lithography and made some modifications so that it is commonly used now for botanical work. It can be made larger or smaller automatically by retracting the brush tip. There should be no more than ten strands of bristle. I prefer the springiest kind made of wolf's hair. Tie it at the base with a very fine thread and insert it into a fine bamboo so that it is possible to extend it or retract it. The advantage of this brush is that the artist has control over the thickness of the line by the way pressure is applied while drawing. This is not possible with the hard, firm tip of a pencil or pen. (Feng C. 1959, 12–13)

When the Japanese occupation forced the Fan Memorial Institute to close in 1941, Feng left the city to return home to Yixing. In 1943, Hu Xiansu and Bing Zhi helped him rent two rooms, which he called the Jiangnan Arts Academy School of Biological Illustration (Jiangnan Meishu Zhuanmen Xuexiao Shengwuhua Zhuanxiuke). In Beiping, he had already taught a group of illustrators. At his academy in Yixing, he gathered some twenty students, including several of his own family members, his younger brother, his son, and several nephews, who all became noted botanical illustrators in their own right after the war.

The Fan Memorial Institute returned to Beiping after the Japanese surrender, and Feng continued to work with Hu, with whom he enjoyed a close personal and professional friendship. In 1948, his drawing illustrated the first published botanical description of the dawn redwood (*Metasequoia glyptostroboides*) (*shui sha*) in the *Bulletin of the Fan Memorial Institute of Biology* (Hu and Cheng 1948, 153–63).[28] After the Fan Memorial Institute had been absorbed into the Chinese Academy of Sciences, Feng and Hu continued their partnership. When Hu was attacked during the Cultural Revolution, his friends and colleagues suffered by association. After Hu died in 1968, hearing that a group of Red Guards was coming from Beijing to subject him to what would be a harsh "criticism meeting," Feng committed suicide with an overdose of sleeping pills (Hai 2017).

KUANG KEREN, 1914–1977

The war years brought institutional upheavals and shortages of basic supplies, such as ink and paper, that reduced activity in botanical illustration to a minimum. Only two illustrated flora were completed and published between 1937 and 1948, both under the direction of Kuang Keren, a young botanist who returned to China in 1937 from Hokkaido Imperial University in Japan. Described as a "brilliant taxonomic botanist and lithographer" by Joseph Needham, who met him in Kunming in February 1943 and again in August 1944,[29] Kuang first taught in Guizhou then accepted a position in 1941 as a researcher at the Yunnan Institute of Agriculture, Forestry, and Botany. He was a specialist in the taxonomy of Juglandaceae (the walnut family), and he took on any assignment that the institute had identified as being necessary for the war effort.[30]

Between 1942 and 1945, Kuang took part in a botanical survey of Emei Shan with Sichuan University. In addition to his contributions as a taxonomist, Kuang proved to be a talented self-taught artist, drawing fifty of the plates in the published flora, *Icones plantarum Omeiensium* (Emei zhiwu tuzhi) (Fang W. 1946). The second project was commissioned by the National Institute of Chinese Medicine (Guoli Guoyi Yanjiusuo) in an attempt to supplement the meager supplies of medicines available in the unoccupied areas of the country. Yunnan had long been known for its diversity of medicinal plants and the many traditions of herbal medicine used by different ethnic groups in the province. Kuang was one of a team of four who compiled the first volume of what was planned to be a series on the medicinal plants of Yunnan.

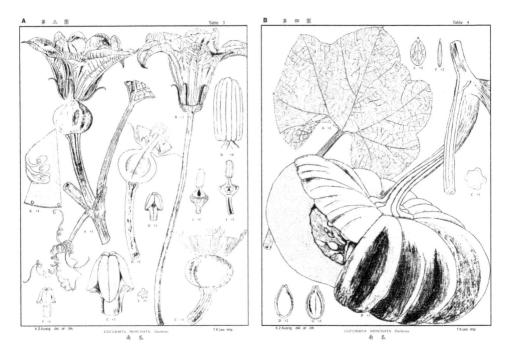

FIGURE 7.7 Kuang Keren. Pumpkin (*Nan gua*) *Cucurbita moschata* Duchesne. Plates 3 and 4 in Jing et al. 1945. *Illustrated Materia Medica of Southern Yunnan* (Diannan bencao tupu) was one of only two botanical works produced during the war, due to shortages of paper, ink, and other publishing materials. Courtesy of Kuang Boli.

With paper and ink in short supply, the institute could only print the first volume of *Illustrated Materia Medica of Southern Yunnan* (Dian nan bencao tupu) (Jing et al. 1945). Each of the twenty-six lithographed plates is accompanied by a lengthy text, beginning with references from the classical literature, followed by a full botanical description, a review of the Western botanical literature on the species, and information collected from herbal medicinal practitioners around the province. Kuang was responsible for eight of the illustrations, with the rest drawn by other artistically gifted taxonomists at the institute. His work is notable for its striking compositions filling the page, arranged to show all the diagnostic parts of the plants, spread over two pages if necessary when the original is very large—such as plates three and four, the pumpkin (*Cucurbita moschata* Duchesne) (*nan gua*).[31] Kuang and his colleagues had compiled one of the first scientific compendia of the Chinese materia medica that successfully integrated the legacy of the great classical works with the science of botany, the conventions of botanical illustration, and a respectful study of indigenous traditions.

None of the four continued their careers as botanical artists, however, joining the *Flora of China* project as taxonomists after 1949. By this time, the younger illustrators Feng Chengru had trained were taking their place in the academies, institutes, and colleges involved in botanical research. As Mu Yu and Wang Zhao wrote in an essay for the catalog for a botanical art exhibition celebrating the nineteenth International Botanical Congress in Shenzhen in 2017, "Botanical illustration, originally no more than simple depictions of plants, has developed into a unique, seamless integration of science and art" (Mu and Wang 2017, 319).

CHAPTER EIGHT

Spaces for Communicating and Informing

> We are going to organize a research association to shed light on all aspects of the natural sciences all over the country. This will advance scientific knowledge, improve teaching materials, develop enterprises, and rebuild after all this time. This magazine will be the messenger proclaiming these goals, the bell that signals the way forward. It will be a medium for the exchange of knowledge, it will gather and collect ideas to magnify their efficacy.
>
> —WU JIAXU, PREFACE TO THE *JOURNAL OF NATURAL HISTORY* ("BOWUXUE ZAZHI XULI")

Botanists collect in the field and then hunch over specimens in laboratories to identify and classify new species, making their discoveries known in the expectation that they will contribute to improving the human condition. Their observations, experimental results, and conclusions circulate through formal and informal exchanges within networks of peers that encourage synergies and communication. Science benefits from collegial spaces where professionals can come together to share their enhanced understanding of the world and from which they can disseminate the lessons of their work to the public (Golinski 1990, 494–95).

The early phase of the encounter with Western botany in China, the years during which natural studies evolved to become botany, drew to a close with the May 4th Movement of 1919. By the 1920s and 1930s, there was an active community of scientists who identified themselves as biologists or botanists, with a voice in public debates on modernization, science, democracy, and China's place in the world. After 1927, with a central government in place, there was sufficient stability to take stock of the status of botany in China, to mobilize resources to build institutions for research and training, and to reach out to inform the public about new discoveries. This brief interlude was harshly interrupted by the Japanese invasion of 1937. The ten years that followed, though, showed that the scientific community and institutions were now strong enough to function even when displaced geographically. During these decades, botanists and their peers in other fields of the sciences completed the transition from being intermediaries in the introduction of new learning to China to being integral parts of a national effort to shape the future (Wang Hui 2011, 57–60).[1]

SPACES FOR COLLABORATION AND COMMUNICATION: THE WRITTEN WORD

It was the city of Shanghai that first made space for learning and exchanges in the sciences. Inkstone Press was more than just a publisher and a bookshop run by missionaries. It invited scholars such as Xu Shou and his friends to pursue their interest in the new learning, equipping them for their role as translators at the Jiangnan Arsenal. They benefited from the press as a place to learn about Western science while remaining deeply embedded in the existing social and political structures (Zhang Jian 2003, 65).[2] In 1874, some years after they began work at the Arsenal, Xu Shou and John Fryer supported a proposal from Walter Medhurst, the British consul in Shanghai, to establish the Shanghai Polytechnic and Reading Room (Shanghai Gezhi Shuyuan) in order to offer technological training and a library to interested Chinese residents of the city. In a letter to Li Hongzhang, the governor of Zhili, who was responsible for relations with Westerners in the Treaty Ports, Xu pointed out that the science that was used and taught at the Arsenal had no impact on society outside: "If we want people to become enlightened and not to lose what they have learned, then there must be a place where people can come together to listen to lectures, where thoughtful people who love learning can come, where they can study after the meetings, and where they can ask for materials to read" (Hao 2003, 86).[3]

Li responded favorably. He secured approval for the project from the authorities in Beijing, waived taxes on imported equipment, and raised funds from his business contacts in Tianjin and Shanghai. At the opening ceremony on June 22, 1876, Li even donated an inscription with the institution's name in his own calligraphy. Shanghai Polytechnic served multiple purposes, with classrooms, a lecture theater, a reading room, and an exhibition hall. It also published the *Chinese Scientific and Industrial Magazine* (Gezhi huibian), coedited by Fryer and Xu (Hao 2003; Elman 2005, 308–10).

As Chinese schools began to teach science and Chinese students began to study abroad, there was less interest in Shanghai Polytechnic's facilities (D. Wright 1996, 11). The magazine, on the other hand, appeared monthly from February 1876 to January 1882, then quarterly from spring 1890 to winter 1892. It described itself on the cover page as "a monthly journal of popular information in the Chinese language relating to the sciences, arts and manufactures of the West." Its contents were eclectic, if always science based. The forty-six pages of the November 1876 issue (vol. 1, no. 11), for example, offered articles that were mostly translations from Western sources on zoology, how to save a drowning person, the mechanical loom, and topics in mathematics. Each issue also included "Miscellaneous Notes" and a section answering readers' questions. The magazine's intended readers were educated literati, so it was written in "the dignified classical, or *wen-li* style" (Peake 1934, 185). Shanghai Polytechnic, together with the *Chinese Scientific and Industrial Magazine*, set a precedent in China for the Western model of a scientific institution with an associated professional journal.

By the turn of the century, Shanghai was home to a group of collaborating institutions that were involved in education, publishing, and the popularization of science. The Agricultural Association, founded in 1896, began publishing *Agriculture News* (Nongxuebao) the following year, with practical advice on all things agricultural. In 1899, Du Yaquan initiated the beginnings of a recognizable scientific community in Shanghai when he published *Yaquan Journal* (Yaquan zazhi) in 1899 as an arm of the Yaquan Academy that he had founded to introduce the public to the sciences. Ten issues of the magazine appeared between 1900 and 1901 before it closed down due to a lack of funds.[4] Du, Zhong Guanguang, and Yu Heqin then opened the Shanghai Scientific Instruments Factory in December 1901 to generate some income. Finding that many of their clients were not sure how to use the equipment, they began to offer lectures on science. The background notes for the lectures became the original materials for their new journal, *Science World* (Kexue shijie), launched in 1903. The following year, Du became the first editor-in-chief of *Eastern Miscellany* (Dongfang zazhi), also dedicated to promoting research and reporting by Chinese scientists. While Shanghai was perhaps the most active center of scientific publishing, in 1910, the Geographical Society (Dixue Hui), founded the previous year in Tianjin, launched *Earth Sciences Journal* (Dixue zazhi). It concentrated on physical geography but made an important contribution to the field of botany between 1921 and 1922, when it published Zhong Guanguang's reports from his collecting expedition in the southwest (Wang Zhenru, Liang, and Wang 1994, 125).[5]

The year 1914 saw a transition from journals with their roots in the traditional framework of natural studies (*gezhixue*) or the broad learning of things (*bowuxue*) to what is now the accepted epistemology of the sciences (*kexue*). In Shanghai, the first issue of the *Journal of Natural History* (Bowuxue zazhi) appeared in 1914, with Wu Jiaxu as editor. Botany was an important theme, and the journal published some of the earliest accounts of botanical field collecting.[6] In the United States, the Science Society of China launched its journal *Science* (Kexue) in 1915.[7] Readers and contributors to both journals noted the significance of the different terminology in their titles. Wu's editorial for the first issue of the *Journal of Natural History* (using the pen name Wu Bingxin) was written in the literary language and style of past scholarship. Philosophically, he situated the natural sciences within a broader learning that seeks to understand the place of humanity in the universal order of things and in relation to other beings: "Humans are born into this universe, not even as significant as a grain of millet in a granary.... What the ears hear, what the eyes see, what the mouth consumes, what the nose smells, whatever touches the skin, clearly there is not a single one of these that is not an object. The form of an object, and its origins all bear some relation to human lives" (Wu Jiaxu 1914a, 3).

In *Science*, the botanist Qian Chongshu (who was studying taxonomy at Harvard) politely praised Wu for paying attention to botany at a time when the experimental sciences such as chemistry and physics were more valued for their applications in engineering and technology. He was unsparing, though, in his criticism of Wu and his journal for being hopelessly tied to tradition:

> China has a rich history of the broad learning of things [*bowu*] going back to ancient times, but it has not budged since the Han dynasty. You can read *Treatise on the Mushrooms of the Lower Yangzi* [Wu xun pu, ca. 1700] and see that [the author] Wu Lin was very learned about fungi. It is all very interesting as a text, but the way past scholars wrote about things does not accord with modern scientific principles; it is no more than a narrative. Modern scientific discoveries have taken place in recent times. We want to be sure that what scholars of the past have done is known, but it is our duty to modernize it and to shed light on it. (Qian 1915, 606)[8]

Despite Qian's condescending view, the *Journal of Natural History* is respected in China today for its role in publishing the work of scientists in China rather than the work of Chinese students overseas. It is also only fair to point out, too, that the journal did move from its traditional frame of reference toward a more contemporary conception of science as the years went by. It is sadly ironic that Chinese scientists at the time who vocally resented foreign domination of China were themselves quite patronizing toward scientists trained in China, who had not had the privilege of studying abroad.

SPACES FOR COLLABORATION AND CONVENING: SOCIETIES AND ASSOCIATIONS

In the summer of 1936, the Science Society of China called a joint meeting of science associations, hosted by the three main universities in Beiping. It was the third such meeting and perhaps the largest scientific gathering held in China before 1949.[9] For three days, August 17–21, over a thousand people, including numerous dignitaries and invited guests, heard keynote addresses and lectures, visited the sights of the capital, and even had an audience on August 19 with the First Lady, Song Meiling, in the government compound of Zhongnanhai. There were so many participants that the group picture taken after the opening ceremony required a panoramic camera and the print was five feet long (Liu Xian 1936, 878). The meeting was an occasion for scientists from the Science Society and six other organizations to combine their efforts, to build synergies among scientists, and to mobilize science.[10]

Addressing the opening ceremony, Jiang Menglin, the president of Beiping University (now Peking University), declared that holding a meeting on such a large scale proved that China was no longer a backward country lagging behind the rest of the world (Liu Xian 1936, 873). In the same vein, Ren Hongjun, the president of the Science Society of China, emphasized that the future of science in China would be in cross-disciplinary collaborations cultivated during meetings such as this one. He expressed the hope that at such a dangerous time, in the face of the threat from Japan, their efforts would contribute to strengthening the nation (Liu Xian 1936, 877–78).

In July the following year, Japanese forces crossed Lugou Qiao (the Marco Polo Bridge) and advanced on Beiping. The threat Ren had alluded to turned into war and occupation. Universities, colleges, and research institutions retreated ahead of the Japanese forces, relocating to the unoccupied southwest. The Beiping meeting had proved to be a high point and a turning point at the same time. The next twelve years of the Anti-Japanese War and the civil war meant that the potential and the hopes of 1936 were not realized until after the founding of the People's Republic of China.

In his report on the 1936 meeting, Zeng Zhaolun stated that by 1935, forty organizations with some scientific or educational character had registered with the Ministry of Education and that there were also many unregistered local or regional associations (Zeng 1936, 807–10). His report began with the claim that the Science Society was the first truly scientific organization in China (Zeng 1936, 798). In his opening address to the meeting, Ren also gently but firmly reminded those present that as the oldest scientific society, it was the *da gege* (older brother) and the other six participating associations were the *xiao didi* (little brothers), "and as the older brother of the seven associations at this meeting, it is quite appropriate for the representative of the Science Society to take on the assignment of representing all of them with some opening words" (Liu Xian 1936, 876). Ren and Zeng may have been factually correct, but at the same time, their words gave credence to the criticism that the Science Society was rather arrogant about its seniority and its origins.

The students who founded the Science Society (later the Science Society of China) at Cornell in 1914 were among the first to receive fellowships to study science and technology in the United States, using funds that had been released by the US Congress from the indemnities demanded by Western powers after the Boxer Rebellion in 1900.[11] They were not the first Chinese students to study in the United States. The first Chinese student known to have studied at an American university graduated from Yale in 1854. In 1872, a group of thirty students left China for the United States, the first of a total of 120 selected for "the Chinese Educational Mission," a project approved by the Qing court. A second wave of students went to Japan in the first decade of the twentieth century, with over 38,000 known to be there by 1911.[12] Official support for sending students abroad was indicative of the profound shift in the goals of higher education at the turn of the century, from training officials to the pursuit of knowledge and expertise in the drive for modernization (Chen and Huang 2013, 94–95).

From the outset, members of the Science Society saw themselves as a force for change, not just a club for a group of students with shared interests. The society invested in the organizing, fundraising, and communications that enabled it to act on its ambitions (Zhang Jian 2002, 2003). It collected dues from members and solicited funds from patrons, including businesses and entrepreneurs. It negotiated contracts for services, such as the selection and translation of scientific books for universities, and it secured grants from government agencies (Wang Zuoyue 2002, 313; Ke and Li 2016, 21–25). In the long run, though, the society was taken to task for being a closed group, with ties dating to their college days and a leadership that resisted change. It has been described as occupying a point between tradition and modernity that prevented it from fully reaching its goals (Zhang Jian 2002, 83).

There was some rivalry between societies, a part of which was related to the country where a group's founders had studied. In 1917, Chinese students in Japan founded Wissen und Wissenschaft (Knowledge and science) (Zhonghua Xueyi She), which deliberately set itself apart from the Science Society of China by using German rather than English for its non-Chinese title. Its journal, *Wissen und Wissenschaft* (Xueyi she), first appeared in Tokyo in 1917 as a quarterly, becoming a monthly when the society moved to Shanghai later that year (Zhang Jian 2002, 91).

In September 1927, teachers and natural scientists who had studied and trained in China met in Sichuan to organize the Natural Science Society of China (Zhonghua Ziran Kexue She). The founders and members, who were mostly teachers trained in China, felt that the societies in Shanghai, with their foreign connections, did not serve the needs of the inland regions. The society revived and relaunched *Science World* (Kexue shijie), the journal that had originally been founded by Du Yaquan and his colleagues in 1903 (Zhang Jian 2002, 91; Fan and Han 2012).

For a long time, botany did not have a dedicated association. Finally, on August 20, 1933, the inaugural meeting of the Botanical Society of China (Zhongguo Zhiwuxue Hui) took place at Beibei, in the suburbs of Chongqing. The 105 people attending approved a constitution, appointed a board of nine, and agreed to publish a new quarterly journal, the *Chinese Journal of Botany* (Zhongguo Zhiwuxue Zazhi), with Hu Xiansu as editor.[13] At the society's second annual meeting at Lushan in August 1934, members approved the launch of the *Bulletin of the Chinese Botanical Society* (Zhongguo Zhiwuxue Huibao), an English-language journal to inform the world about botany in China. The *Bulletin* published reports on botanical research in China together with abstracts in English (and occasionally in French, German, or Japanese) of papers and theses originally written in Chinese. The decision to publish the *Bulletin* was a response to the persistent issue of language. The problem was no longer how to express scientific ideas in Chinese; it had become a matter of how to inform colleagues around the world about botany in China. The solution for those associations that had the resources was to produce two publications, one in Chinese for domestic readers and one, usually in English, for an international audience. In a disciplinary quirk dictated by the International Code of Botanical Nomenclature, botanical descriptions in journals for both domestic and international audiences had to be in Latin. Chinese or English (or any other language) could accompany or supplement the Latin but could not be recognized as diagnostic under the code.[14]

SPACES FOR COLLABORATION AND COOPERATION: RESEARCH AND TRAINING

Late in life, reminiscing about his experience as president of the Science Society of China, Ren Hongjun wrote that a society cannot thrive on the enthusiasm of its members alone. It also needs a space, a physical structure in which members can pursue the interests that first brought them together (Ren 1983, 4). The Biological Laboratory of the Science Society of China, inaugurated in Nanjing on August 18, 1922, was the embodiment of Ren's sentiment in bricks and mortar. The society's decision to focus on biology had been pragmatic. Much of the subject material—plants and animals—could be found locally, and botany and zoology required less equipment than physics, chemistry, or any other branch of the sciences. The laboratory organized a permanent display of zoological and botanical specimens in a public area at the ground level of the building, which became the Nanjing Natural History Museum, attracting visitors and building public support (Ke and Li 2016, 26). Although the laboratory had very basic facilities, it was the only place in the country at the time where botanists and zoologists could get together to exchange ideas and to collaborate on research. Besides, as Wu Weishi, the director of the Jiangsu Entomological Bureau and one of the dignitaries at the opening ceremony, said optimistically, "You should not be disheartened by the lack of equipment or books. When I first began my own research, I encountered the same problem. What I did was to look for new problems that no one else had worked on, so there were no books on those subjects anyway! . . . I hope that you, too, will be able to find new problems to which you will certainly find answers that will be talked about in research institutions around the world" (*Kexue* 1922, 848)

The laboratory's Department of Botany devoted much of its energy to collecting and preparing specimens for a new herbarium, aiming to compile the first complete flora of China. Together with their colleagues from the Department of Zoology, they also wrote manuals and textbooks for schools. With these two tracks, the laboratory proposed to implement the goals that Bing Zhi, the chair of the society's fundraising committee, had identified at the inauguration ceremony: to plumb the depths of research problems and to popularize and disseminate scientific achievements (*Kexue* 1922, 846).

The laboratory was careful to cultivate good relations with the Jiangsu and Sichuan provincial authorities, who responded with modest financial support to help cover its operating expenses (Jiang L. 2016, 155). More substantial funding came from the China Foundation for the Promotion of Education and Culture (Zhonghua Jiaoyu Wenhua Jijinhui) (referred to hereafter as the Foundation), which managed the second remission of Boxer Rebellion indemnity payments from the United States.

The indemnity payments legislation, approved by Congress and signed into law by President Calvin Coolidge in May 1924, specified that the funds were to be used for the promotion of Chinese education and culture, disbursed by a philanthropic foundation modeled on the Carnegie Foundation and the Rockefeller Foundation's China Medical Board. The Foundation was to be devoted to the development of scientific knowledge and to the advancement of cultural enterprises such as libraries (Cowdry 1927, 150). The board of trustees, composed of ten Chinese and five American members, officially inaugurated the Foundation during its first meeting on September 18, 1925, and it approved its first grant, to the National Library in Beijing, in November. The Foundation promoted what it referred to as "local" sciences: the field sciences of biology and geology, which it believed were best suited to this stage in China's development. The Biological Laboratory clearly met these conditions and received several grants, beginning in 1926, to cover some of its operating expenses, part of the costs of a building in Nanjing, and a research professorship for Bing Zhi. In the absence of reliable government support before 1949, grants from the Foundation were one of the most significant sources of funding for research and training in the fields of botany and zoology.[15]

Hu Xiansu reported in March 1934 that there were now four research institutes in China dedicated to taxonomy and many universities teaching botany, all equipped with facilities such as herbaria and libraries (Hu X. 1934, 1). They shared the goals of carrying out research and improving society and the nation, but each of them owed its existence to a different configuration of organizational support, funding, and leadership—although they all benefited to some extent from grants from the Foundation. The Biological Laboratory in Nanjing was the research arm of the Science Society of China. Sun Yat-sen University's Institute of Agriculture, Forestry, and Botany in Guangzhou, founded in 1928, was to a large extent an initiative of its founder, Chen Huanyong, who was also well connected with the provincial government. The Fan Memorial Institute of Biology, founded in Beiping in 1928, was a private institution supported almost entirely by grants from philanthropic foundations and other nongovernmental sources. The Institute of Botany of the National Academy of Beiping (Beiping Yanjiuyuan Zhiwu Yanjiusuo), founded in 1929 as the northern arm of the Academia Sinica, was the only one of the four that was a government body with financial support from the government.

Sun Yat-sen University's Institute of Agriculture, Forestry, and Botany was first organized in 1928 by Chen Huanyong as a laboratory in the university's Department of Forestry to study the vegetation and to compile a flora of Guangdong in order to assist in planning the future of agriculture and forestry in the province.[16] After just a year, the laboratory had been so successful that the president of the university approved upgrading it to an institute, which opened in April 1930. It organized a collecting program in southern and southwestern China, sending ten expeditions between 1928 and 1932, three of which were to Hainan Island alone. The specimens and cuttings they brought back to Guangzhou added to the institute's growing herbarium, a second area of its work. Exchanges with herbaria around the world were important enough that Chen listed them in his annual reports as a third area of work in their own right. A table in the 1933 report shows 18,707 mounted vouchers exchanged since 1928 with thirteen institutions in nine countries or regions.[17] The fourth area of work was to enrich the herbarium with more specimens and a growing portfolio of drawings for a planned flora of Guangdong. Finally, and of growing importance, was the Specimen Garden (Biaoben Yuan), which was to become today's Huanan Botanical Garden (Huanan Zhiwuyuan). Chen reported that it included a covered area for orchids collected in Hainan and a section for rare plants from Guangdong. The garden was open to the public and quickly became an attraction for residents of the city. The institute took advantage, too, of opportunities to foster an interest in plants among the public, taking part, for example, in an exhibition organized by the city of Guangzhou in 1933, featuring a display of dried specimens displayed and labeled according to their different uses, together with photographs of botanical gardens in other countries.[18]

Meanwhile in Beiping, in December 1927, the Shang Zhi Society (Shang Zhi Xuehui) a network of Chinese students who had studied in Japan, created a fund in memory of one of their members, Fan Yuanlian (1876–1927), who had just died unexpectedly in Tianjin shortly after making a substantial donation to the society. Fan had studied natural sciences in Tokyo, had been president of Tsinghua and National Peking Normal Universities, and had served as minister of education in the northern military government after 1911. Most importantly, he had been the first director of the China Foundation for the Promotion of Education and Culture. The Shang Zhi Society proposed that the funds should be used to found a research institute in Beiping in his name, the Fan Memorial Institute of Biology (Jingsheng Shengwu Diaochasuo).[19] The association presented a proposal to the Foundation, which agreed in March 1928 to add to the fund and to contribute its expertise and guidance in planning the new institute. The Foundation helped to design a management structure that it hoped would become a model for scientific research institutions in the future. The institute had a management committee and a board with representatives from the Foundation, government agencies, higher education, and business. Specialized committees supervised different aspects of the institute's work, each with written rules relating to the selection and election of members, term limits, and job descriptions (Jiang Y. 2005, 295–96). Bing Zhi was appointed director of the institute and the Department of Zoology, with Hu Xiansu as the director of the Department of Botany.[20]

The committee accepted the donation by Fan Jingsheng's brother Fan Xu of his former residence in Beiping to house the institute and announced that it would open on October 1. At the opening ceremony, attended by more than fifty guests, including several foreign dignitaries, Bing Zhi announced that the institute would work mostly on the flora and fauna of northern China, complementing the work of the Science Society's Nanjing laboratory in the south and the Institute of Agriculture, Forestry, and Botany at Sun Yat-sen University, which covered the subtropical and tropical zones.[21] The regional coverage of the country and collaboration between institutions was an explicit strategy of the Foundation to concentrate its funding on a limited number of institutions to develop a cluster of internationally respected research centers in China (Yang T. 1991, 55–91, 127–39).

The scientists of the Fan Memorial Institute believed that an inventory of the nation's resources should be the foundation of the biological sciences in China. The institute opened in October 1928 as winter approached, limiting the opportunities for fieldwork. Nevertheless, it sent two groups to collect close to home, in the mountains to the north and east of Beiping, adding to its herbarium and its collection of wood samples (Hu Z. 2005a, 63). Over the years, the institute became more expansive in its regional focus. As director after 1932, Hu extended field collecting to the southwest, which he knew had the richest botanical diversity in the country. In 1932, he sent Cai Xitao and two volunteers on what became a three-year expedition to Yunnan.[22] Hu also collaborated with other research institutions on ambitious projects, such as a joint survey in 1932 with the West China Academy of Sciences in Chongqing of the Sichuan-Tibet-Yunnan borderlands, led by Yu Dejun. The most elaborate such expedition was a survey of Hainan, conducted jointly in 1934 with the Science Society of China and three other institutions, involving separate but coordinated teams working on the coast and inland (Jiang Y. 2005, 293–305).

The core of the Fan Memorial Institute's work was the identification and classification of the specimens it received from its collecting expeditions and exchanges with other herbaria in China and around the world. Its work on the ferns quickly made an impression internationally with the publication in 1930 of the first fascicle of *Icones filicum Sinicarum* (Zhongguo juelei tupu) (Illustrated manual of the cryptogams of China) by Qin Renchang and Hu Xiansu.[23] A fellowship from the Foundation allowed Qin to travel to Europe to pursue his research on ferns, examining specimens from China held in collections there (Hu Z. 2005a, 36). He arrived in Denmark in June 1930 to study with Carl Christensen, the curator of the botanical museum at the University of Copenhagen, then went on to London to visit the herbarium at Kew. Confronted with the sheer quantity of material from China in their collections, he embarked on a project to take photographs of as many specimens as possible so as to make them available to Chinese botanists.[24] He returned to China in 1932 with more than 18,000 negatives, which when printed became an invaluable tool for botanists all over the country and are still considered to be one of the most important legacies of the Fan Memorial Institute.

The institute moved into new premises in the spring of 1931. It added a laboratory for wood sciences (wood anatomy, physics, mechanics, and chemistry), with a collection of over 1,800 samples. It launched a publishing and printing venture run by the botanical illustrator Feng Chengru,[25] and it converted its earlier offices into a science museum, which attracted over eight thousand visitors in the first two months after it opened. In December 1933, Hu Xiansu began to work with the Jiangxi Provincial College of Agriculture to plan the Lushan Arboretum and Botanical Garden, which opened on August 20, 1934. It was by far the largest in China at the time, the first that was not located on a university campus and the first to be designed for in situ botanical research, as well as a space open to the public for education and recreation. It is the only physical part of the Fan Memorial Institute that still exists today.[26]

The Biological Laboratory of the Science Society of China, Sun Yat-sen University's Institute of Agriculture, Forestry, and Botany, and the Fan Memorial Institute all came into being as independent research centers before the Guomindang government had consolidated its authority over the country. Creating the Academia Sinica in June 1928 in Nanjing was a key element in the new regime's moves to bring all social activities, including education and science, under the oversight of the state. In September 1929, the Academia Sinica established a regional National Academy of Beiping to serve the north—and to act as a state-sanctioned counterweight to the Fan Memorial Institute.

Li Yuying, the first president of the Beiping Academy, had an ambitious vision of an institution with seven departments, each with several research institutes. With very limited funding, though, he was only able to set up an Institute of Physics and Chemistry and an Institute of Biology, under which there was a Department of Botany and a Department of Zoology. Liu Shen'e, the director of the institute, had a doctorate in botany from the University of Paris, and he successfully marshaled his contacts there to obtain support for the institute and its new building, which was completed in 1934 (Hong and Blackmore 2015, 227). The Department of Botany undertook some forty collecting expeditions between 1930 and 1936, mostly in northern and northwestern China. Liu himself took part in a joint Franco-Chinese scientific expedition to western China and Xinjiang in 1931. He then continued on his own for nearly two years, on an extraordinary odyssey that took him across the Karakoram range to Ladakh, Kashmir, northern India, Assam, and Calcutta, from where he traveled by ship back to Shanghai.[27] He had almost been given up for dead when he finally returned to Beijing, but he resumed his position as director, leading several more collecting expeditions over the next four years (Hu Z. 2006). The department became known for its work on biogeography and its journal, *Contributions from the Institute of Botany* (Beiping Yanjiuyuan Zhiwuxue Yanjiu congkan), published from 1931 to 1937. Responding to a call from the government in 1934 to develop the northwest, the academy entered into a partnership with the Northwestern College of Agriculture at Wugong in northern Shaanxi. The partnership led to the establishment in November 1936 of the Northwest China Botanical Institute (Zhongguo Xibei Zhiwu Diaochasuo), dedicated to surveying and developing the region's biological resources (Jiang Y. 2003, 34–41).

THE WAR YEARS, 1937–49

The 1936 Beiping meeting of seven scientific organizations celebrated a vibrant scientific community that had taken shape over the previous twenty years. Participants looked forward to the future with optimism.

The Japanese invasion in 1937 destroyed that vision overnight. More than 125 universities, colleges, and research institutions left the occupied areas of northern China and relocated to the south and southwest. Some institutions had to relocate a second time in 1941 when the Japanese advanced south, taking Guangzhou and Hong Kong. Some even had to move a third time in 1944, when Japanese troops entered Guangxi, large parts of Sichuan, and Indochina in an attempt to cut off supply routes to China (Xu Guoli 1998). The Japanese authorities considered the Fan Memorial Institute, funded by the US-based China Foundation, to be an American institution and allowed it to continue working under the protection of the American embassy until the attack on Pearl Harbor and the outbreak of the Pacific War, when it was seized, looted, and turned into Military Hospital 151.[28]

The dispersal immediately following the occupation triggered a spate of reorganization and regrouping in which some institutions formed branch universities or colleges, others merged to form new universities, and some developed collaborative arrangements with the local authorities where they had relocated. A cluster of universities and research centers moved to the wartime capital of Chongqing, where an academic district took shape in the suburb of Beibei, which had been selected before the outbreak of war by the central government as a site to resettle higher education institutions if necessary.

Beibei was already the home of the West China Academy of Sciences (Zhongguo Xibu Kexueyuan). Lu Zuofu (1893–1952) had founded the academy in 1930 with funds from the Minsheng Corporation, China's largest private shipping company, which he had founded in 1925. Lu also received donations from the Sichuan warlords Liu Xiang and Liu Wenhui, the Bank of Szechuan, and various industrialists and financiers with the support of academic and political leaders in Beiping and Shanghai. The impetus behind the academy was the belief among the military and political authorities in Sichuan as well as planners inside and outside China that the western provinces would not only serve as a defensive barrier in the event of invasion from the north but also had the potential to match the northeastern provinces in economic importance. The academy's mission was to "engage in scientific research in order to develop the region's natural wealth and enrich the people" (Yang T. 1991, 255–57). It grew out of Lu's interest in preparing young people to play their part in modernization, with facilities for training and research, as well as a natural history museum that is still considered to be one of the best in China. The academy offered logistical assistance to the research centers in Beiping and Nanjing and assigned students to accompany their botanical expeditions in the southwest. Lu had also sent specimens collected by students on their own field trips to his colleagues at the Biological Laboratory of the Science Society or the Fan Memorial Institute for identification and to add to their collections.[29] When those same colleagues arrived in Chongqing at the end of their forced journeys south, the West China Academy was a welcoming host, putting its equipment and premises at their disposition (Pan and Peng 2007).

Beibei was a pleasant area, with easy road and river access to Chongqing, which together with the presence of the West China Academy of Sciences made it a conducive location for a scholarly enclave devoted to research. The Biological Laboratory of the Science Society of China and the Fan Memorial Institute arrived in late 1937, followed by Academia Sinica's Institute of Biology in winter 1940 (Jiang and Zhang 2002, 24; Pan and Peng 2007). More than twenty scientific and research institutions eventually settled in Beibei, fostering interest in new areas of research with work on genetics, plant fibers, and the processing of plant materials as substitutes for scarce materials, such as plastics and textiles (Jiang and Zhang 2002, 24). Joseph Needham, the director of the British Council's Cultural Scientific Mission in China, wrote in a report on the Institute of Biology that, despite makeshift facilities and limited equipment, "a visitor . . . viewing the work going on in it, feels that it has the true research atmosphere of the world's best laboratories" (Needham 1943, 65). By July 1943, the scientific community was able to organize a three-day meeting of six scientific societies at Beibei with 240 members attending, at which more than three hundred papers were read (*Science* 1943).

Beibei was collegial if rustic. Some institutions, though, were in more remote locations, where the intellectual elite of the country suddenly found themselves rubbing shoulders with rural communities. At the Northwest China Botanical Institute, interest moved from collecting and classification to a focus on inventories of lesser-known but potentially valuable local varieties of vegetables, fruits, and medicinal herbs. The institute also collected seeds for propagation and carried out applied research on pest control and on improving yields of horticultural crops.

The shift to economic botany was especially pronounced in the very different ecological and political environment of the "liberated areas" of northwestern and central China under communist control. Some scientists had gone to Yan'an, the seat of the communist government, inspired by the vision of a new, more egalitarian society. Under the supervision of the Communist Party, they established the Yan'an Institute of Natural Sciences

(Yan'an Ziran Kexueyuan) and the Yan'an Biological Research Institute (Yan'an Shengwu Yanjiusuo) and designed a program of surveys and research on economic botany (Wang Zhenru, Liang, and Wang 1994, 140). Beginning in 1939, academicians carried out at least five inventories of the semiarid region's little-known flora and forest resources, on the basis of which the Ministry of Economics ordered the creation of a Bureau of Forestry to implement its recommendations for forest management (Wu H. 1987). After 1950, the institutes gave the specimens they had collected during these surveys to the Chinese Academy of Sciences. Le Tianyu and Xu Weiying, two of the botanists who had been responsible for the surveys, assembled these materials in 1957 to compile what became the standard reference work on the region, *The Flora of the Shaan[xi] Gan[su] Ning[xia] Basin* (Shaan Gan Ning pendi zhiwu zhi) (Le and Xu 1957).

In Beiping, Hu Xiansu was painfully aware that the Fan Memorial Institute's reprieve as an institution under American protection could not last long and the time would come when it, too, would have to leave Beiping. He had been in contact with Gong Zizhi, the director of the Yunnan Province Bureau of Education, since Cai Xitao's first collecting expedition to Yunnan. Gong had been very supportive of the institute's work in the province and had sent two of his staff to Beiping in 1934 to study botany. In 1937, Hu sent Cai with a proposal to establish a joint botanical research institute in Kunming. Gong responded enthusiastically and offered some funds from the province, as well as a plot of land to build on.[30]

Things moved rapidly. Cai visited Kunming again at the end of a collecting expedition in 1938 and drew up an agreement between the Fan Memorial Institute, the Yunnan Provincial Economic Commission, and the Yunnan Province Bureau of Education, which established the Yunnan Institute of Agriculture, Forestry, and Botany (Yunnan Nonglin Zhiwu Yanjiusuo) at Heilong Tan in the northern suburbs of Kunming. On July 1, 1938, the provincial government issued a proclamation approving the new institute, with Hu Xiansu as its first director, although he had to divide his time between Beiping, where the Fan Memorial Institute was still active, and Kunming. He finally left Beiping for Kunming in the spring of 1940, and the remaining staff moved south after the Japanese authorities closed it in December 1941. Hu left Kunming in early 1941 to take up a position as president of National Chung Cheng University in Jiangxi, handing the Yunnan Institute to Deng Wanjun, who had recently returned from France.

The Yunnan Institute of Agriculture, Forestry, and Botany continued and completed botanical surveys that the Fan Memorial Institute had already started. In accordance with the agreement that set out the terms of the partnership, the new institute committed to taking on any research project needed by the Bureau of Education or the Economic Commission. Projects included an investigation of the distribution, production, and sales of the medicinal tuber *san qi* (*Panax notoginseng*), a study of the distribution and ecology of the lacquer tree (*Toxicodendron vernicifluum*), and research into the sources of fiber used in indigenous paper production (Ding and Chen 2013, 99). In the field of forestry, the institute carried out surveys of forest cover and distribution, as well as programs on wood anatomy, the chemical and physical properties of wood, and the management of native species. Only the first volume of a planned comprehensive *Illustrated Materia Medica of Southern Yunnan* was completed, though, published in 1945 with a very small print run, a victim to the wartime scarcity of printing supplies (Jing et al. 1945).[31]

On August 15, 1945, Emperor Hirohito announced Japan's unconditional surrender. By the time the war ended, all research and educational institutions, no matter where they were or how well connected, were in very difficult straits. Laboratory-based research groups were unable to find the equipment and supplies they needed. Inflation was reducing their already meager salaries to worthless paper. The Lijiang Research Station sold pine resin and propagated high-elevation plants, such as rhododendrons and azaleas, to sell in Kunming and abroad (Hu Z. 2018, 156). In Kunming, Cai Xitao began experimenting in 1946 with tobacco production, testing varieties that he had obtained through the institute's international contacts. When he found a successful variety (at the time, the soils and climate of Yunnan were thought to be unsuitable for tobacco), he and Hu Xiansu contracted with the tobacco company to lease more land, from which they harvested enough to earn a useful profit to boost dwindling salaries—and laying the foundation for what was to become by the 1970s the mainstay of the provincial economy (Hu Z. 2000, 58; Chinese Academy of Sciences, Kunming Institute of Botany Editorial Committee 2008, 99–106).

Recovery and reconstruction were not easy. The process of restoring buildings that had been taken over during the occupation was very slow. The Academia Sinica's Institute of Biology and the Fan Memorial Institute were not able to move back into their former premises until the first half of 1948. When civil war broke out, inflation became hyperinflation, so the rare grants or budget disbursements they might receive barely covered basic needs (Jiang Y. 2003, 43–45; Hu Z. 2000, 57–59). Botanical institutions that had been collecting specimens during the war years were able to continue working on identification and classification, which required little funding or equipment. A rare breakthrough, such as the discovery of the dawn redwood (*Metasequoia glyptostroboides*) showed that Chinese botanists were making exciting contributions to the field, but with the country at war and a collapsing economy, there was no chance of realizing the hopes of the 1936 meeting of scientific associations in Beiping.[32] There was little resistance in December 1949 when Guo Moruo, the president of the Chinese Academy of Sciences (CAS), proposed a restructuring of scientific research and education (Liu Xiao 2013). In January 1950, the Fan Memorial Institute and the Beiping Academy's Institute of Botany merged to form the CAS Institute of Taxonomy (Zhongguo Kexueyuan Zhiwu Fenlei Yanjiusuo). In September, the Institute of Agriculture, Forestry, and Botany in Kunming became the CAS Kunming Taxonomy Research Station (Zhongguo Kexueyuan Zhiwu Fenlei Yanjiuso Kunming Gongzuozhan). In November, the Lushan Arboretum and Botanical Garden became the CAS Institute of Botany Lushan Botanical Garden (Zhongguo Kexueyuan Zhiwu Yanjiusuo Lushan Zhiwuyuan).

The Science Society of China had planned to continue functioning as a nongovernmental institution promoting the sciences, based in Shanghai. With no constituency for what was no longer an urgent message, though, the society slowly wound down. The journal *Science* ceased publication in 1951. In 1954, the society gave its herbarium and equipment to the Chinese Academy of Sciences. The Shanghai city library accepted the donation of the society's library in 1956. The Science Press of China took over the society's printing and publishing venture in 1957. Finally, on May 4, 1959, at the offices of the Shanghai Science Commission, the remaining members of the society signed the papers that dissolved the Science Society of China. Ren Hongjun, who had been the society's president from 1914 to 1923, wrote: "At this point we announced that the Science Society of China had gloriously accomplished its mission" (Ren 1983, 13).

CHAPTER NINE

Museums, Exhibitions, and Botanical Gardens

> Every day since we established the garden during the winter of 1931, we have been increasing the number of specimens collected in the wild, including those brought back from Hainan, which has been an excellent way to expand our collection. We have built five sheltered beds, a dormitory for the laborers, and a covered area for orchids from Hainan as well as rare plants from Guangdong. We receive a lot of visitors whenever they are in bloom. They are all delighted at what they see and our visitor numbers are increasing.
>
> —CHEN HUANYONG, A REPORT ON RECENT WORK OF THE INSTITUTE FOR AGRICULTURE, FORESTRY, AND BOTANY ("NONGLIN ZHIWU YANJIUSUO ZUI JINZHI GONGZUO BAOGAO")

FIELD botanists and their colleagues in the laboratory report on their work in scientific journals and the popular press. The record of a collecting expedition might appear in a professional journal with lists of botanical names and data on soils, temperature, humidity, and other ecological parameters. A more popular geographical periodical might publish an account of the same expedition with less scientific jargon and more description of landscapes, the customs of ethnic minority groups encountered on the way, or the dangers of bandits in the forests. In recent decades, film and electronic media have become the dominant channel through which to reach the public, but during the decades of interest here, there were few alternatives to print. Zhong Guanguang is reported to have broadcast a paper he wrote, "The Different Interests of Chinese and Japanese Botanists," on Central Radio in 1933, but radio and nonprint media did not have a wide audience until the latter half of the twentieth century.[1] Information circulated primarily through written texts.

More powerful than the written word, though, is a visual presentation or display, especially where literacy is a luxury enjoyed by the few. Even today, with access to a huge range of audio and visual media, the popularity of special exhibitions and museums, with their opportunities to see objects, plants, or other materials related to the sciences, suggests that a direct, face-to-face experience is a highly effective way to engage the public. Botany invites sensory interaction. Plants are a part of everyday life. Color, shape, fragrance, and texture are not only familiar to everyone, botanists also use them in identifying and classifying plant life. Close observation is a vital skill in fieldwork, so inviting the public to observe plant life for themselves is an attractive strategy to introduce a nonprofessional audience to botany. When the Science Society of China opened the Biological Laboratory in Nanjing in 1922, it used one of its two buildings for a museum of zoology and botany, where it displayed specimens from its collections. Botanical gardens represent a much greater, longer-term planning effort and commitment of resources. Even so, the first large-scale botanical garden in China designed for research as well as for public education and recreation opened at Lushan in 1934, less than a century after Li Shanlan and his colleagues coined the neologism *zhiwuxue* for the new science of botany.

MUSEUMS

At different scales, museums, exhibitions and botanical gardens all display plant materials in a way that embodies scientific principles. In one sense, museums are gatekeepers, controlling access to the objects they store. At the same time, they are places where collaboration and learning take place (Spary 2000, 51). The first botanists in China were cut off from basic research tools, such as type specimens, vouchers of pressed plants, and the field notes that situated them in their ecological context, which were stored "safely" in European herbaria. Two museums in the foreign concessions in Shanghai played a part in beginning to redress the balance, building collections and publishing monographs on the flora and fauna of China, and offering physical spaces "where expert science meets public cultures" (Kohler 2002, 189).

In 1868, the French Jesuit Père Pierre-Marie Heude founded the Musée de Zikawei near the Jesuit mission in Shanghai. It was the first museum in China to collect natural history specimens, which were sent there from the order's inland missions and from other travelers.[2] The collections were mainly zoological, but the museum also housed over 50,000 plant specimens, which had been studied and classified with assistance from the Muséum National d'Histoire Naturelle in Paris. The collection was not open to the public until 1932, after it had moved to the campus of Aurora University, where it was renamed the Musée Heude. In its new home, the museum became more involved in teaching, and it collaborated with Chinese scientists, who used its collections and published in its journals (Dai 2013, 350–53).

Close on the heels of the Zikawei museum, the North China Branch of the Royal Asiatic Society opened the Shanghai Museum in 1874 in one room in the society's building in the British Concession. It was originally dedicated to the natural history of China and Japan, but it soon began to add archeological and ethnographic objects, which became the heart of its collections. The museum was open to the public, but although most visitors were Chinese, it remained closely associated with British interests in Shanghai. As foreign residents left China after 1949, its finances became more and more precarious, and it closed in 1952, giving its collections to the new government of the People's Republic of China (Lu T. 2014, ch. 3).

Museums were associated with modernity and progress, in explicit contrast to "traditional" or "superstitious" ways of thinking. Outside Shanghai, other museums, founded by Western missionaries in Shandong during the 1890s and in 1905 by the modernizing Chinese entrepreneur Zhang Jian in his model municipality of Nantong in Jiangsu, assembled collections of rocks, animal skeletons, stuffed birds, or living plants, arranged to convey a vision of a natural world made orderly through scientific investigation.[3] By the time the Science Society of China opened to the public in 1922, museums were already familiar instruments in the project of modernization through education—with the added benefit for the society that they could generate some income from ticket sales.

The Science Society was not alone in its efforts to educate and enlighten. Planning began in January 1929 for a natural history museum under the Academia Sinica in Nanjing. The academy had sent an expedition to Guangxi in April 1928 to collect zoological and botanical specimens. It had also purchased ten crates of paleontological and paleobotanical specimens from France, which arrived at the same time that the expedition returned. The academy appointed a planning committee to organize a museum that would be a space for research as well as a venue to display their new acquisitions. The Academia Sinica National Natural History Museum (Guoli Zhongyang Yanjiuyuan Ziran Lishi Bowuguan) opened its doors in January 1930. In July 1934, though, the academy restructured its institutes, and the museum stepped back from its public, educational role to become a research institute, the Zoological and Botanical Institute of the Academia Sinica (Jiang and Zhang 2002, 18–20).

EXHIBITIONS

Museums and permanent collections, however well displayed, do not generate the same level of excitement as a well-publicized, once-only special exhibition. Exhibitions offer a spectacle that both diverts and engages the visitor. At its grandest, an event like Prince Albert's Great Exhibition of 1851 generated enthusiasm for the transformations that science and the Industrial Revolution were bringing to people's lives. China's participation in this and other world's fairs and international expositions during the late nineteenth and early twentieth centuries showed the persuasive power of an exhibition that used visual display to a pedagogical end. During the last days of the Qing dynasty, the hugely popular Nanyang Exposition of 1910 in Nanjing projected the message that China was a dynamic, modernizing nation confidently facing the future (Fernsebner 2006). All around the country, smaller regional or local exhibitions also took place at this time, celebrating symbols of modernity such as education, agriculture, or industry.

In the spring of 1933, in the midst of an existential crisis triggered by the worldwide economic depression, Japanese expansionism, and tensions between Nationalist and Communist forces, the city of Guangzhou organized a grand City Exposition (Shi Zhanlanhui). The event portrayed Guangzhou as a dynamic, forward-looking metropolitan region, forging a path that kept its distance from the increasingly corrupt and repressive Nanjing regime. Science and education were shining examples of this imagined future (Ding L. 2017, 206–34). In a report on the activities of Sun Yat-sen University's Institute for Agriculture, Forestry, and Botany, Chen Huanyong wrote that the institute had played its part with a display of dried, pressed plants, as well as specimens preserved in spirits, illustrated books, and other exhibits. The dried specimens were mostly of useful plants, grouped and displayed according to different uses, each with an explanatory label. The institute also submitted photographs of plants and botanical gardens in other countries to the photography exhibition. Chen was confident that visitors would "develop a deeper knowledge about plants, to encourage them to develop a powerful love for plants so they will want to protect them" (Chen Huanyong 1933a, 119).[4]

More ambitious still was a large natural history exhibition organized by the Biological Laboratory of the Science Society of China in Nanjing in January 1934. When a scientific survey team returned from Hainan in January 1933, the popular press published such lurid, exoticized accounts of their adventures that representatives of the city's schools and colleges asked the laboratory to organize an exhibition on the flora and fauna of China to teach schoolchildren, students, and interested residents of the city what natural history was really about. The leadership of the laboratory immediately agreed, seeing an opportunity to take advantage of a wave of interest in science to raise awareness of biology and to encourage young people to consider becoming involved in research (Zhang M. 1934a, 550).[5]

The exhibition opened on January 29, 1934. Scheduled to last ten weekdays, it was extended to include Sunday, when more working people could attend, then extended again to accommodate popular demand. In the end, more than 10,000 visitors came over a period of sixteen days (Zhang M. 1934a, 551). They were treated to a panorama of life-forms spread over four rooms, beginning with parasites, then moving on to taxonomy, flora in the third room, and finally fauna in the fourth and final section, where displays of stuffed animals in a darkened room were so lifelike that some children jumped in fear (Zhang M. 1934a, 561). Visitors learned about climate zones and vegetation from maps, with examples of the trees, plants, and fruits from each zone. They could peer down microscopes to study cellular structures, and there were demonstrations of how to make slides from cross-sections of tissue, with enlarged drawings pointing out key features of plant anatomy (Zhang M. 1934a, 562). Throughout the exhibition, the underlying narrative was that there was an order to the extraordinary profusion of life-forms and that evolution explained both their commonalities and their diversity.

Pictures of the exhibition show shelves of specimens arranged in neat rows that do not seem very imaginative by contemporary standards (Zhang M. 1934b). The report of the exhibition makes it clear, though, that the sheer variety of objects on view and the opportunities for hands-on experiences, such as using microscopes, handling the tools and materials used in taxidermy, or preparing vouchers of pressed plants for a herbarium, all had a powerful impact, particularly on the intended audience of teachers and students. Visitors' most common complaint was not that the presentation was dull but that the labeling was not very helpful—and that there were too many technical terms, especially scientific names in English and Latin (Zhang M. 1934a, 563). Visitors were eager to learn but responded more readily to what they could look at and touch than to written explanations of what they were seeing.

BOTANICAL GARDENS

Historically, gardens in China were usually private places for rest, contemplation, and leisure. Lists of plants recorded in a wide range of sources suggest that some gardeners, at least, not only had an interest in creating a pleasing assembly of trees and blooms but also wanted to keep a record of what they were collecting. The Song scholar Sima Guang (1019–1086) divided his Garden of Solitary Pleasures (Dule Yuan) into plots dedicated to categories of plants, including medicinal herbs, peonies, and ornamental flowers, with at least two of each variety, a design that could be said to prefigure the different sections of a modern botanical garden.[6] The purpose of such careful cataloging, though, was not to educate the owner or the visitor about the names and relationships between plants in the natural world, it was the familiar Confucian imperative of knowing the correct names of the myriad things encountered in the classics.

There is evidence that not all gardens in China were the exclusive retreats of a scholarly elite. There are records that by the end of the Ming dynasty, many of the great gardens in a city like Suzhou were accessible to the public (Clunas 1996, 94). Even if these gardens did not exclude the world, though, they were built according to the personal tastes of their owners for their own pleasure. The visitor strolling through a Ming scholar's garden might come away with a better appreciation of the importance of naming things correctly, but it was not created for the purpose of education, unlike its near-contemporaries in Europe, the medicinal gardens that became today's botanical gardens.

The concept of a botanical garden as a place devoted to the collection, cultivation, study, and display of plants has its origins in the shift in Europe from the inquiry into plants as materia medica to research on plants as one piece in the complicated puzzle of how the natural world operates. The ancestor of the gardens attached to universities for the purpose of learning was the monastic garden, dedicated to growing medicinal herbs or vegetables for the community. As the study of plants extended beyond the medicinal, university gardens included specimens cultivated to teach students about classification and how to distinguish between species. The university gardens are notable for their longevity. The botanical gardens of the universities of Pisa (founded in 1543), Padua (1545), and Florence (1545) still exist today.[7] Belonging to established educational institutions, they were not dependent on the changing whims and fortunes of an individual or a family.

The introduction of exotics, initially through exchanges between networks of institutions and individuals, made it possible to learn about the diversity of nature without the need for students to travel. Later, as exploration and commerce expanded, companies such as the Dutch East India Company brought plants and seeds home from their distant trading posts to botanical gardens that were fast becoming repositories for exotic species. In the Netherlands, the University of Leiden received permission as early as 1594 to establish a *hortus botanicus* to display its collection of plants from around the world (Egmond 2010, ch. 9, ch. 12). Commercial networks paved the way in turn for empire-building. The colonial powers established new botanical gardens in their capitals as testing grounds for economic species, from which they were redistributed around their far-flung territories as plantation stock. The result was that gardens such as Kew Gardens, the Real Jardín Botánico in Madrid, and the Jardin du Roi (later the Jardin des Plantes) in Paris became, in the words of a later assistant superintendent of Kew Gardens, "a sort of botanical clearing-house or exchange for the Empire" (Thiselton-Dyer 1880, 6).[8]

Hong Kong became a crown colony on June 26, 1843, two years after British troops had taken possession of it during the First Opium War. As early as 1842, notices began to appear in the horticultural press in London about the flora of the colony, with suggestions that there should be a botanical garden dedicated to the study of the island's plants and to material now flowing there from China.[9] On June 4, 1861, the governor approved nearly £7,000 "for the formation of Public Gardens." (Griffiths and Lau 1986, 60). A curator was appointed in October 1861, and work began on the Hong Kong Botanical Gardens, which partially opened to the public on August 6, 1864, and was completed and fully opened in 1871. The first superintendent, Charles Ford, had been selected in 1871 by none other than Joseph Hooker, the director of Kew Gardens, but the small size of the colony meant that it never had the importance or the prestige of the Calcutta Botanical Garden or the Royal Botanical Gardens at Peradeniya (Ceylon, now Sri Lanka). The initiative had come from the colonial authorities, British residents of Hong Kong, and the scientific establishment in London, but geographically, this was the first modern botanical garden on Chinese territory.

Despite its foreign genealogy, the Hong Kong Botanical Gardens played a part in the development of botany in China. In the course of its brief history, Hong Kong has been a hub of exchanges of all kinds between China and the West, including scientific and educational ones. Botany is no

FIGURE 9.1   The Hong Kong Botanical Gardens in 1873, two years after they opened to the public. Although the gardens were in the British crown colony, they were the first on Chinese territory and played an important role as a point of exchange between Western and Chinese botanists well into the twentieth century. Photograph by John Thomson, 1873. Wellcome Library no. 18674i, https://catalogue.wellcomelibrary.org/record=b1176552.

exception. Charles Ford went on several collecting expeditions in South China (Griffiths 1988, 193). Cai Shou, who briefly left his mark on the field of botanical illustration, spent a year in Hong Kong in 1908 when he was forced to leave Shanghai because of his political activities (see chapter 7). When Chen Huanyong visited in October 1919, J. Tutcher, the superintendent of the Botanical Gardens, helped him plan his collecting expedition to Hainan. He spent several months there again in 1927, studying the Chinese plants in the gardens' herbarium and working with Qin Renchang, who used the herbarium's collection of ferns from China to refine his ideas about the taxonomy of cryptogams.[10] Most importantly, when Chinese botanists began to think about establishing botanical gardens in China, they had a well-established, well-respected example close at hand in Hong Kong.

Hu Xiansu had studied botany at the Arnold Arboretum. Qin Renchang spent several months in 1930 at the botanical gardens in Copenhagen and Kew after attending the Fifth International Botanical Congress in Cambridge, which itself has a noted botanical garden established between 1760 and 1763 (Hu Z. 2014, 9–10; Lu D. 2014). Other Chinese botanists studied at Edinburgh, Berlin, and Vienna, where they saw Chinese plants displayed as the pride of the gardens. It is not surprising that they would feel there was a pressing need to create botanical gardens in China.

There is some debate about when and where the first botanical garden in China was established.[11] Chen Rong (1883–1971) was professor of forestry and president of the Jiangsu Agricultural Academy (Jiangsu Jiazhong Nongye Xuexiao) in Nanjing when he set up an arboretum on the campus for teaching purposes in 1915.[12] The academy moved several times, though so there is no longer any trace of his creation (Wang Zhenru, Liang, and Wang 1994, 136). Zhong Guanguang is usually credited with establishing the first educational botanical garden in 1927 at what was then the Third National Sun Yat-sen University in Hangzhou. The College of Agriculture gave him a plot of 50 *mu* (3.3 hectares) and a modest budget. In January 1929, the plot officially became the Botanical Garden of the College of Agriculture (Fan and Chen, 1990). The garden was in use until 1969, when it was destroyed during the Cultural Revolution. Other short-lived botanical gardens were created in 1929 on the grounds of the Beiping Natural History Museum (Beiping Tianran Bowuguan) and surrounding Sun Yat-sen's mausoleum in Nanjing. Both were destroyed during the Japanese occupation (Wang Zhenru, Liang, and Wang 1994, 136).

The oldest botanical garden that still exists today started as the Specimen Garden at Sun Yat-sen University's Institute for Agriculture, Forestry, and Botany in Guangzhou.[13] In his annual report for 1932, Chen Huanyong, the director, wrote that one of the garden's main goals was to conserve threatened species, an important expansion from the focus on teaching in the earlier gardens:

> Year after year the institute has sent out collecting expeditions. Everywhere they go they come across terrible examples of burned mountainsides, and it is getting worse by the day. The original natural vegetation will be completely destroyed. We had to think of how we could transplant the plants to protect them.

> So we came up with the plan to build a Specimen Garden. . . . There are now 15,000 to 16,000 plants, which were all collected in their original habitat. After going through all kinds of difficulties, we have now brought them safely into the garden. After identifying each plant correctly, we label it with the botanical name and its place of origin so that people can tell what it is. (Chen Huanyong 1933b, 174–75)

The Specimen Garden was quite small, with a total area of just 10 *mu* (two-thirds of a hectare). It was open to the public and very popular with visitors, especially when the orchids from Hainan and rare plants from remote parts of Guangdong were in bloom. It survived the Japanese occupation and eventually became the Huanan Botanical Garden under the Chinese Academy of Sciences. It is still a popular attraction in the city of Guangzhou.

Less than a decade after Zhong Guanguang had established the botanical garden at Zhejiang Agricultural College, Hu Xiansu presented a proposal to a board meeting of the Jiangxi College of Agriculture in December 1933 for a botanical garden on the site of the Demonstration Forest of the Jiangxi Provincial Forestry School (Jiangxi Sheng Linxiao Yanxi Linchang) (Hu Z. 2014, 7). The Lushan Arboretum and Botanical Garden opened to the public on August 20, 1934, and by 1936, its second annual report could already announce that it had welcomed several thousand visitors that summer (Hu Z. 2014, 117).

Lushan is now designated as Lushan National Park, which consists of the Lushan UNESCO World Heritage site, together with the 5,000 *mu* (333 hectares) of the Chinese Academy of Sciences' Lushan Botanical Garden. The UNESCO designation cites Lushan for its "strikingly beautiful landscape, which has inspired countless artists who developed the aesthetic approach to nature found in Chinese culture." The Botanical Garden, for its part, prides itself on being "scientific in content, beautiful in appearance, resting on a bedrock of culture."[14] Between them, these descriptions capture the features of the mountain that prompted Hu Xiansu to propose it as the site for a botanical garden (Xu et al. 2009; Liu and Wang 2014, 205).

When Hu Xiansu returned to China in 1916 with a BA in botany from the University of California, Berkeley, he was appointed deputy director of the Lushan Bureau of Forestry in his home province of Jiangxi. He was concerned by the extent of deforestation on the mountain and wrote of his hopes that reforestation and professional forest management could play a role in solving poverty in China (Wang G. 1986, 31, 32; Liu and Wang 2014,

204). In 1926, when Hu was studying dendrology at the Arnold Arboretum, his mentor Charles Sprague Sargent proposed a partnership between Harvard and a botanical garden in East Asia in order to establish formal channels for regular exchanges and training. Hu was receptive but pointed out that the time was not yet right in China. The military campaigns against the warlords were at their height, and the country was too insecure to risk such an initiative (Hu Z. 2014, 4).

While serving as director of the Botanical Department of the Fan Memorial Institute of Biology, Hu raised the possibility of creating a botanical garden, initially in the western hills outside Beijing. At the October 1930 meeting of the institute's management committee, he presented detailed plans, with cost estimates and the possibility of a donation of land from a local timber company. The plan fell through, but Hu continued to pursue his idea.

In 1931, Hu and two colleagues from the institute visited Lushan, accepting an invitation to write the section on botany and zoology for a planned new gazetteer of the mountain. With the help of the director of the forestry station, they collected over three hundred plant specimens and eleven wood sections for the herbarium at the Fan Memorial Institute. Most importantly, he was reminded of the botanical riches—and the extent of deforestation—of Lushan, an isolated mountain rising more than a thousand meters above the Yangzi River and Poyang Lake with a remarkable variety of microclimates and soil conditions. The terrain varied from gentle slopes around the settlement of Guling (known at the time to Westerners as Kuling), to precipitous cliffs, waterfalls, and steep ravines. Having failed in Beijing, Hu was encouraged by the support he received during his survey, and he began to consider locating the botanical garden on Lushan instead of in the north (Hu Z. 2014, 6–7).

During the 1920s and 1930s, the settlement of Guling on Lushan had become a popular summer resort for missionaries and other foreigners working in the hot and humid plains of the lower Yangzi. As the Nanjing government consolidated its authority, it became a popular retreat for officials of the regime, including Chiang Kai-shek himself. Hu believed that the only way to get the province to pass and to enforce laws to protect the remaining forest flora and fauna was to designate an officially protected area and that the mountain's de facto if unofficial status as China's summer capital would help him in his efforts to get approval for an arboretum or a botanical garden.[15]

The governor of Jiangxi, Xiong Shiyao, a personal friend of Hu's, chaired the first meeting of the board of the Jiangxi College of Agriculture in

December 1933. He presented the proposal for an arboretum and botanical garden, to be managed jointly by the Fan Memorial Institute and the college (Liu and Wang 2014, 206). The proposal was easily adopted and forwarded first to the Fan Memorial Institute, then to the China Foundation for the Promotion of Education and Culture with a request for funding. In March 1934, the Foundation approved a grant of 120,000 yuan annually for three years, to be divided equally between the two partner institutions. A week later, the college accepted the management plans Hu had drawn up and approved a further expenditure of 6,000 yuan for the running costs. In April, the college decided to offer its Hanpokou Forestry Station (Hanpokou Linchang) as the headquarters of the new Lushan Arboretum and Botanical Garden. It was a welcome offer, since there was already a variety of planted trees, including some exotics, available for research, as well as several buildings, at least one of which was in good enough condition to be used immediately as an office (Hu Z. 2009, 186–87).

Qin Renchang, the Fan Memorial Institute's specialist on ferns, accepted the planning committee's offer of an appointment as the first director, and he moved to the mountain with his wife in July. The first order of business was to plan for the official opening in late August, scheduled to coincide with the Science Society of China's nineteenth annual meeting, its first joint meeting with other scientific societies. The opening ceremony took place at 3:00 p.m. on August 20, 1934, with participants of the Science Society meeting attending, as well as members of the Foundation, officials from the provincial government, and even a representative sent by Chiang Kai-shek. Later that day, the management committee held its first formal meeting, during which it adopted a seven-point statement to guide its planning and for use in fundraising. The fourth point spelled out the garden's mission, which was to combine pure and applied botanical research. Pure research meant to identify and to classify plants, to plant them, and to grow them for the purposes of scholarly research. Applied research meant to study how to grow and to propagate plants and to investigate their uses in order to improve agriculture and forestry in China (Hu Z. 2014, 14).[16]

In 1936, Lushan was well on the way to realizing its founders' dream to create a world-class botanical garden in China. That year, the garden hired its first horticultural technician, Chen Fenghuai (1900–1993), who had just returned from two years study at the Royal Botanic Garden in Edinburgh. Construction of the 4,419 *mu* (295 hectare) garden was now complete, with boundary stones in place to mark the newly surveyed perimeter. Teams from the garden were actively collecting on the mountain, as well as in several other provinces. By this time the displays of flowering trees and ornamentals were well established, attracting several thousand visitors, with an additional three hundred delegates attending a meeting of the Chinese Society for Children's Education—an opportunity to encourage schools to teach botany in their science curriculum (Hu Z. 2014, 128).

The outbreak of war in 1937 did not immediately affect Lushan. In September 1937, Qin instructed staff to begin preparations in case they would have to evacuate, but during that summer, the garden had carried on international seed exchanges and collecting expeditions uninterrupted. By June 1938, Japanese forces had advanced to the Yangzi, and on July 24, Qin was on his way to Changsha when he received instructions from the Guomindang military authorities to evacuate Lushan. He told staff to leave, storing 120 crates of equipment and documents at the American School in Guling, where staff cared for it until the outbreak of the Pacific War in December 1941. Japanese troops then took over the school, seizing the crates of botanical material. Some of the contents were sent to Japan; the rest went to the Fan Memorial Institute in Beiping, where it stayed until the Japanese surrender (Hu Z. 2014, 23–24).

FIGURE 9.2 A photograph of Lushan Botanical Garden in the snow, taken in 1937, a year after it opened to the public. Photograph by Ren Hongjun. Courtesy of Hu Zonggang, Lushan Botanical Garden.

From 1938 until 1945, the story of the Lushan Arboretum and Botanical Gardens merges with the story of the Fan Memorial Institute. Qin Renchang, Chen Fenghuai, and most of the staff made their way to Kunming, where they joined the Yunnan Institute of Agriculture, Forestry, and Botany at Heilong Tan. By this time, institute staff from Beiping had already reached there, and there was no room for the new arrivals. Since the Lushan Botanical Garden had been working on montane flora, Qin decided to open a research station in Lijiang, moving there in December 1938. After the Japanese surrender, Qin stayed in Kunming. Chen was appointed director of the Lushan Arboretum and Botanical Garden, returning there in the summer of 1946 to find that it had been completely destroyed during the Japanese occupation. A few of the former staff were still in the Guling area, and he hired them to begin to rebuild what they could while the civil war and hyperinflation ravaged the surrounding area. Lushan was liberated in 1949, and in 1950 the gardens became the Lushan Botanical Garden, joining other research institutions under the Chinese Academy of Sciences (Wang G. 1986, 33–35; Hu Z. 2014, 28–31).

Today, there are three graves in a quiet grove near the administrative offices of the Lushan Botanical Garden, where the three men most closely associated with founding the garden now rest. The tombstones bear the names Chen Fenghuai, Hu Xiansu, and Qin Renchang. A brief walk from there is a plaque with a picture of Mao Zedong to commemorate the Eighth Plenary Meeting of the Central Committee Political Bureau of the Communist Party of China that took place at Lushan from July 2 to August 16, 1959. The meeting was an attack on those who doubted the wisdom of the Great Leap Forward, and the antirightist movement that followed set a pattern of harsh criticism of critics of the party line, which led to the excesses of the Cultural Revolution.[17] It was the unrelenting attacks of the Red Guards that led to Hu Xiansu's death and to the suicide of his friend, the botanical illustrator Feng Chengru, in 1968.

When he died, Hu's family was not allowed to bury his ashes at Beijing's Babao Shan Cemetery. It was only in 1979 that he was officially rehabilitated, with a memorial service organized by the Academy of Sciences. In 1982, the leadership of the Botanical Garden initiated a lengthy bureaucratic process to get permission to bury his ashes at Lushan, as he would have wished. The ceremony finally took place in 1983, in preparation for the fiftieth anniversary of the garden's opening. In due course, Qin Renchang and Chen Fenghuai joined him in his resting place, where the three are now honored as the fathers of the Botanical Garden. In a sad footnote, though, in an essay honoring Hu Xiansu, his son Hu Dekun noted that his father's tombstone simply remembers him as the "founder of the Lushan Botanical Garden." Qin Renchang, on the other hand, is honored with the title "academician of the Chinese Academy of Sciences and founder of the Lushan Botanical Garden."[18] Hu Dekun questioned why, during the celebrations in 1983 of the eightieth anniversary of the Botanical Society of China, there was no mention at all of Hu Xiansu, a founding member who served for many years as the society's president. "I hope and I trust," he wrote, "that the way Hu Xiansu is remembered will slowly come closer to the truth."

CHAPTER TEN

Metasequoia glyptostroboides, the Dawn Redwood

> The Pliocene period is generally accepted as having begun some 7 or 8 million years ago. Thus it may be appreciated that accounts of this new conifer which appeared in the daily press, hailing it as "a tree believed extinct for 100,000,000 years," may be commended for their enthusiasm but not their accuracy!
>
> —HENRY N. ANDREWS, "*METASEQUOIA* AND THE LIVING FOSSILS"

IN 1948, nearly three years after the Japanese surrender, the Fan Memorial Institute of Biology had finally reclaimed its building in Beiping. Staff had moved back from their wartime refuge in Kunming. Hu Xiansu, the director, was struggling to find funds to reactivate a research program that had been severely curtailed during the war years. In May, the institute was able, after a lapse of many years, to republish its respected journal, the *Bulletin of the Fan Memorial Institute of Biology*, now described as the new series. Only one issue of the new series ever appeared, but it made its mark and is remembered for a paper by Hu Xiansu and Zheng Wanjun, "On the new family Metasequoiaceae and on *Metasequoia glyptostroboides,* a living species of the genus *Metasequoia* found in Szechuan and Hupeh" (Hu and Zheng 1948). It was one of four papers in the same issue describing newly discovered species, but it was *Metasequoia*— described as a "living fossil" and given the evocative name of "dawn redwood" by the *San Francisco Chronicle*—that captured the imagination of botanists and the general public around the world. The story of its discovery

FIGURE 10.1 The first published drawing of the dawn redwood (*Metasequoia glyptostroboides*), by Feng Chengru in the *Bulletin of the Fan Memorial Institute of Biology*, new series (Hu and Zheng 1948, 153). © President and Fellows of Harvard College. Arnold Arboretum Archives.

and its introduction into urban landscapes, parks, and gardens around the world incorporates many of the elements that marked the story of botany in China over the previous century.

There are many accounts of the discovery and identification of *Metasequoia glyptostroboides*, with conflicting claims about the roles of the actors involved. The United States Department of Agriculture's Plant Inventory Catalog for 1955 indicates receiving seeds of *Metasequoia glyptostroboides* from China on January 28, 1948, several months before Hu and Zheng's paper (United States Department of Agriculture 1955, 4). It had in fact been some years since the first trees were found in the village of Modao Xi in Wanxian County, Sichuan (now within the administrative boundary of Lichuan County, Hubei).[1]

Hu and Zheng reported in their paper that Wu Zhonglun, professor of forest ecology at National Central University in Chongqing, was visiting Wang Zhan, a colleague at the National Bureau of Forest Research, in the summer of 1945. Wang showed him some specimens of an unusual deciduous conifer he had collected during a recent visit to Modao Xi (Hu and Zheng 1948, 154). Wang had been staying with a friend there who asked for help in identifying a strange tree that local people called *shui sha*, or "water fir." From the samples he and his friend were able to collect, Wang had provisionally identified it as the Chinese swamp cypress (*Glyptostrobus pensilis*), widely distributed in southern China but not previously reported in this central part of the country.

Wu Zhonglun gave one of Wang's specimens to Zheng Wanjun, a leading authority on the taxonomy of the gymnosperms at National Central University in Nanjing. Zheng thought it was probably a new species, not *G. pensilis*. The sample was too small to make an identification, however, so he sent his student Xue Jiru (1921–1999) in February and again in May 1946 on the arduous 120 km trek to Modao Xi to collect more specimens for a correct identification (Hsueh 1985). By March or April, before Xue's second visit, Zheng had enough material to confirm that this was a new species, but he also sent some specimen vouchers to Hu Xiansu in Beiping for assistance in making a firm identification.

Hu had been working on a special issue of the *Chinese Journal of Botany* on paleobotany (Hu X. 1950, 9). Among the materials he was reviewing was a paper published in 1941 by the Japanese paleobotanist Miki Shigeru (1901–1974) of Kyoto University identifying a new genus *Metasequoia* from Pliocene fossils found near Tokyo and in Manchuria (Hu X. 1948, 202; Tsukagoshi, Momohara, and Minaki 2011). Hu determined that the conifer from Modao Xi was a living member of this genus otherwise known only from the fossil record. He had also been writing a paper for the *Bulletin of the Geological Society of China* on Miki's work and took the opportunity to announce that a living relative of *Metasequoia* might have been found in Wanxian County (Hu X. 1948, 202; 1950).

Zheng and Hu were eager to inform their international contacts about *Metasequoia*. They both told their mentor E. D. Merrill at the Arnold Arboretum in April 1946. In May, Hu also sent the news to Ralph Chaney, professor of paleobotany at the University of California, Berkeley. Merrill received a specimen in May 1947. By June, both Chaney and Merrill had sent Hu some funds to collect seed, and Zheng sent a packet of seeds labeled *Metasequoia glyptostroboides* to the Arnold Arboretum in December (Ma J. 2003, 11). Zheng and Hu also sent seeds to at least seven botanical gardens in the United States, Europe, and India, and to the US Department of Agriculture.

The living fossil from China quickly became a public sensation. In February 1948, Milton Silverman, the science correspondent of the *San Francisco Chronicle*, accompanied Chaney to China. His first report from Wanxian appeared on March 25 under the eye-catching headline "Science Makes a Spectacular Discovery. 100,000,000-year-old race of redwoods. Story of a tree whose family lived with the dinosaurs" (Silverman 1948). It seems that even before leaving for China, Silverman and his editor had coined the name "dawn redwood," alluding to the dawn of time, a far more catchy name than the botanical *Metasequoia glyptostroboides* (Silverman 1990).

Over the next two years, there were further expeditions to the *Metasequoia* area, studying the distribution and ecology of the species, its conservation status, the cultural history of the area, and the species' ability to reproduce in its natural habitat (Gressitt 1953; Chu and Cooper 1950). The seeds sent to the various botanical gardens flourished, and the once-rare dawn redwood eventually became a popular, widely distributed species, although the naturally occurring population is still considered endangered in its original habitat.

After 1950, there were no direct contacts between Western botanists and their colleagues in the People's Republic of China until the 1980s. During this time, the dawn redwood, the living fossil *Metasequoia glyptostroboides*, acquired a mystique fed by disputes about who had discovered it, who first obtained and distributed seeds outside China, and speculation about its possible spiritual significance.

In the United States, there were reports of a bitter feud between Merrill and Chaney over who first brought seeds from China. They both always gave full credit to their Chinese colleagues, but the perception still lingers that Western botanists saved *Metasequoia* from extinction, and there have even been suggestions that Chaney discovered the first tree in Modao Xi. Regrettably, this popular memory of the story in the West, with the prominence given to Chaney, Merrill, Gressitt, and others, carries uncomfortable echoes of the heroic representation of earlier Western plant hunters.[2]

In China, too, there were arguments about who should be credited with the discovery. Gan Duo, professor of forestry at National Central University, told Hu and Zheng that he had passed through Modao Xi and seen the tree as early as 1941, but he had not been able to take any specimens because it was winter and the tree had no leaves. Others were unhappy that Wang Zhan, who had verifiably collected the first specimens, had received so little credit for the discovery (Ma J. and Shao 2003; Kyna 2016). There have even, very occasionally, been questions about the "prescientific life" of *Metasequoia*. A blog post from 2007 by two researchers at Beijing Forestry University asks, "Who discovered it? Before it was officially discovered and named, it was known by ordinary people, the Tujia people who live there, who have known it as a sacred tree and planted and tended it for 470 years" (Tie and Wang 2007).

Hu Xiansu himself considered the identification of the *Metasequoia* to be one of the most important achievements of his career. His last professional publication was "Metasequoia Poem" (Shui sha ge), published in the *People's Daily* on February 17, 1962, which he translated himself for the Hong Kong periodical *Eastern Horizon* in October 1966.[3] Unusually for Hu, the poem was written in modern Chinese, not in his preferred medium of classical verse. It follows the geological transformations of the earth from the cretaceous era to the present, pausing at the Ice Age to describe the extraordinary survival of just a small group of *Metasequoia*:

> Metasequoias, mighty kings of old, found their last refuge in a tiny spot in central China.
> Miki first studied their fossil remains, Hu and Zheng continued the search,
> Miraculously some descendants of the Herculean giants have been preserved!
> (Hu X. 1966, 27)

The poem continues with praise for Chinese scholars of old and today, ending with the patriotic and politically correct phrase for the time from a speech by Mao Zedong: "The East wind will undoubtedly surpass the West wind." It was just two years later that Hu died, hounded by Red Guards fired into action by that very phrase.

The story of *Metasequoia* is a suitable place to conclude this study. Chronologically it was the last botanical discovery and identification to involve a collaboration between a Chinese team and a group of Western colleagues before the resumption of cooperative research programs with the People's Republic in the 1980s. Hu and Zheng's paper, with a finely rendered illustration by Feng Chengru, appeared exactly one century after Wu Qijun published the *Research on the Illustrations, Realities, and Names of Plants*. Wu had provided brief descriptions and even illustrations of some plants that were not found in the classical corpus, but he had simply listed them as *"wuming"* (no name). His purpose was not to look for affinities or differences with other plants that would allow him to bring the unknown into a comprehensive system of classification. With the publication of *Metasequoia* coming just ninety years after Li Shanlan's translation of *Elements of Botany*, the transition from traditional knowledge to scientific botany would appear to be complete. The richness of China's flora was recognized worldwide as part of a global heritage, not just as a commercial opportunity for seed companies and nurseries in Europe and the United States. Chinese scientists were acknowledged, their names now associated with the plants they collected. They were no longer just the anonymous "Native Collector" on the labels on specimen vouchers stored in herbaria in foreign institutions.

Hu and Zheng's paper and the others in the same issue of the *Bulletin of the Fan Memorial Institute of Biology* all used a recognized, standardized Chinese terminology in their descriptions of new species. They included the scientific Latin or internationally sanctioned term to guard against possible misinterpretations of Chinese terms. The language issue had been addressed as Bing Zhi, Zhong Guanguang, and others had recommended twenty to thirty years earlier, with mechanisms in place to generate new terminology as needed.

The discovery and identification of *Metasequoia glyptostroboides* had involved not only the acknowledged founders of the field of botany in China who had been trained abroad but also a new generation of botanists who had studied in China, such as Wang Zhan and Xue Jiru, the "young men who have recently graduated from the universities ... full of enthusiasm and energy" that Hu spoke about in his 1938 address to the Royal Botanical Society. There was, not surprisingly, some discomfort at the institutional hierarchy that seemed to give more credit for the discoveries to the senior scientists in Nanjing and Beiping than to Wang Zhan, Xue Jiru, and others who had actually made their way to the stands of *Metasequoia* in the Sichuan-Hubei borderlands to collect specimens and seed. Such personal and institutional rivalries are of course in no way confined to China, as can be seen from the reported dispute (exaggerated in the press) between Chaney and Merrill over who first received and distributed seed from China.

This book is framed as the transition from traditional to scientific, from one epistemology to another. The discovery and identification of *Metasequoia glyptostroboides* followed a pattern of practice and research that conforms to the "scientific" norm. It is not difficult to see from their treatment of new species that Hu Xiansu and Wu Qijun a century before him were working in different intellectual spheres, asking quite different questions. It is not unreasonable to ask, though, whether there is or was any connection between them. Was there a transition from one to the other, or was there a sharp, decisive break? The orthodox view in China is clear. *The History of Botany in China*, compiled under the auspices of the Chinese Academy of Sciences Institute of Botany, states that there was a complete break. The first section of the history celebrates the achievements of traditional botany using the term *gudian* (classical) where I have used the word traditional (Wang Zhenru, Liang, and Wang 1994, 3–119). The next section, on the modern history of botany, begins with the unequivocal statement that "China's accomplishments in classical botany are glorious and brilliant, with many achievements that were far ahead of the rest of the world. But our modern botany has come to us from the West" (Wang Zhenru, Liang, and Wang 1994, 121).

Reading the words of Zhong Guanguang, Hu Xiansu, their contemporaries, and the generation of botanists that followed them, such as Cai Xitao and Xue Jiru, I would argue together with David Wright that "traditional Chinese science did not suddenly die; nor did modern Western science suddenly arise in China to take its place. The process was, rather, a gradual transformation" (D. Wright 2000, 24). I would also agree with Jing Tsu and Benjamin Elman who warn against reducing the history of science in twentieth-century China to a "simple reception history," suggesting instead that "the adapted uses of scientific knowledge range from creative appropriation to disarticulated, small-scale efforts" (Tsu and Elman 2014, 2). The past was very much a part of the way the early Chinese botanists negotiated the present and faced the future. Hu Xiansu wrote classical poetry about the plants he was studying. Cai Xitao wrote allegorical fables about the cycles of change in nature. As he was preparing a botanical guide to the camellias of Yunnan in 1947, Yu Dejun had no hesitation in deferring to the old-fashioned, anachronistic monograph *A Brief Account of the Camellias of Southern Yunnan* by Fang Shumei. When Zhong Guanguang dedicated the latter part of his life to matching plants in the classics with their botanical counterparts, he was not simply engaged in a practical task of translation. He was deeply embedded in a system of thought, expounded in the works of the classical thinkers such as Xunzi, in which correctly matching names to things was a prerequisite for attaining a desirable degree of order in the world. For Zhong, a scientific name and a name recorded in the past were different, but related. One did not exclude the other. It was his responsibility to determine how one mapped onto the other.

In botany, taxonomy was especially conducive to the creative adaptation of the traditional body of knowledge to develop and absorb Western scientific botany. Plants had names. The concept of ordering and organizing was well entrenched. There was a rich body of existing literature and scholarship about plants. It is not unreasonable to conclude that there was a transition from one mode of classification to another, rather than a break.

One area where there was a break, though, was in the collection of materials to study, the field research that is arguably the most essential practice of botany.

Travel writing and writing about landscapes and vegetation encountered while traveling was a respectable literary genre. The close study of plants with reference to existing sources of information to identify them and to classify them was a known practice, although the goal was to correct names rather than to build new knowledge. But going into wild, untraveled, even dangerous places to collect unknown plants of uncertain or quite possibly of no utility was new. Doing science in this way was an unprecedented practice, and botanists responded to the experience by opening themselves to everything they encountered on the way. They inquired into rural life. They reported on local economies, crops, land uses, and natural resources. They developed an affinity for landscapes and plants that fostered an appreciation and pride in what had until then been unseen and uncared for.

The connection between nationalism and botany—and other fields of science—is a theme that has recurred in almost every chapter here. The connection could take the form of anger against Western dominance and control over knowledge, or it could be expressed as pride in China's bountiful resources and centuries of scholarly learning about plants. Constant reminders that foreign naturalists knew more than Chinese botanists did about their own country's flora and fauna inflected their pride with nationalism. To conduct an inventory of the nation's plant life was to reclaim a natural world that explorers and plant hunters had appropriated, and to extend China's own knowledge of itself. Whether it was explicit, as in the calls for Mr. Science to save China and the founding of Academia Sinica to make science a part of national development, or whether it was unspoken but still central, as in the exploration of remote border regions such as Hainan and Yunnan, science, botany, and nationalism have been closely entwined from the Opium Wars to the founding of the People's Republic and into the present.

In the field of botany, at least, the evidence seems to confirm the proposition that the dichotomy between modern science and "prescientific" or traditional knowledge is not an unbridgeable chasm. To insist on the distance between two worlds is less enlightening than to look for the intersections between them or to explore the pathways that lead from one to the other.[4] Zhong Guanguang, working to align vernacular, common names with scientific nomenclature had no doubt that the traditional and the scientific needed each other in a new global and connected world.

> Common names and scientific names are like the wings on a bird or wheels on a cart—they are both necessary and they must work together to move forward. If one is missing, then both will fail. Universities, museums, and private collections all contain specimens of plants. If there is no record of the scientific name, then the classification will not be clear and the information will be cut off from the body of international culture. The specimens will be no more than kindling for a fire. If there is no record of the common name, it will be impossible to communicate the information. The specimens will be cut off from our Chinese culture and they will be no more than shiny baubles. (Zhong 1932a, 2)

GLOSSARY

PERSONAL NAMES

Ai Yuese (Joseph Edkins) 艾約瑟
An Ji 安吉

Bing Zhi 秉志

Cai Shou 蔡守
Cai Xitao 蔡希陶
Cai Xun 蔡珣
Cai Yuanpei 蔡元培
Chen Fenghuai 陳封懷
Chen Haozi 陳淏子
Chen Huanyong 陳煥鏞
Chen Rong 陳嶸
Chen Zhen 陳禎
Cheng Yaotian 程瑤田
Ci Xi 慈禧

Deng Mei 鄧渼
Deng Wanjun 鄧萬均
Deng Zhizhi 鄧直指
Ding Wenjiang 丁文江
Du Yaquan 杜亞泉

Fan Chengda 范成大
Fan Diji 範迪吉
Fan Jingsheng 范靜生

Fan Xu 范旭
Fan Yuanlian 范源濂
Fang Shumei 方樹梅
Feng Chengru 馮澄如
Feng Shike 馮時可
Feng Youlan 馮友蘭

Gan Duo 干鐸
Gong Lixian 龔禮賢
Gong Zizhi 龔自知
Gu Jiegang 顧頡剛
Guo Moruo 郭沫若
Guo Pu 郭璞

Han Guojun 韓國鈞
Hong Xiuquan 洪秀全
Hu Buzeng 胡步曾
Hu Dekun 胡德焜
Hu Xiansu 胡先驌
Hua Hengfang 華衡方
Huang Yiren 黃以仁
Huang Zong 黃宗

Jiang Jieshi (Chiang Kai-shek) 蔣介石
Jiang Menglin 蔣夢麟
Jiang Tingxi 蔣廷錫
Jiang Zong 江總

183

Jiayepo (Sanskrit: Kāśyapa) 加葉波
Jingwen　靜聞

Kang Youwei　康有為
Kuang Keren　匡可任

Lang Shining (Giuseppe Castiglione)　郎世寧
Le Tianyu　樂天宇
Li Gefei　李格非
Li Hongzhang　李鴻章
Li Huilin　李慧林
Li Shanlan　李善蘭
Li Shizhen　李時珍
Li Yuying　李煜瀛
Li Zhongli　李中立
Lin Zexu　林則徐
Liu Dayou　劉大猷
Liu Shen'e　劉慎諤
Liu Shoudan　劉壽珊
Liu Wenhui　劉文輝
Liu Xian　劉咸
Liu Xiang　劉湘
Liu Youtang　劉幼堂
Lu Yinggu　陸應穀
Lu Zuofu　盧作孚

Mai Xiu (Māāk or Mauk Sow-u) 麥秀
Mao Zedong (Mao Tse-tung) 毛澤東
Matsuda Sadahisa　松田定久
Matsumura Jinzō　松村任三
Miki Shigeru　三木茂
Mile (Sanskrit: Maitreya)　彌勒
Mu Yu　穆宇

Panlong Shan Ren　盤龍山人
Pei Shengji　裴盛基

Pu Jingzi　撲靜子
Pu Yi　溥儀

Qian Chongshu　錢崇澍
Qin Renchang　秦仁昌

Ren Hongjun　任鴻雋

Shen Gua　沈括
Shen Huanzhang　沈煥章
Sima Guang　司馬光
Song Boren　松伯仁
Su Song　蘇頌

Takano Chōei　高野長英
Tan Yuese　談月色
Tao Bi　陶弼
Tao Hongjing　陶弘景
Thopa Xin　拓拔欣

Wang Ang　汪昂
Wang Hanchen　王漢臣
Wang Jie　王介
Wang Qi　王圻
Wang Ren'an　汪訒庵
Wang Rongbao　汪榮寶
Wang Shimao　王世懋
Wang Siyi　王思義
Wang Xiangjin　王象晉
Wang Zhan　王戰
Wang Zhao　王釗
Wang Zhonglang　王仲朗
Wei Lianchen (Alexander Williamson)　威廉臣
Wu Bingxin　吳冰心
Wu Jiaxu　吳家煦
Wu Lin　吳林
Wu Qijun　吳其濬
Wu Weishi　吳偉士
Wu Yuandi　吳元滌

Wu Zhonglun　吳中倫
Wu Zixiu　吳子修

Xie Zhaozhe　謝肇淛
Xing Shuzhi　幸樹幟
Xiong Shiyao　熊式耀
Xu Jianyin　徐建寅
Xu Shou　徐壽
Xu Weiying　徐維英
Xu Xiake　徐霞客
Xue Jiru　薛紀如
Xun Kuang　荀況
Xunzi　荀子

Yan Fu　嚴復
Yang Jiankun　楊建昆
Yang Xingfo　楊杏佛
Ye Lan　葉瀾
Ye Qizhen　葉其楨
Yu Dejun　俞德浚
Yu Heqin　虞和欽

Yu Heyin　虞和寅
Yuan Shikai　袁世凱

Zeng Guofan　曾國藩
Zeng Zhaolun　曾昭掄
Zhang Hua　張華
Zhang Jian　張謇
Zhang Jingyue　張景鉞
Zhang Zongxu　張宗緒
Zhao Chengzhang　趙成章
Zheng Qiao　鄭樵
Zheng Wanjun　鄭萬鈞
Zhong Guanguang　鐘觀光
Zhong Xianchang　鍾憲鬯
Zhong Xinxuan　鐘心煊
Zhu Su　朱橚
Zhu Yuanzhang　朱元璋
Zhuangzi　莊子
Zou Bingwen　鄒秉文
Zou Yigui　鄒一桂
Zuo Jinglie　佐景烈

OTHER TERMS

151 Bing Zhan Yiyuan　151 兵站醫院　Military Hospital 151

bai he hua　百合花　Liliales
Bao Zhu　寶珠　Precious Pearl (camellia variety)
Beibei　北碚　a suburb of Chongqing; the wartime home of universities and research centers displaced from Japanese-occupied territory
Beiping Tianran Bowuguan　北平天然博物館　Beiping Natural History Museum
Beiping Yanjiuyuan Zhiwu Yanjiusuo　北平研究院植物研究所　Institute of Botany of the National Academy of Beiping
Beiping Yanjiuyuan zhiwuxue yanjiu congkan　北平研究院植物學研究叢刊　*Contributions from the Institute of Botany*
beizi zhiwu　被子植物　angiosperms
Bencao gangmu　本草綱目　*Classification of Materia Medica*
Bencao tujing　本草圖經　*Illustrated Pharmacopeia*
Bencao yuanshi　本草原始　*Sources for Materia Medica*

Bianyi putong jiaoyu baike quan shu 編譯普通教育百科全書 *The Compiled and Translated Encyclopedia for General Education*
biaoben yuan 標本園 specimen garden
boluomi 波羅蜜 jackfruit (*Artocarpus heterophyllus*)
bowu 博物 the broad range of things; natural history
bowu tuhua 博物圖畫 natural history drawings
Bowu zhi 博物志 *Treatise on the Broad Learning of Things*
bowuxue 博物學 broad learning of things; natural history
Bowuxue zazhi 博物學雜誌 *Journal of Natural History*
bu 部 section or class

cai 菜 vegetables
Cai Zhai Zhen 蔡宅鎮 a market town in Zhejiang; birthplace of Cai Xitao
cao 草 grass
cha 茶 tea
cha hua 茶花 Camellia
Cha hua pu 茶花譜 *A Compendium of Camellias*
chang ye 長葉 *magnifolia*
Chang'an 長安 a former capital of China (now Xi'an)
chi 尺 one Chinese foot (approximately 30 cm)
chong 蟲 insects
chu 畜 livestock
cihua 雌花 female inflorescence
cirui 雌蕊 pistil
congwu 叢物 plants that grow in clumps

Da Situ 大司徒 The Great Director of the Multitudes
Da xue 大學 *The Great Learning*
Dali 大理 a city in northern Yunnan
dan ziye 單子葉 monocotyledons
danbaguke 淡巴苽科 tobacco (*Nicotiana tabacum*) (translation used in *Zhiwuxue* [Botany] (1858), no longer in use today)
Daxue Yuan 大學院 University Council
Dian 滇 Yunnan
Dian nan bencao tupu 滇南本草圖譜 *Illustrated Materia Medica of Southern Yunnan*
Dian nan cha hua xiao zhi 滇南茶花小志 *A Brief Account of the Camellias of Southern Yunnan*
Dian zhong cha hua ji 滇中茶花記 *An Account of the Camellias of Central Yunnan*

Dian zhong hua mu ji 滇中花木記 *A Note about the Plants of Central Yunnan*
Diao zhong hua 弔鍾花 hanging bell flower (*Enkianthus quinqueflorus* Lour.)
Diexi 疊溪 a village in Sichuan, site of a major earthquake in 1933
Dinghu Shan 鼎湖山 a mountain in Zhaoqing Prefecture, near Guangzhou
Dixue Hui 地學會 Geographical Society
Dixue zazhi 地學雜誌 *Earth Sciences Journal* (later *The Geographical Journal*)
Dongfang zazhi 東方雜誌 *Eastern Miscellany*
du cao 毒草 poisonous grasses
Dule Yuan 獨樂園 Garden of Solitary Pleasures

Emei Shan 峨眉山 Mt. Emei in Sichuan, one of four Buddhist sacred mountains in China
Emei zhiwu tuzhi 峨眉植物圖志 *Icones plantarum Omeiensium* (Flora of Mt. Emei)
Er ya 爾雅 *The Literary Expositor*
Erhai 洱海 Er Lake in northern Yunnan

feng shui 風水 geomancy
fenke 分科 taxonomy (no longer in use)
fenleixue 分類學 taxonomy

Gao deng zhiwuxue 高等植物學 *Advanced Botany*
Gao Song ju pu 高松菊谱 *Gao Song's Compendium on the Chrysanthemum*
gaowu 橐物 plants with a quiver-like protective covering over their buds
ge 茖 *Allium victorialis* (victory onion or Alpine broad-leaf allium)
gewu 格物 the investigation of things
gewu zhizhi 格物致知 the investigation of things and extension of knowledge
Gezhi huibian 格致彙編 *Chinese Scientific and Industrial Magazine*
Gezhi Shuyuan 格致書院 Shanghai Polytechnic Institution and Reading Room
gezhixue 格致學 the extension of knowledge; natural studies
gongbi hua 工筆畫 fine-brush style (of painting)
gu 穀 grains

Gu jin tu shu ji cheng 古今圖書集成 *The Imperially Commissioned Compendium of Literature and Illustrations, Ancient and Modern*
Guang Hua Daxue 光華大學 Guang Hua University
guanhua 官話 official language (as opposed to classical or literary Chinese)
gudian 古典 classical or classics
Gui 桂 a prefecture in today's Guangxi
Guling 牯嶺 a summer resort on Lushan
guo 菓 fruit
Guocui xuebao 國粹學報 *Journal of National Esssence*
guojia xue 國家學 theory of the state
Guoli Bianyiguan 國立編譯館 National Institute for Compilation
Guoli Di San Zhongshan Daxue 國立第三中山大學 Third National Sun Yat-sen University
Guoli Dongnan Daxue 國立東南大學 National Southeastern University
Guoli Guoyi Yanjiusuo 國立國醫研究所 National Institute of Chinese Medicine
Guoli Zhong Zheng Daxue 國立中正大學 National Zhong Zheng (Chung Cheng) University
Guoli Zhongyang Yanjiuyuan Ziran Lishi Bowuguan 國立中央研究自然歷史博物館 Academia Sinica National Natural History Museum
Guoli Zhongyang Yanjiuyuan Ziran Lishi Bowuguan tekan 國立中央研究院自然歷史博物館特刊 *Sinensia*
Guoxue Baocun Hui 國學保存會 Society for Preserving National Learning
Guxue huibao 古學匯報 *Journal of Ancient Learning*

Hai shiliu 海石榴 sea pomegranate (early term for the camellia)
Hai song 海松 dead man's fingers (*Codium fragile*)
Hanpokou Linchang 含鄱口林場 Hanpokou Forestry Station (on Lushan)
Heilong Tan 黑龍潭 a suburb of Kunming
hewu 覈物 plants with a kernel in the fruit
hua 花 flower
Hua jing 花鏡 *Mirror of Flowers*
Hua shi 花史 *History of Flowers*
Huanan Zhiwuyuan 華南植物園 Huanan (South China) Botanical Garden
Hubei Ziqiang Xuetang 湖北自強學堂 Hubei Self-Strengthening School
hui 卉 ornamental flowers
Hui Tong Si Yi Guan 會同四譯館 Combined Four Translations Office
Huiwen Xueshe 會文學社 Literary Academy Society

Jiangnan Meishu Zhuanmen Xuexiao Shengwuhua Zhuanxiuke 江南美術專門學校生物畫專修科 Jiangnan Arts Academy School of Biological Illustration

Jiangsu Jiazhong Nongyexuexiao 江蘇甲種農業學校 Jiangsu Agricultural Academy

Jiangsu Sheng Jiaoyu Hui 江蘇省教育會 Jiangsu Provincial Educational Association

Jiangxi Sheng Linxiao Yanxi Linchang 江西省林校演習林場 Demonstration Forest of the Jiangxi Provincial Forestry School

Jiangyin 江陰 a city in Jiangsu; birthplace of Xu Xiake

jiawu 莢物 plants with a seedpod

jing 經 one of the classic texts; warp (in weaving)

Jing Shi Daxue 京師大學 Imperial University

Jing Shi Yixue Guan 京師譯學館 Beijing Teachers' College

Jingsheng Shengwu Diaochasuo 靜生生物調查所 Fan Memorial Institute of Biology

jinjina 金雞納 *Cinchona*

jiu 韭 Chinese chive (*Allium tuberosum*)

Jiuhuang bencao 救荒本草 *Treatise on Plants for Use in Emergency*

Jizu Shan 雞足山 Mt. Jizu (Chickenfoot Mountain) in northern Yunnan, a Buddhist pilgrimage site

jun zao 菌藻 fungi and algae

ke 科 family (in botanical taxonomy)

Kexue 科學 *Science*

Kexue huabao 科學畫報 *Popular Science*

Kexue Mingci Shencha Hui 科學名詞審查會 Joint Scientific Terminology Commission

Kexue She 科學社 Science Society

Kexue shijie 科學世界 *Science World*

Kexuede Zhongguo 科學的中國 *Scientific China*

ku zhu 苦櫧 *Castanopsis sclerophylla*

lei 類 category

li 理 order or pattern (within the changes of the *Dao*); principles

Lichuan 利川 a county in Hubei

Lijiang 麗江 a city in northern Yunnan

Lingnan Daxue 嶺南大學 Lingnan University (formerly Canton Christian College)

Longjiang　龍江　a township in Shunde County, Guangdong; birthplace of Cai Shou

Lü chanyan bencao　履巉嚴本草　*The Cliff Walker's Herbal*

Lugou Qiao　盧溝橋　Lugou Bridge or Marco Polo Bridge (western entry to Beijing)

luo zi　裸子　gymnosperms

Luoyang ming yuan ji　洛陽名園記　*Record of the Celebrated Gardens of Luoyang*

Lushan　廬山　a mountain in Jiangsu, site of the Lushan Botanical Garden

Lushan Senlin Zhiwuyuan　廬山森林植物園　Lushan Arboretum and Botanical Garden

mantuoluo　曼陀羅　alternative name for the camellia in *The Imperially Commissioned Compendium of Literature and Illustrations, Ancient and Modern*

Mao Song　懋松　a town in northwestern Sichuan, in today's Aba (Ngawa) Tibetan and Qiang Autonomous Prefecture

Meihua xishen pu　梅花喜神譜　*Spirit of Joy: A Collection of Portraits of the Apricot Flower*

Mengxi Bitan　夢溪筆談　*Dream Pool Essays*

Mengxue bao　蒙學報　*Journal of Pedagogy*

milu pi　秘魯皮　Peruvian bark, *Cinchona* (in a 1904 glossary of technical terminology)

ming　茗　tea leaves picked late in the season

Mingci Luntan　名詞論壇　Terminology Forum

Minsheng Gongsi　民生公司　Minsheng Corporation

Modao Xi　磨刀溪　a village, formerly in Sichuan, now in Hubei; site of the discovery of the dawn redwood

Mohai Shuguan　墨海書館　Inkstone Press

mu　木　trees, wood

mu　畝　a unit of surface area ($^1/_{15}$ hectare)

Mudan Hong　牡丹紅　Peony Red (camellia variety)

nan gua　南瓜　pumpkin (*Cucurbita moschata* Duchesne)

nan shan cha hua　南山茶花　southern mountain camellia (*Camellia reticulata*)

Nanjing Gaodeng Shifan Xuexiao　南京高等師範學校　Nanjing Higher Education Normal School

Nanjing Jiangsu Jia Zhong Nongye Xuexiao　南京江蘇甲種農業學校　Nanjing Jiangsu Higher Agricultural College

Nanmu 楠木 nanmu or Chinese cedarwood (*Phoebe nanmu*)
Nantong 南通 a model municipality in Jiangsu
niao 鳥 birds
ningmeng 檸檬 lemon
Nong Gong Shang Bu Nongshi Shiyan Chang 農工商部農事試驗場 Agricultural Experimental Station of the Ministry of Agriculture, Trade, and Industry
Nongxue Hui 農學會 Agricultural Association
Nongxuebao 農學報 *Agriculture News*

penjing 盆景 miniature landscape (Chinese equivalent of Japanese bonsai)
pin 品 grade
ping cao 瓶草 pitcher plant (*Nepenthes* sp.)
pingguo 蘋菓 apple
Pu gongying 蒲公英 *The Dandelion*

qi 氣 qi (primordial matter)
qin 禽 birds
Qing Li 青櫟 ring-cupped oak (*Quercus glauca*)
Qing Yi bao 清議報 *Qing Yi News*
Qun fang pu 群芳譜 *Assembly of Perfumes*

rui 蕊 in literary Chinese: the yellow parts near the center of a flower; now used in the binomials *xiongrui* (stamen) and *cirui* (pistil)

Sai xiansheng 賽先生 Mr. Science
San cai tu hui 三彩 圖會 *Universal Encyclopedia*
san qi 三七 pseudoginseng root (*Panax notoginseng*)
sanxingke 傘形科 Umbelliferae
Shaan Gan Ning Pendi zhiwu zhi 陕甘宁盆地植物志 *Flora of the Shaan[xi] Gan[su] Ning[xia] Basin*
shan cao 山草 mountain grasses
shan cha hua 山茶花 camellia
shan cha ke 山茶科 Theaceae
shan cha shu 山茶屬 *Camellia* (genus)
Shang Zhi Xuehui 尚志學會 Shang Zhi Society
Shangguan 上關 a town by Erhai (Er Lake), Yunnan
Shanghai Bing Gongchang 上海兵工廠 Shanghai Military Works
Shanghai Gezhi Shuyuan 上海格致書院 Shanghai Polytechnic and Reading Room

Shanghai Kexue Yiqi Guan　上海科學儀器館　Shanghai Scientific Instruments Factory

Shanyin　山陰縣　a county in Shaoxing Prefecture, Zhejiang; birthplace of Du Yaquan

Shending Mingci Guan　審定名詞館　Office for the Standardization of Terminology

shengwu huitu　生物繪圖　biological illustration

Shennong bencao jing　神農本草經　*Classical Pharmacopoeia of the Heavenly Husbandman*

Shennong bencao jing jizhu　神農本草經集注　*Collected Commentaries on the Bencao Jing*

shi　石　rocks or minerals

shi cao　石草　grasses from rocky areas

Shi ci ming ju　詩詞名句　*Famous Poetry and Verse*

Shi Zhanlan Hui　市展覽會　(Guangzhou) City Exposition

Shi zhiwu　釋植物　*On Plants*

Shijing　詩經　*Book of Poems* or *Book of Odes*

shou　獸　wild animals, beasts

shu　屬　genus

shu　蔬　vegetables

Shu　蜀　Sichuan; also a camellia variety

shuang ziye　雙子葉　dicotyledons

shuimo hua　水墨畫　ink-wash style of painting

shui sha　水杉　dawn redwood (*Metasequoia glyptostroboides*)

Shui sha ge　水杉歌　*Metasequoia Poem*

Shumu tushuo　樹木圖說　*Illustrated Manual of Trees*

Shunde　順德　a city in the Foshan district of Guangdong

Shuo fu　說郛　*Environs of Fiction*

Shuo wen jie zi　說文解字　*Analytical Dictionary of Characters*

Si Yi Guan　四譯館　Office for Four Translations (Qing translation office)

Si Yi Guan　四夷館　Office of the Four Barbarian Languages (Ming translation office)

Taihua Shan　太華山　Mt. Taihua, on the shores of Dianchi Lake, near Kunming, Yunnan

Taiping Tian Guo　太平天國　Kingdom of Heavenly Peace (Taiping)

Taiyuan　太原　the capital city of Shanxi

teng　藤　vines, lianas

Tengyüeh (Tengchong)　騰越 (騰沖)　a city in southwestern Yunnan

Tiantai Shan 天台山 Mt. Tiantai, Zhejiang
Tong yi lu 通藝錄 *Notes on All the Arts*
Tongwen Guan 同文館 School of Combined Learning (Imperial School for Interpreters)
Tupu lüe 圖譜略 *Brief Account of the Graphic Arts*

Wan guo gongbao 萬國公報 *The Global Magazine* (also *A Review of the Times*)
wan wu 萬物 the myriad things
Wan Xian 萬縣 a county in Sichuan
Wang Zhi 王制 Institutions of the Sovereign
wei 緯 weft
Wei Wang hua mu zhi 魏王花木志 *Flowers and Trees of the King of Wei*
Wenxue 文學 Literature
wu wei 無為 "nonwillful action" (in Daoism)
wu xing 五行 the five phases (earth, fire, metal, water, and wood)
Wu xun pu 吳蕈譜 *Treatise on the Mushrooms of the Lower Yangzi*
Wu za zu 五雜俎 *The Fivefold Miscellany*
Wugong 武功 a city in Shaanxi
wuming 無名 nameless
Wuxi Di San Shifan Xuexiao 無錫第三師範學校 Third Normal College (Teachers' Training College) Wuxi

Xiamen (Amoy) 廈門 a port city in southern Fujian
xiantai 蘚苔 mosses
xibao 細胞 cell
Xichang 西昌 a city in southern Sichuan
Xin er ya 新爾雅 *New Literary Expositor*
Xinjian 新建 a district in the Nanchang metropolitan area, Jiangxi; birthplace of Hu Xiansu
xionghua 雄花 male inflorescence
xiongrui 雄蕊 stamen
xiucai 秀才 xiucai degree (the first level in the civil service examinations)
Xiuqiuke 繡球科 Hydrangeaceae
Xu Xiake youji 徐霞客遊記 *Travels of Xu Xiake*
Xueheng 學衡 *Critical Review*
Xuepu za shu 學圃雜疏 *Miscellaneous Gleanings from the Nursery of Learning*
Xujiahui (Zikawei) 徐家匯 a district of Shanghai

Yan'an 延安 a city in northern Shaanxi, headquarters of the Communist-held areas during the Anti-Japanese War

Yan'an Shengwu Yanjiusuo 延安生物研究所 Yan'an Biological Research Institute

Yan'an Ziran Kexueyuan 延安自然科學院 Yan'an Institute of Natural Sciences

yancao 煙草 tobacco (*Nicotiana tabacum*)

Yandang Shan 雁蕩山 Yandang Mountains (literally "Wild Goose Pond Mountains"), Zhejiang

yang 陽 yang

Yang Fei 楊妃 Concubine Yang (camellia variety)

yangchi 羊齒 brackens and ferns

Yantai 煙台 a port city in Shandong

Yao Jiang An 姚江岸 a village in Zhejiang; birthplace of Zhong Guanguang

Yaquan Xueguan 亞泉學館 Yaquan Academy

Yaquan zazhi 亞泉雜誌 *Yaquan Journal*

Yiming Tongyi Weiyuanhui 譯名統一委員會 Translated Terminology Standardization Committee

yin 陰 yin

Yixing 宜興 a city in Jiangsu; birthplace of Feng Chengru

Yizhi Shu Hui 益智書會 Educational Association of China

yu 魚 fish (and other scaly animals)

Yunnan Nonglin Zhiwu Yanjiusuo 雲南農林植物研究所 Yunnan Institute of Agriculture, Forestry, and Botany

Yunnan xueshu piping chu zhoukan 雲南學術批評處週刊 *Yunnan Scholarly and Critical Weekly*

Yunnan zong bu yiwen 雲南總部藝文 Literary Works on Yunnan

zaowu 皁物 plants with acorns

zhang 丈 a measure of ten Chinese feet (*chi*) equivalent to approximately three meters

Zhangzhou 漳州 a city in Fujian

Zhaoqing 肇慶 a city in Guangdong

zhen li 貞理 true principles

Zhendan Daxue 震旦大學 Zhendan (Aurora) University

Zhili 直隸 a northern administrative region directly ruled by the imperial government from the Ming Dynasty until 1911, renamed Hebei Province in 1928

zhiwu 植物 plants

Zhiwu mingshi tukao　植物名實圖考　Research on the Illustrations, Realities, and Names of Plants
Zhiwu shoujing shuo　植物受精說　On Plant Pollination
Zhiwu tushuo　植物圖說　Illustrated Botany
Zhiwu xu zhi　植物須知　Essential Knowledge about Plants
zhiwuxue　植物學　botany
Zhiwuxue　植物學　Botany (translation of John Lindley's *Elements of Botany*)
Zhiwuxue da cidian　植物學大辭典　Dictionary of Botany
Zhiwuxue gelüe　植物學歌略　A Verse Primer of Botany
Zhiwuxue jiaoke shu　植物學教科書　Textbook of Botany
Zhiwuxue qimeng　植物學啟蒙　Primer of Botany
zhizhi zai gewu　致知在格物　extension of wisdom in the investigation of things
zhong　種　species, kind
Zhong xi jian wen lu　中西見聞錄　Peking Magazine
Zhongguo jingji shumu　中國經濟樹木　Economic Trees of China
Zhongguo jue lei tupu　中國蕨類圖譜　Icones filicum Sinicarum (Illustrated manual of the cryptogams of China)
Zhongguo Kexue She　中國科學社　Science Society of China
Zhongguo Kexue She Shengwu Yanjiusuo　中國科學社生物研究所　Biological Laboratory of the Science Society of China
Zhongguo Kexue She Shengwu Yanjiusuo Zhiwubu lunwen congkan　中國科學社生物研究所植物部論文叢刊　Contributions from the Biological Laboratory of the Science Society of China, Botanical Series
Zhongguo Kexue She Zhiwu Yanjiusuo　中國科學社植物研究所　Institute of Botany of the Science Society of China
Zhongguo Kexueyuan　中國科學院　Chinese Academy of Sciences (CAS)
Zhongguo Kexueyuan Zhiwu Fenlei Yanjiusuo　中國科學院植物分類研究所　CAS Institute of Taxonomy
Zhongguo Kexueyuan Zhiwu Fenlei Yanjiusuo Kunming Gongzuozhan　中國科學院植物分類研究所昆明工作站　CAS Kunming Taxonomy Research Station
Zhongguo Kexueyuan Zhiwu Yanjiusuo Lushan Zhiwuyuan　中國科學院植物研究所廬山植物園　CAS Institute of Botany, Lushan Botanical Garden
Zhongguo Xibei Zhiwu Diaochasuo　中國西北植物調查所　Northwest China Botanical Institute
Zhongguo Xibu Kexueyuan　中國西部科學院　West China Academy of Science

Zhongguo zhiwu tupu　中國植物圖譜　*Flora of China*
Zhongguo zhiwu zhi　中國植物志　*Flora Reipublicae Popularis Sinicae*
Zhongguo Zhiwuxue Hui　中國植物學會　Botanical Society of China
Zhongguo zhiwuxue huibao　中國植物學匯報　*Bulletin of the Botanical Society of China*
Zhongguo zhiwuxue zazhi　中國植物學雜誌　*Chinese Journal of Botany*
Zhonghua Jiaoyu Wenhua Jijinhui　中華教育文化基金會　China Foundation for the Promotion of Education and Culture
Zhonghua xuesheng jie　中華學生界　*Chinese Students' World*
Zhonghua Xueyi She　中華學藝社　Knowledge and Science (Wissen und Wissenschaft)
Zhonghua Ziran Kexue She　中華自然科學社　Natural Science Society of China
Zhongshan Daxue　中山大學　Sun Yat-sen University
Zhongshan Daxue Nong Lin Zhiwu Yanjiusuo　中山大學農林植物研究所　Sun Yat-sen University Institute of Agriculture, Forestry, and Botany
Zhongyang Yanjiuyuan　中央研究院　Academia Sinica
Zhongyang Zhong Zheng Daxue　中央中正大學　Central Zhong Zheng University
Zhou li　周禮　*Rites of Zhou*
Zhoushan (Chusan)　舟山　a port city in Zhejiang
zhu　竹　bamboo
zifang　子房　ovary
ziran　自然　nature
ziran kexue　自然科學　natural sciences
ziranzhi li　自然之理　self-generation's pattern
zu　租　tribe (in Li Shizhen's classification system)

NOTES

INTRODUCTION

1. Between 1928 and 1949, under the Nationalist government, Beijing was not the capital of China and was named Beiping.
2. On George Forrest, see Cowan (1952) and McLean (2004). On Joseph Rock, see Mueggler (2011).
3. This argument is closer to Michel Foucault's concept of "codes of knowledge" (1994, ix). See Pollini (2013) for a critical review of essentialist and antiessentialist arguments on nature and science.
4. On John Ray, see Raven (2009).
5. Foucault (1994, 96–110) argues that language is fundamental to the process of ordering, that its very structure creates the discourse that frames the way the world is seen and understood.
6. Li Nanqiu 1999; Métailié 2001b; Wen 2005; Luesink 2012.

CHAPTER ONE: HOW THE SOUTHERN MOUNTAIN TEA FLOWER BECAME *CAMELLIA RETICULATA*

Unless stated otherwise, all translations are mine.

1. Yu Dejun's research on camellias in the gardens of Kunming is discussed later in this chapter.
2. Petiver 1702a, 1702b; Kilpatrick 2007; Jarvis and Oswald 2015. Cuninghame's term *swatea* for the species, apparently derived from the local pronunciation of the Standard Chinese *shan cha*, did not include a rendition of the word flower (*hua* in Standard Chinese), which would normally distinguish the camellia from the tea bush.
3. Issues of concern to Chinese scholars today include: the role of tea in Chinese culture (Shen D. 2007); the philology of words used for tea (Chen Chuan 1991; Wang Zhenru, Liang, and Wang 1994); and

archaeological evidence for the earliest cultivation of tea or use of tea as a beverage (Lu et al. 2016; Zhu Jifa 2016).

4　The text used here is in *Environs of Fiction* (Shuo fu) (ca. 1368), ch. 108, 66–71. The camellia is on p. 67. Accessed July 11, 2017, through the Chinese Text Project, http://ctext.org/library.pl?if=gb&file=66723&page=67.

5　In *Complete Works of Jiang Zong* (Jiang Zong quan ji). Accessed July 11, 2017, at *Famous Poetry and Verse* (Shi ci ming ju), www.shicimingju.com/chaxun/zuozhe/386.html.

6　The author and date of *The History of Flowers* are not known. Internal textual evidence in the only extant copy, in the National Library in Beijing, is said to be from the Wanli era of the Ming dynasty (approximately 1600). See Li and Wright (2016, 148).

7　In the Shanghai, 1934 edition, the camellia is in vol. (*ce*) 554, 41a. The camellia image is on 43b.

8　Feng's *Account* is in the *Compendium* (Shanghai, 1934 edition), juan 1456, pt. 2 of *Literary Works on Yunnan* (Yunnan zong bu yiwen). Accessed July 18, 2017, at the Chinese Text Project, http://ctext.org/library.pl?if=gb&file=91562&page=6.

9　*Classification of Materia Medica* was first published in 1596. References here are to the Shunzhi edition (1645).

10　On Wu Qijun in relation to botanical illustration, see chapter 7. On the place of *Research on the Illustrations, Realities, and Names of Plants* in "traditional" botany, see Métailié (2015, 659–69). For other studies, see Chen Chongming (1980); Wang Zhenru, Liang, and Wang (1994, 102–13); Haudricourt and Métailié (1994, 406–9); Sun, Ma, and Qin (2018, 155); Li and Chen (2015, 168–69); and Zheng Jinsheng (2018, 155).

11　Szczesniak 1955; Boutan 1993; Camus 2007; Kilpatrick 2014. On the Jesuit scientific missions to China during the sixteenth and seventeenth centuries, see chapter 2.

12　Valder 1999; Fan F. 2004a; Mueggler 2005; Kilpatrick 2007, 2014; Glover et al. 2011. On Western plant hunters in China, see chapter 5.

13　Bretschneider was one of the rare Western scholars to show an interest in Chinese written sources about plants. He published two seminal texts on what he was prepared to call "Chinese botanical works," one in 1871, the other in 1895 (Bretschneider 1871, 1895).

14　This account of *C. reticulata* is based mostly on Bartholomew (1986, 8–10); Kilpatrick (2007, 193–96); and F. Griffin (1964, 27–108). The outlines of the story can be found in a number of other works on the camellia and on garden plants from China (Trehane 1998; Macoboy 1981; Valder 1999; Durrant 1982).

15　I am grateful to Professor Pei Shengji and Professor Wang Zhonglang, both of the Kunming Institute of Botany, for taking the time in 2018 to tell me the story of Liu Youtang, Yu Dejun, and *Camellia reticulata*.

16 There were some sporadic exchanges in the 1960s with collectors in Japan and New Zealand, but partnerships and exchanges with non-Communist countries did not resume until the 1980s (F. Griffin 1964; Bartholomew 1986).

17 Durrant 1982, 86; Bartholomew 1986, 9; Valder 1999, 146.

CHAPTER TWO: THE HISTORICAL CONTEXT OF AN EPISTEMIC TRANSITION

1 Gordin et al. 2002; Schneider 2003, 2012; Hu D. 2005; Schmalzer 2007; Wang Zuoyue 2007, 2015.

2 On Western plant hunters and collectors in China, see chapter 5.

3 Boym 1656; Szczesniak 1955; Camus 2007, 6; A. Griffin 2016.

4 Camus 2007, 13; Franchet 1882; Cao and Xu 2016, 616. On Boym and d'Incarville, see chapter 5. A digitized copy of Boym's *Flora sinensis* is available through the Biodiversity Heritage Library at www.biodiversitylibrary.org/bibliography/123322#/summary.

5 The emperor's edict expelled missionaries, most of whom went to Macao. A small number of Jesuits remained at the Bureau of Astronomy in Beijing and in some other official functions, so for many, the Pope's suppression of the order was the final blow to the mission. See Spence (1980, 1984); Elman (2005, ch. 2); Camus (2007); and Hsia (2009).

6 Le Rougetel 1982; Valder 1999, 63–70; Fan F. 2003a, 2003b; Chan Y. 2015. On these paintings in the context of botanical illustration in China, see chapter 7. See also the papers in *Curtis's Botanical Magazine* 34, no. 4 (Crane 2017).

7 On the Opium Wars and the Treaty of Nanking, see Waley (1958); Fay (1975); Wakeman (1978); Zhang T. (2004); van Dyke (2007); and Lovell (2011).

8 Linnean Society of London 1943; Cotton 1944, 5; Coates 1969, 101–11; Kilpatrick 2007, chs. 16 and 17.

9 On *Research on the Illustrations, Realities, and Names of Plants*, see chapter 1.

10 On the Inkstone Press, Li Shanlan, and the translation of scientific terminology, see chapter 4, as well as Luo G. (1987); Masini (1993, 36–40); D. Wright (1995, 55–61; 1998); Vittinghoff (2004, 89–92); and Sun and Ma (2017).

11 The British general Charles Gordon earned his nickname "Chinese Gordon" for his role in assisting the imperial armies during the Taiping Rebellion. For an eyewitness account, see Mossman (1893). For more scholarly studies of the Taiping Rebellion, see Teng (1971); Spence (1996); Platt (2012); and Reilly (2014).

12 Pope Pius VII had restored the order in 1814.

13 Bretschneider 1898; Boutan 1993; Kilpatrick 2014; T. Lu 2014, ch. 2; Dai 2013, 2017. On museums and botany in China, see chapter 9.
14 On the Second Opium War and the Self-Strengthening Movement, see Kwong (2000); Philips (2011); Lung (2016); and Greve and Levy (2018, 164–66).
15 On the Jiangnan Arsenal and scientific translation, see chapter 4. See also Wang Yangzong (1988); Masini (1993, 44); Fan Z. (2011); and Huang and Zhu (2012, 26–27).
16 On the Shanghai Polytechnic, the *Chinese Scientific and Industrial Magazine*, and the role of the printed media and journals, see chapter 8, as well as Biggerstaff (1956); D. Wright (1995, 1996); and Wang Hui (2011, 51, 57).
17 On the Sino-Japanese War of 1894–95, see Lone (1994); Fung A. (1996); Paine (2003); Zong (2012); and Greve and Levy (2018).
18 In 1886, Robert Hart and Joseph Edkins had published *A Primer of Botany* (Zhiwuxue qimeng), an edited volume used for teaching purposes. Fryer's translation of Balfour was notable as a translation of a complete work and because the illustrations were linked directly to the text, making it especially useful as a teaching aide.
19 In 1901, Ye Lan was one of the first Chinese students to go to Japan to study botany. I am grateful to Professor Ching May-bo of the City University of Hong Kong and her student Zhang Wenyang of Sun Yat-sen University in Guangzhou for locating a digitized copy of the primer in the Guangdong Provincial Library.
20 On the Hundred Days' Reform, see Kwong (2000); Karl and Zarrow (2002); and Weston (2002).
21 Separate courses in chemistry and geology entered the curriculum in 1910, with physics added in 1912. The Imperial University became the Government University of Peking in 1912 (Tsu and Elman 2014, 30).
22 On the Boxer Rebellion, see Esherick (1987); Elliott (2002); Xiang (2003); Bickers and Tiedemann (2007); and Silbey (2012).
23 Peake 1934; Fan F. 2004b, 413–14; Jin and Liu 2005. Data from the Qing archives relating to the civil service exams and from the records of early revolutionary groups make it feasible to quantify the linkages between the loss of possible careers in the bureaucracy and recruitment to anti-Qing organizations. An analysis has shown a positive and significant correlation between the two (Bai and Jia 2016).
24 Du Yaquan was later to be the editor-in-chief of the first dictionary of botany in Chinese, published in 1918. There is a brief biography of Du by Cai Yuanpei in Wang Yangzong (2009, vol. 2, pt. 4, 470–73). See also Elman (2005, 340); Chen, Yang, and Gu (2011, 9–10); Wang X. (2013, 104–6); and Liu Yongli (2017).

25 On the warlords, see Perry (1980); Billingsley (1988); McCord (1993); and Lary (2010). On Yuan Shikai, see MacKinnon (1992) and Shan (2018). On the 1911 Revolution, see Dingle (1912); Esherick (1976); M. C. Wright (1978); and Etō and Schiffrin (2008).
26 The first university herbarium opened in 1916 at Guangzhou's Lingnan University (Hong and Blackmore 2015, 228). Haas (1988b, 42–45) lists the higher education institutions teaching botany before 1930. On botanical gardens, see chapter 9.
27 The May 4th Movement launched the New Culture Movement, a period of intellectual and cultural change with a strong element of nationalism. On the New Culture Movement, see Weston (1998); on nationalism in the sciences in republican China, see Grace Yen Shen (2009).
28 Li Hui-lin (1911–2002) was born in Suzhou. He earned degrees in biology from Suzhou University and Yanjing University (now merged with Peking University). He went to the United States in 1940 to study for his PhD at Harvard, then spent the rest of his career in the United States.
29 *Earth Sciences Journal* was later renamed the *Geographical Journal* in English. On Zhong Guanguang's collecting expedition, see chapter 5.
30 Chen Shiwei 1997, 25–32; Wakeman 1997; Jiang and Zhang 2002, 18–20; Jiang Y. 2003, 34–36; Zhang Jian 2003, 72.
31 For biographical sketches of Hu XIansu and Chen Huanyong, see chapter 5.
32 Peake 1934, 200–203; Schneider 1982; Yang T. 1991. Chapter 8 discusses the funding agencies and research institutes. The two journals were the *Bulletin of the Fan Memorial Institute of Biology, Botanical Series*, published from 1929 to 1942, and *Sunyatsenia*, published from 1930 to 1948.
33 On the institute during the war years, see chapter 8.

CHAPTER THREE: NATURE, THE MYRIAD THINGS, AND THEIR INVESTIGATION

1 Accounts of this world view can be found in Schafer (1965); Sivin (1973); Elman (2002); Schäfer (2011, 52–55); Fan H. (2016); and Jalais (2018). On the concept of *ziran* in its classical and present meanings, see Fung Y. (1922); Van Houten (1988); Elvin (2004, ch. 10); Kim (2004); Liu Xiaogan (2004); Perkins (2005); Harbsmeier (2010); and Roetz (2010).
2 For a biography of Cai Xitao, see chapter 5.
3 Chen Siqing (1987, 66–68) interprets this as a fable about Darwinian processes of evolution with a Marxist dialectical dynamic, praising the spirit of resistance of the oppressed in their fight against the powerful. It is questionable whether at this stage in his career, Cai had absorbed Marxist literary theory to this depth.

4 This discussion of the concept of nature only covers Daoism and Confucianism, the two major indigenous schools of Chinese thought. Buddhism, introduced to China during the second and third centuries CE, offers a path to personal enlightenment and eliminating suffering from a world in which, ideally, humans and other living beings coexist in harmony. For deeper analysis in English of nature and the environment in Chinese philosophies, see Chan W. (1963); Creel (1970); Schwartz (1985); and Callicott and Ames (1989). On Daoism, see Robinet (1997) and Kirkland (2004). On Confucianism, see Nivison and van Norden (1996) and the essays in Tucker and Berthrong (1998).
5 On the publication history of the *Journal of Natural History*, see chapter 8.
6 On Du Yaquan, see chapter 2.
7 On the classification of plants, see chapter 6, as well as Pan J. (1984); Needham, Lu, and Huang (1986, 142–81); Wang Zhenru, Liang, and Wang (1994, 3–15); Mittag (2010); and Métailié (1989; 2015, 32–118).
8 On botanical terminology in the translation of *Elements of Botany*, see also chapter 4.
9 *The Treatise on Broad Learning of Things* (Bowu zhi) by Zhang Hua (late third century CE) was the earliest compilation to use the expression. It covered thirty-nine topics, including geography, unusual birds, and unusual plants.
10 Zhu F. 2003; Fan F. 2004b; Elman 2005, 43–46; Jiang L. 2016, 164–65.
11 Needham, Lu, and Huang 1986, 214–16; Métailié 1989, 2001b; Elman 2002, 2004; Zhu F. 2003; Nappi 2009b; Wang Hui 2011, 53–56.
12 Textual research indicates that between 1902 and 1905, the two terms were used almost equally frequently, followed by a period of transition, when *kexue* began to replace *gezhi*. After 1905, *kexue* became more common and after 1906, *gezhi* disappeared almost entirely (Elman 2002; Zhou and Ji 2009; Feng T. 2008; Zhang Fan 2009, 2016; Chen, Yang, and Gu 2011).
13 Métailié 2002; Feng T. 2008; Zhang F. 2009; Zhou and Ji 2009; Shen Guowei 2009, 147; 2014. The school at Deshima was shut down by the Japanese authorities in 1828. Takano was forced to flee and settled in Edo, where he was arrested in 1850 and was either beaten to death or used a dagger to stab himself in the neck. He had been accused of encouraging Western learning, a crime for a samurai. On the history of natural history in Japan, see Marcon (2015).
14 For studies of Yan Fu's use of the term *kexue* in his translations and the evolution of his understanding of science in contrast to the investigation of things, see Elman (2004); Feng T. (2008); Huang K. (2008); Zhang Fan (2009); Wang F. (2009); and Shen Guowei (2009, 2014).

15 Named after *The Literary Expositor* (Er ya), an early dictionary regarded as a classic, probably from the late Zhou to early Qin period (approximately 200 BCE), which had a commentary by Guo Pu, itself considered a classic, appended in 300 CE. On taxonomies in the original *Literary Expositor*, see chapter 6.
16 See also Brockway (2002); Schiebinger (2004); Thomas (2006); and Batsaki, Cahalan, and Tchikine (2016, 1–34).
17 Jardine, Secord, and Spray 1996; Outram 1996; Farber 2000; Daston 2008; Endersby 2008; Magnin-Gonze 2009.
18 Fan H. 1989; Wang Zuoyue 2002; Fan F. 2004b, 2008; Elman 2014; Zhang Fan 2016.
19 The entry in *Flora of China* continues over several pages, with descriptions of the many species within the genus *Camellia*.
20 On the new school curriculum after the abolition of the civil service examination system, see chapter 4.
21 On "The Inventory of Nature," see chapter 6.
22 Yu Heyin (1879–1944), also known as Yu Heqin, studied engineering in Japan and became one of China's first mining engineers (Wang X. 2013). He was a close friend and collaborator of Du Yaquan and Zhong Guanguang.
23 On the dictionary and the part it played in standardizing the vocabulary of botany, see chapter 4.

CHAPTER FOUR: A NEW LANGUAGE TO NAME AND DESCRIBE PLANTS

1 D. Wright 1998, 654–55; Alleton 2001, 26; Zürcher 2007, ch. 2.
2 Jami 1999; Elman 2002, 211; Cao and Xu 2016, 64. On the Jesuit scientific mission in China, see chapter 2.
3 Honour 1962; Jacobson 1993; Rinaldi 2016.
4 Li Hongqi 1991; Roberts 1997; Gao Y. 2017.
5 Masini 1993, 68–69; Wang Hongkai 1998, 79–80; Wang Hongxia 2009.
6 On science in the curriculum in the late Qing educational reforms, see Zhang Y. (2005). The page reference here is to the location in the original article in *Eastern Miscellany*, followed by its location in the digitized text at https://ctext.org/library.pl?if=en&file=96754&page=120 (accessed October 15, 2018).
7 On the role of *Eastern Miscellany* in popularizing science, see chapter 8.
8 On the May 4th Movement, see chapter 2.
9 Masini treats *baihua* as "another literary language ... used to compose narrative works that were to become very popular in China." He translates *baihua* as "lingua clara," which is perhaps a more appropriate rendition than "plain language" (Masini 1993, 3).

10 On translation and interpreters during the Opium Wars, see Wong (2007).
11 On the scientific translation programs of the late Qing and early republican periods, see D. Wright (2000); Li Nanqiu (1996); and Ma Z. (1998).
12 Fryer 1880, 2–4; D. Wright 1995; Elman 2005, 356–60. Fryer was one of at least six British and American translators working at the Arsenal (Lung 2016, 46–47; Wang Yangzong 1988, 71). He served as head of the Translation Department from 1868 to 1896, when he left to go to Berkeley. He continued translating for the Arsenal from America until 1903 (D. Wright 1996; Wong 2004, 241–42; Wang Hongxia 2009).
13 The numbers are from a report Fryer wrote for the *North China Herald* in 1880. The report also listed 45 unpublished translations and 13 in the course of translation (Fryer 1880, 20).
14 On the Shanghai Polytechnic and Reading Room and the *Chinese Scientific Magazine*, see chapter 8.
15 Li N. 1999; D. Wright 2001, 236; Zhang and Zhao 2004; Huang K. 2008, 30–31; Fan X. 2007; Fu Lei 2014, 52–54. On standardization and the institutionalization of botany, see chapter 8.
16 Masini 1993, 138; Alleton 2001; T'sou 2001. There are numerous tables showing the vowels and syllables of modern standard Chinese. The Chinese Language Institute of Guangxi Normal University has posted a useful chart at https://assets.studycli.org/CLI/wp-content/uploads/20180518223229/pinyin-chart.pdf (accessed August 2, 2018). For a study of Chinese phonology, see San (2007). On loanwords in Chinese, see Novotná (1967).
17 Nappi (2009a, 743–45) suggests that transliteration was not unusual in medicinal texts.
18 Luo G. 1987; Elman 2005, 327–30; Sun and Ma 2017.
19 Luo G. 1987; Métailié 2002; Sun and Ma 2017. On classification systems and taxonomy, see chapter 6.
20 On the *New Literary Expositor*, see chapter 3; on its literary ancestor, *The Literary Expositor* (Er ya), see chapter 6.
21 On the possible Japanese origin of the term *kexue* (science), see chapter 3. On Yan Fu and his approach to translation, see D. Wright (2001); Huang K. (2008); Wang F. (2009); Shen Guowei (2014); and Zhu Jiachun 2016. On the Chinese or Japanese origins of neologisms in the sciences, see Pan J. (1984); Lippert 2001; Métailié 2001b; Zhao (2006); Huang K. (2008); and Zhou C. (2009).
22 The present term is *fenleixue*.
23 On the history of what is now known as the International Code of Nomenclature for Algae, Fungi, and Plants, see Nicolson (1991).

24 On the use of traditional groupings of plants in the table, see chapter 6.
25 On the assignation of correct botanical names to plants in the dictionary, see chapter 6.

CHAPTER FIVE: OBSERVING NATURE, PRACTICING SCIENCE

1 Other terms for the field sciences include inventory sciences (Zeller 1987), survey sciences (Kohler 2007, 435), and open-air sciences (Raj 2007). In addition to botany and geology, the category includes archaeology, paleontology, anthropology, ethnology, and zoology.
2 Outram 1996; Farber 2000; Schiebinger 2004, ch. 1; Piementel 2009; Bleichmar 2011; Batsaki, Cahalan, and Tchikine 2016, 1–34.
3 Cannon, Jones, and Paton 1975; Banks et al. 2000; Chambers 2007, 21–49; Endersby 2008, ch. 3.
4 On Cuninghame, see chapter 1. Bretschneider 1898, 21–28 is an account of the early trading visits and pages 29–44 list all the species Cuninghame sent to London. See also Petiver 1702a and b; Jarvis and Oswald 2015; Kilpatrick 2007, 34–48.
5 In his report to the Société Botanique de France, Franchet said that there was another herbarium probably sent by d'Incarville with 144 species from Macao. See also Bretschneider (1898, 46–56); Camus (2007, 13); and Kilpatrick (2007, 61–70).
6 On the Swedish East India Company, see Bretschneider (1898, 58–113). See also Peter Osbeck's account of his voyage to China (Osbeck 1771).
7 Wilkinson 2006; Kilpatrick 2007, 71–110, 200–237; Easterby-Smith 2018.
8 See Fortune (1847); Wilson and Sargent (1913); Farrer (1926); and Handel-Mazzetti (1927). For biographies of plant hunters, see Bretschneider (1898); Cox (1945); Coates (1969); Pim (1966); Boutan (1993); R. Briggs (1993); Kilpatrick (2014); O'Brien (2011); and Holway (2018). Critical studies include Luo G. (2002); Fan F. (2003a, 2004a); Mueggler (2005, 2011); Glover et al. (2011); and Long C. (2016).
9 This does not include Japanese botanists in Taiwan, which was a Japanese colony at the time.
10 Matsuda is honored in the name *Salix matsudana*, the weeping willow of northern China.
11 See Wan (2020) for a popularized biography of Zhong Guanguang, who also had the name Zhong Xianchang and was known as K. K. Tsoong in early Western materials.
12 On Du Yaquan and *Science World*, see chapter 2.
13 George Weidman Groff (1884–1954) helped establish Lingnan University in Canton. Between 1927 and 1929, he also collaborated with Chinese researchers studying Chinese plants at the University of California, Berkeley. He left China in 1940.

14　Wu Zhengyi 1953, 13; Fan and Chen 1990, 451; Zhu and Liang 2005; Chen Jinzheng and Zhong 2008.
15　Zhong 1932a and 1932b; Jiang W. 1941, 59; Chen Jinzheng and Zhong 2008.
16　Most of the information here about Hu's early life comes from a biography by Hu Zonggang (2005a) and from Yu Dejun's article in Tan (1985). Hu Zonggang had access to Hu Xiansu's correspondence and archival materials, which as of 2018 were no longer accessible even to Chinese scholars (Hu Zonggang, pers. comm., December 3, 2018).
17　On Academia Sinica and the Fan Memorial Institute of Biology, see chapter 8. On the Lushan Arboretum and Botanical Garden, see chapter 9.
18　Chen was referred to in English language publications as Chun Woon-Young or W. Y. Chun, the Cantonese pronunciation of his name.
19　Haas 1988b, 13; Wu D. 2008, 2; Hu Z. 2013a, 12–16.
20　Twenty-six issues of *Sunyatsenia* appeared between 1930 and 1948. It was the first journal about Chinese botany written by Chinese botanists in English (Wu D. 2008, 2).
21　Using archival material, Hu Zonggang (2013c) concludes that Chen did not initiate the overtures to the puppet government. He did not reject the opportunity offered, though, seeing cooperation as leading to a better outcome than having the whole herbarium and library fall into Japanese hands.
22　On *The Dandelion*, see chapter 3.
23　On the institute in Kunming, see chapter 8.
24　Wu Yuandi 1914, 1918; Wu Zixiu 1924; Yang G. 1928.
25　Fang and Zhang 1928; Xing 1928, 123; Zuo J. 1934. Augustine Henry was one of the few Western collectors who worked in this area (Fan F. 2004a).
26　*Castanopsis tibetana* Hance is in the original text. I have added other botanical names in parentheses. See Jiang L. (2016, 171–73) on Hu's 1920 collecting expedition.
27　I have used the 1996 reprint of a 1928 edition by Ding Wenjiang of Xu Xiake's travels. References are to the page number in that edition, abbreviated to XXKYJ (Xu Xiake you ji). Unless stated otherwise, translations are from Ward (2001).
28　Fan C. 1692; Ward 2001, 174; Kindall 2012, 421–22.
29　Translations from Handel-Mazzetti are from Winstanley (1996). The Handel-Mazzetti China Collection in the herbarium of the University of Vienna is accessible online at https://herbarium.univie.ac.at/database/results.php.

CHAPTER SIX: THE INVENTORY OF NATURE

For references (including this chapter's epigraph) to Xunzi from the original Chinese, I have used the juan number, together with the page number in the digitized text of the *Xunzi Siku quan shu* (SKQS) edition available at the China-America Digital Academic Library (CADAL), https://archive.org/details/cadal?and%5B%5D= 荀子&sin= (accessed December 20, 2018).

1. In the same vein, Foucault (1994, 96–110) argued a millennium and a half later that language is fundamental to the process of ordering, that the very structure of grammar creates the discourse that frames the way the world is seen and understood.
2. Zhang Fan has reprinted several of these contributions in a study of the popularization of science during the early decades of the twentieth century (Zhang Fan 2016, 131–33).
3. On the concept of *ziran* (nature) and *ziranzhi li* (self-generation's pattern), see chapter 3.
4. A more fundamental classification is the identification of a category of thing that is named "plant" (*zhiwu*). See chapter 3, "Plants: Things that Grow."
5. On Caesalpino's contributions to plant taxonomy, see Ogilvie (2006, 222–26) and Pavord (2005, 228–41).
6. Studies of the classical philosophers and their relevance to modern biological taxonomy include Ryle (1938); Atran (1985); Carpenter (2010); and Lennox and Bolton (2010). Marder (2014) reviews the treatment of plants by philosophers from Plato to Derrida.
7. For a study of the power of naming the natural world in the French colonies, see Bonneuil (1997), especially ch. 1, 78–96. On the Royal Society, Kew Gardens, and the British Museum, see also Endersby (2008).
8. The most recent congress was held in Shenzhen in 2018.On the different iterations of the code and Chinese participation in the fifth congress in Cambridge, see chapter 4.
9. On Li Shanlan's translations of botanical terminology and nomenclature, see chapter 4.
10. On misrepresentations of Chinese systems of classification, see Zhang L. (1988) and Nappi (2009b, 1–7 and 148–49). For a survey of the tradition of encyclopedias, collectanea, compilations, and other works with reference to botany, see Needham, Lu, and Huang (1986, 182–220); Drège (2007); Elman (2007); and Jiang S. 2007.
11. Scholars disagree on the exact number of plants that are named in *The Book of Poems*, but it is generally accepted that the collection names over one hundred plants. Wang Zhenru, Liang, and Wang (1994, 6–7) give a list of 130 plants. It is not always clear whether a character refers to a plant at all, and if it does, it is can be very difficult to identify it. See also Keng (1974).
12. *Cao* is more accurately translated as "grasses." With some exceptions, it is the generic term used for plants in compendia, encyclopedias, and other reference works before the mid-nineteenth century.

13 On classification systems in materia medicas, see Métailié (2015, 32–118) and Chen Jiarui (1978, 108–9).
14 The materia medicas (*bencao*) were not limited to plants. They included animals, minerals, mythical beings (dragons and dragon parts), and even human parts and excretions that were used for medicinal purposes. See Nappi (2009b, 50–68) on the different materials in Li Shizhen's compilation.
15 *Sources for Materia Medica* is notable for its 380 illustrations, said to have been drawn from examples in pharmacies that Li visited. The illustrations include annotations indicating how to identify the plant and how to ensure that the product on sale is genuine. See chapter 7 and figure 7.1.
16 Francesca Bray (1988, 17) points out that within the wider ordering of nature as seen in the organization of materia medicas, agricultural manuals, encyclopedias, and other reference works, plants are placed after minerals and before animals.
17 The author of the article is not named but was probably not Chinese. Much of the content of the magazine was translated from Western languages or written by a foreign resident in China and translated by Chinese staff. On the *Chinese Scientific and Industrial Magazine* and scientific journals, see chapter 4 and chapter 8.
18 Superseded by the Translated Terminology Standardization Committee (Yiming Tongyi Weiyuanhui) in 1927 and renamed the National Institute for Compilation (Guoli Bianyiguan) in 1932 (also translated as the General Committee on Scientific Terminology). In the ten years from 1932 to 1942, when wartime conditions forced it to close down, it only produced forty-five publications, a small proportion of the estimated 9,671 scientific works published during the same period (Li Nanqiu 1999, 43).
19 On nationalism and science during the republican period, see Chen Shiwei (1997); Fan F. (2004b, 2007); Luo G. (2006); Yen (2015); and Jiang L. (2016, 157–61).

CHAPTER SEVEN: BOTANICAL ILLUSTRATION

1 For surveys in English of the history of Chinese painting see Barnhart (1997); Clunas (1997); and Cahill (2010).
2 Drawing and painting are still more widely used than photography, because the artist has the flexibility to select, to highlight, and to synthesize. It is likely that the manipulation of images digital photography allows will add to the ways of recording plant materials rather than replace existing techniques entirely. For recent reviews of technological

innovation in botanical illustration, see Simpson and Barnes (2008) and Bridson (2010).
3. Blunt and Stearn 1994, 26; Sun, Ma, and Qin 2008, 772; Fu Liangyu 2013, 79.
4. Zheng's essay can be dated to 1150 CE. The translation here is by Mu and Wang (2017, 284). Page references are to the *Siku quan shu* edition.
5. Haudricourt and Métailié (1994); Métailié (2007); Li and Chen (2015); Métailié (2015, 182–238); Mu and Wang (2017).
6. On the history of printing and publishing in China, see Twitchett (1983); Chow (2004); Brokaw and Chow (2005); Bussotti (2007); and Mun (2013). On printing technologies and early scientific illustration in China, as distinct from artistic paintings of plants, see Métailié (2007) and Pang (1997, 152–62).
7. Li and Chen 2015, 165; Métailié 2015, 161–63; Zheng J. 2018, 152–53.
8. The Ming copy is kept at the National Library in Beijing. A facsimile edition in black and white has been published, and a digitized copy from Peking University Library is available online through the Internet Archive at https://archive.org/search.php?query=creator%3A%22%E9%A1%8C%28%E5%AE%8B%29%E7%8E%8B%E4%BB%8B%E6%92%B0%22. On his project to reconstruct the body of illustrated materia medica from the Ming Dynasty, see Cao H. (2018).
9. Wang Yonghou 1994; Wang Zhenru, Liang, and Wang 1994, 66–68; Li and Chen 2015, 166; Métailié 2015, 66–70; Mu and Wang 2017, 290–92; Zheng J. 2018, 154.
10. Zheng J. 2003; Li and Wright 2018; Cao H. 2018.
11. Haudricourt and Métailié 1994, 394–96; Li and Chen 2005, 166; Métailié 2015, 192; Zheng J. 2018, 156.
12. On the format and placement of illustrations in *Sources for Materia Medica*, see Zheng J. (2018, 154–55) and Li and Chen (2015, 170). For more about Li Zhongli and the compilation of *Sources for Materia Medica*, see Needham, Lu, and Huang (1986, 321–323); Haudricourt and Métailié (1994, 339); Zhang and Zhang (2010, 2); and Zhang Fang 2015.
13. On horticultural and gardening manuals, see Métailié (2015, 416–47) and Wang Zhenru, Liang, and Wang (1994, 82–93).
14. On *The History of Flowers*, see chapter 1.
15. Haudricourt and Métailié (1995, 35) reach a similar conclusion after a careful study of the much older *Spirit of Joy* (Meihua xishen pu), a collection of portraits of the apricot flower (*Prunus mume* Sieb. et Zucc.) by Song Boren, published in 1231: "This work . . . is a fine example of a minute level of observation, but realized in a very subjective way."
16. On *Research on the Illustrations, Realities, and Names of Plants*, see chapter 1.

17 Lang Shining in Chinese. On Western artists in China, the Guangzhou art market, and artistic exchanges between China and the West during the Qing era, see Chu and Ning (2015); Koon (2014); and Musillo (2017).
18 On the plants Cuninghame collected in Chusan (Zhoushan), see chapter 1. None of the paintings in the Brewster portfolio has the Chinese characters for "Dr. Bun-Ko." I am grateful to Sir Peter Crane of the Oak Spring Garden Foundation (OSGF) for information about and access to both the Brewster collection and the Blake archive (see below).
19 In the Oak Spring Garden Library, John Bradby Blake archive, record no. M-152, PDF no. 9. On John Bradby Blake, see the special issue of *Curtis's Botanical Magazine* (Crane 2017). On Blake's life and work, see also Goodman and Crane (2017). See figure 2.1 for a painting of a litchee in the collection.
20 On the paintings commissioned by Reeves, see Fan F. (2003a, 68–69); Kilpatrick (2007, 200–205); and Magee (2011, 10–11).
21 Clunas 1984; Saunders 1995, 73–81; Fan F. 2004a, ch. 2; Cheng 2007, 45–47; Mu and Wang 2017, 304–6.
22 Guo Yanbing (2013, 59) says that Cai went to study at Zhendan (Aurora) University when he was seventeen, which would have been in 1896. Zhendan was only founded in 1903, so this is not possible. Guo used the memoirs of Cai's second wife, Tan Yuese (1891–1976), as a source. Tan was thirty-one when she left a nunnery to marry Cai in 1922. Cheng Meibao (2009, 103) argues that Tan's account of Cai's earlier years may not be very reliable, since it is based on her memories of what her husband had told her about events that had happened before they met. Cheng makes more use of Cai's own writings than other scholars, and I have followed her narrative here.
23 Fan F. 2004b, 415; Cheng 2006, 20–21; Cheng 2007, 31–35; Ding L. 2017, 43–55.
24 The flower is identified by Fèvre and Métailié (2005) as *Enkianthus quinqueflorus* Lour. It is a low bush found in the hills and mountains of Guangdong, Guangxi, Guizhou, and Yunnan and flowers around the New Year. Dinghu Shan is 18 km northeast of today's Zhaoqing City, Guangdong.
25 Cheng 2006; 2007, 39, 40; 2009, 97–102.
26 On Chen Zhen's project, see Jiang L. 2016, 177–89. Chen's work appeared with illustrations by Feng Chengru in the first issue (1925) of *Contributions from the Biological Laboratory of the Science Society of China*.
27 Wang Zhenru, Liang, and Wang 1994, 367–68; Hu Z. 2005a, 49–55; Sun, Ma, and Qin 2008, 777; Sun et al. 2010, 153; Mu and Wang 2017, 306.
28 On the discovery of *Metasequoia glyptostroboides*, see chapter 10.
29 The Needham Research Institute in Cambridge has compiled his notes on the people he met during his visits to Chinese scientific institutions

during the war as a spreadsheet available online at www.nri.cam.ac.uk/JN_wartime.htm. Kuang Keren is on row 2366.

30 Very little has been published about Kuang Keren. The information here comes from a biographical sketch on the Chinese Academy of Sciences Institute of Botany's website, www.ibcas.ac.cn/90zhounian/Academic_source/201808/t20180801_5052299.html (accessed May 10, 2019). I also benefitted from an afternoon (January 18, 2018) with Yang Jiankun, a botanical artist at the Kunming Institute of Botany familiar with Kuang's work, who showed me an original copy of *Illustrated Materia Medica of Southern Yunnan* and some of the drawings prepared for it and other publications from the early 1950s.

31 Although the pumpkin is originally from South America, the text notes that it had become an integral part of home gardens and had been adopted in local herbal remedies, and so it was included in the compendium.

CHAPTER EIGHT: SPACES FOR COMMUNICATING AND INFORMING

1 Haas (1988a, 33–38) is a review, focused on republican China, of the networks of people and institutions that were involved in botanical research.
2 On Inkstone Press and the Jiangnan Arsenal, see chapter 4.
3 Hao Bingjian quotes from Xu's letter, originally published in *The Global Magazine* (Wan guo gongbao).
4 Xie Zhensheng 1989; Wang Hui 2011, 51; Liu Yongli 2017; Wang X. 2013, 104–5. On Du Yaquan, see chapter 2. Two-thirds of the articles published in the ten issues of *Yaquan Journal* were on chemistry. The remaining third covered topics that included biology, zoology, and botany (Jiang and Wang 2011, 79).
5 On Zhong Guanguang's field reports, see chapter 5.
6 On the legacy of the broad learning of things in the *Journal of Natural History*, see Li and Yao (2011). The earliest reports of fieldwork in the journal were by Wu Jiaxu (1914b) on his work in Jiangsu, Wu Yuandi (1914) on collecting in the suburbs of Nanjing, and Wu Zixiu (1924) on collecting in Yantai (Shandong).
7 On the Science Society of China, see chapter 2. According to Fan Hongye (2015, 5), the idea for the journal came first, then a society was proposed as a home for the journal.
8 See Métailié (2015, 361–67).
9 The organizers appointed Liu Xian rapporteur for the meeting. His report appeared in *Science* 20, no. 10: 871–83, together with the text of a keynote address by Hu Xiansu on the utilization of China's plant

resources (Hu X. 1936). The same issue opens with a review by Zeng Zhaolun of the scientific associations in China at the time and of the two previous joint meetings of science associations at Lushan (1934) and Nanning (1935) (Zeng Z. 1936, 798–843).

10 The six societies were the Chinese Mathematical Society, the Chinese Physical Society, the Chinese Chemical Society, the Chinese Zoological Society, the Botanical Society of China, and the Geographical Society of China. The societies' names are sometimes translated differently into English. I have used the translations used in Zeng's review of scientific associations in China (Zeng Z. 1936).

11 The first group of scholars left China in 1910. On the Boxer Indemnity fellowships, see Wang Zuoyue (2002, 294–95).

12 Rhoads 2005; Bevis and Lucas 2006; Chen and Huang 2013, 93–94. There were Chinese students in France, the United Kingdom, and Germany, but the First World War disrupted higher education in Europe, making the United States a more attractive destination. There were programs during the war to bring Chinese to France and Britain on work-study programs, where they provided labor to replace men who had been sent to the front. Many of the students in France were politically radicalized and became leading figures in the Chinese Communist Party. Zhou Enlai and Deng Xiaoping were among the best known of this group. See Bailey (1988) on the programs in France.

13 The first issue of the journal appeared in March 1934. The society's bylaws specified that members of its management or organizing committees had to have a degree in botany from a Chinese or foreign university. Zhong Guanguang did not meet this requirement, so he never served on the committees, despite his seniority and accomplishments (Xiao L. 2014, 42).

14 Zeng Z. 1936, 806–7, 829–31; Wang Zhenru, Liang, and Wang 1994, 136–37; Xiao L. 2014. The requirement for a Latin description was in force until the 18th International Botanical Congress in Melbourne in 2011, since when either English or Latin may be used.

15 On the Boxer Rebellion, see chapter 2. On the Rockefeller Foundation in China, see Brown (1980); Schneider (1982); and Zi (1995). The fullest account of the China Foundation for the Promotion of Education and Culture is by Yang Tsui-hua (1991). Peake (1934, 203) has a summary of grants approved between 1926 and 1934. Other sources are Cowdry (1927); Fan Y. (1927); Schneider (1982); and Haas (1988a).

16 For a biography of Chen Huanyong, see chapter 5.

17 The references are from Chen's reports to the Foundation. On Chen Huanyong's correspondence with the foundation requesting financial support, see also Hu Z. (2013a, 76–82).

18 Chen Huanyong 1933a, 118–19; 1933b, 163–65; Hu Z. 2013a; Ding L. 2017, 229–31. On the Guangzhou City Exposition of 1933, see chapter 9.
19 The English name for the institute, The Fan Memorial Institute of Biology, used Fan Yuanliang's surname. The Chinese name, Jingsheng Shengwu Diaochasuo, used his courtesy name or style, Jingsheng. On the Shang Zhi Association, see Hu Z. (2005a, 15–16). My account of the Fan Memorial Institute of Biology relies heavily on the chronology by Wu Jiarui (1989) and the history based on archival materials by Hu Zonggang (Hu Z. 2005a).
20 Bing spent six months each year in Nanjing as director of the Biological Laboratory of the Science Society, traveling to Beijing to spend the next six months as director of the Fan Memorial Institute. In January 1932, he asked to be relieved of his duties as director in Beijing. The board agreed and appointed Hu Xiansu as the full-time director (Wu Jiarui 1989, 28; Hu Z. 2005a, 25–26).
21 Bing's speech at the opening ceremony was published in *Kexue* in 1929 (vol. 13, no. 9). I am using the text as reproduced in Hu Z. (2005a).
22 On Cai Xitao's expedition to Yunnan, see Hu Z. (2018, 29–34). Also see chapter 5.
23 Qin not only was an expert on Chinese ferns but he also played an important part in developing a global taxonomy of the class (Hu Z. 2005a, 35–44).
24 In addition to Copenhagen and Kew, Qin visited herbaria in France, Sweden, Austria, and Germany (Yu D. 1981, 84).
25 On Feng Chengru and botanical illustration, see chapter 7.
26 On the Lushan Botanical Garden, see chapter 9.
27 There is a brief note about Liu's journey in the Beiping Academy's fifth anniversary report (Beiping Yanjiuyuan 1934, 80–81). I have not been able to find any reports in English, French, or Chinese on a "Franco-Chinese scientific expedition" in 1931. Hu Zonggang (2006) believes that it was the "Croisière Jaune," also known as the "Mission Centre-Asie," organized by André Citroën of the Citroën car company. One group set out from Beirut, the other from Beijing, meeting in Ürümqi on October 27. The Chinese government was reluctant to allow the expedition to travel so far west but agreed on the condition that some Chinese scientists travel with them. The expedition was widely publicized, and a film with the title *La Croisière Jaune* was very successful in France in 1934. Studies of the expedition do not give the names of any Chinese scientists (Gourlay, 2004; Audoin-Dubreuil, 2007).
28 Military Hospital 151 is said to have been a research center for biological warfare, but all the documents from the facility were destroyed in 1945, shortly before the Japanese surrender, so it has not been possible to prove what happened there (Wu Jiarui 1989, 31; Hu Z. 2005b, 188).

29 Beginning in 1932, the Foundation made several grants to the West China Academy of Sciences to support collaborative projects with the major research centers (Yang T. 1991, 256).
30 The most complete source for the history of the Fan Memorial Institute in Yunnan is Hu Z. (2018), which includes transcriptions and copies of archival documents.
31 On Kuang Keren and his illustrations for *Illustrated Materia Medica of Southern Yunnan*, see chapter 7.
32 On *Metasequoia glyptostroboides*, see chapter 10.

CHAPTER NINE: MUSEUMS, EXHIBITIONS, AND BOTANICAL GARDENS

1 The text of the broadcast was published in *Scientific China* (Kexuede Zhongguo) in 1933 (Zhong 1933).
2 Borrell 1991; Lu T. 2014, ch. 2; Tai 2017. The mission was located in the Xujiahui district of the city (Zikawei in the French romanization of the time). There had been a small museum in Macao between 1829 and 1834, but the Musée de Zikawei was the first to have a stated educational vocation and a collecting plan (Puga 2012).
3 Ren 1983, 9; Qin S. 2004, 688–90; Claypool 2005, 580–85; Lu T. 2014, chs. 2 and 3.
4 I would like to thank Ching May-Bo of the City University of Hong Kong for bringing my attention to Ding Lei's monograph on museums and exhibitions in republican Guangzhou.
5 On the expedition to Hainan led by Zuo Jinglie, see chapter 5. The Science Society commissioned a report of the exhibition with photographs that was published in both *Science* and *Popular Science* (Kexue huabao) (Zhang M. 1934a, 1934b).
6 Wang G. (1985, 13) makes this tentative suggestion. In a detailed study of the Garden of Solitary Pleasures, Harrist (1993) concludes, though, that the design and the names of the different sites represent the moral world of the Confucian scholar, not a form of botanical identification.
7 On the origins of botanical gardens in Europe, see Ogilvie (2006, 151–64); B. Johnson (2007); N. Johnson (2011); and Rakow and Lee (2015).
8 On imperialism and botany, see chapter 3. Also see McCracken (1997, 85–145); Brockway (2002); and Deleuze and Williams (2011, 161–63).
9 This history of the Hong Kong Botanical Gardens draws mainly on Griffiths and Lau (1986); Griffiths (1988); Fan F. (2003b, 4–7); and McCracken (1997, 29–32).
10 See chapter 5. The Hong Kong Botanical Gardens were noted for their exceptional fernery (Griffiths 1988, 193). On Qin Renchang's work on the cryptogams, see chapter 8.

11   In 1896, the Japanese authorities in Taiwan established the Taibei Botanical Garden, the first on the island. I have not covered the history of botanical gardens on Taiwan during the colonial period in this study.
12   Chen had studied in Hokkaido, Japan, from 1906 to 1913. He was at the academy in Nanjing from 1913 until 1923, when he went to study forest taxonomy at Harvard's Arnold Arboretum, graduating with a master's degree in 1924. He spent a further year in Germany, then returned to take a position as professor of forestry at Jinling University in Nanjing.
13   On the Institute for Agriculture, Forestry, and Botany, see chapter 8.
14   The UNESCO World Heritage Sites' Lushan web page is at https://whc.unesco.org/en/list/778/ (accessed June 23, 2019).
15   Wang G. 1986, 32; Hu Z. 2014, 6; Liu and Wang 2014.
16   Hu Zonggang's history of the Lushan Botanical Garden reprints many of the original documents from this early period that are now held in archives in Nanjing or Beijing (Hu Z. 2014). At the time of writing (2017–20), they were no longer accessible even to Chinese scholars. I am grateful to Professor Hu for having transcribed and reprinted this invaluable material.
17   On the Lushan meeting, see Bernstein (2006).
18   Hu Dekun 2014. Hu Zonggang has included this essay in his history of the Lushan Botanical Garden (Hu Z. 2014, 336–41).

CHAPTER TEN: *METASEQUOIA GLYPTOSTROBOIDES*, THE DAWN REDWOOD

1   The archival research of Ma Jinshuang (Ma J. 2003; Ma J. and Shao, 2003) has been vital in reconstructing the chronology of the story. I also use documents reprinted in a special issue of *Arnoldia* in 1998 (vol. 58 no. 4) celebrating the fiftieth anniversary of Hu and Zheng's paper. I had the extraordinary privilege of working with Professor Xue Jiru on several field trips in Yunnan between 1989 and 1995, and his lively stories about his part in the discovery have helped me navigate the arguments about who did what and when. Other sources are Merrill (1948); Chu and Cooper (1950); Fulling (1970); Silverman (1990); Hu Z. (2000); and Kyna (2016).
2   The Save the Redwood League website has a section on the dawn redwood, which states that "in 1944, a Chinese forester found an enormous dawn redwood in Sichuan Province." The bulk of the website's information, though, is about Chaney's 1948 visit to Wanxian and its significance for the conservation of the species. Accessed July 5, 2019, www.savetheredwoods.org/redwoods/dawn-redwoods/.

3 The Chinese text is also in a recent study of Hu Xiansu's poetry (Liu W. 2012). I would like to thank Guo Lingling, librarian at the University of the West, in Rosemead, CA, for locating a copy of *Eastern Horizon* and forwarding it to me through Zou Xiuying, the East Asia librarian at Connold Mudd Library, Claremont Colleges.
4 See Latour (1986, 1) and Jullien (1990, 9–10).

REFERENCES

Aldrich, Robert, and Kirsten McKenzie, eds. 2014. *The Routledge History of Western Empires*. Abingdon, UK; and New York: Routledge.

Alfonso-Goldfarb, Ana M., Silvia Waisse, and Márcia H. M. Ferraz. 2013. "From Shelves to Cyberspace: Organization of Knowledge and the Complex Identity of History of Science." *Isis* 104, no. 3 (September): 551–60.

Allain, Yves-Marie, Lucile Allorge, and Cécile Aupic, eds. 2008. *Passions botaniques: naturalistes voyageurs au temps des grandes découvertes*. Rennes: Editions Ouest-France.

Alleton, Viviane. 2001. "Chinese Terminologies: On Preconceptions." In *New Terms for New Ideas: Western Knowledge and Lexical Change in Late Imperial China*, edited by Michael Lackner, Iwo Amelung, and Joachim Kurtz, 15–34. Leiden, Netherlands; Boston; and Cologne: Brill.

Amelung, Iwo. 2014. "Historiography of Science and Technology in China." In *Science and Technology in Modern China, 1880s–1940s*, edited by Jing Tsu and Benjamin A. Elman, 39–65. Leiden, Netherlands: Brill.

An Ji 安吉 and Zhang Zongxu 張宗緒. 1920. *Zhiwu minglu shiyi* 植物名錄拾遺 (A repository of plant names). Shanghai: Shangwu Yinshuguan.

Andrews, Henry N. 1948. "*Metasequoia* and the Living Fossils." *Missouri Botanical Garden Bulletin* 36, no. 5 (May): 79–85.

Atran, Scott. 1985. "Pre-theoretical Aspects of Aristotlean Definition and Classification of Animals: The Case for Common Sense." *Studies in History and the Philosophy of Science* 16, no. 2: 113–63.

Audouin-Dubreuil, Ariane. 2007. *Croisière Jaune: sur la Route de la Soie*. Grenoble: Glénat.

Bai, Ying, and Ruixue Jia. 2016. "Elite Recruitment and Political Stability: The Impact of the Abolition of China's Civil Service Exam." *Econometrica* 84, no. 2: 677–733.

Bailey, Paul. 1988. "The Chinese Work-Study Movement in France." *China Quarterly* 115 (September): 441–61.

Balfour, John Hutton. 1851. *A Manual of Botany: Being an Introduction to the Study of the Structure, Physiology, and Classification of Plants*. London: J. J. Griffin.

Banks, R. E. R., B. Elliott, D. King-Hele, and G. L. Lucas. 2000. *Sir Joseph Banks: A Global Perspective*. Kew, UK: Royal Botanic Gardens.

Barnhart, Richard M. 1997. *Three Thousand Years of Chinese Painting*. New Haven, CT: Yale University Press.

Bartholomew, Bruce. 1986. "The Chinese Species of Camellia in Cultivation." *Arnoldia* 46, no. 1: 2–16.

Batsaki, Yota, Sarah Burke Cahalan, and Anatole Tchikine, eds. 2016. *The Botany of Empire in the Long Eighteenth Century*. Washington, DC: Dumbarton Oaks Research Library and Collection.

Behr, Wolfgang. 2004. "'To Translate' Is 'To Exchange' 譯者言易也: Linguistic Diversity and the Terms for Translation in Ancient China." In *Mapping Meanings: The Field of New Learning in Late Qing China*, edited by Michael Lackner and Natascha Vittinghoff, 174–209. Leiden, Netherlands; and Boston: Brill.

Beiping Yanjiuyuan 北平研究院. 1934. *Beiping Yanjiuyuan wu zhou nian gongzuo baogao* 北平研究院五週年工作報告 (Work report on the fifth anniversary of the National Academy of Beiping). Beiping: Beiping Yanjiuyuan.

Berlin, Brent. 1992. *Ethnobiological Classification: Principles of Categorization of Plants and Animals in Traditional Societies*. Princeton, NJ: Princeton University Press.

Berlin, Brent, Dennis E. Breedlove, and Peter H. Raven. 1973. "General Principles of Classification and Nomenclature in Folk Biology." *American Anthropologist*, n.s., 75, no. 1: 214–42.

Bernstein, Thomas P. 2006. "Mao Zedong and the Famine of 1959–1960: A Study in Wilfulness." *China Quarterly* 186: 421–45.

Bevis, Teresa Brawner, and Christopher J. Lucas. 2006. "Chinese Students in US Colleges: The First Hundred Years." *International Educator* 15, no. 6 (November–December): 26–33.

Bickers, Robert A., and R. G. Tiedemann, eds. 2007. *The Boxers, China, and the World*. Lanham, MD: Rowman and Littlefield.

Biggerstaff, Knight. 1956. "Shanghai Polytechnic Institution and Reading Room: An Attempt to Introduce Western Science and Technology to the Chinese." *Pacific Historical Review* 25, no. 2 (May): 127–49.

Billingsley, Phil. 1988. *Bandits in Republican China*. Stanford, CA: Stanford University Press.
Bing Zhi 秉志. 1915. "Shengwuxue gailun 生物學概論" (A brief account of biology). *Kexue* 科學 (Science) 1, no. 1: 78–85.
———. 1926. "Zhongwenzhi shuangming zhi 中文之雙名制" (The binomial system in Chinese). *Kexue* 科學 (Science) 11, no. 10: 1346–50.
Bleichmar, Daniela. 2011. "The Geography of Observation: Distance and Visibility in Eighteenth Century Botanical Travel." In *Histories of Scientific Observation*, edited by Lorraine Daston and Elizabeth Lunbeck, 373–95. Chicago: University of Chicago Press.
———. 2016. "Botanical Conquistadors: The Promises and Challenges of Imperial Botany in the Hispanic Enlightenment." In *The Botany of Empire in the Long Eighteenth Century*, edited by Yota Batsaki, Sarah Burke Cahalan, and Anatole Tchikine, 35–60. Washington, DC: Dumbarton Oaks Research Library and Collection.
Bleichmar, Daniela, Paula de Dos, Kristin Huffine, and Kevin Sheehan. 2009. *Science in the Spanish and Portuguese Empires, 1500–1800*. Stanford, CA: Stanford University Press.
Blunt, Wilfrid, and William T. Stearn. (1950) 1994. *The Art of Botanical Illustration*. London: Antique Collectors Club and Royal Botanic Gardens, Kew.
Bonneuil, Christophe. 1997. "Mettre en ordre et discipliner les tropiques: les sciences du végétal dans l'empire français, 1870–1941." PhD diss., Université Paris-Diderot (Paris VII).
Borrell, Octavius William. 1991. "A Short History of the Heude Museum, "Musée Heude," 1858–1952: Its Botanist and Plant Collectors." *Journal of the Hong Kong Branch of the Royal Asiatic Society* 31: 183–91.
Boutan, E. 1993. *Le nuage et la vitrine: une vie de Monsieur David*. Bayonne, France: Editions Chabaud.
Bowers, John Z., William Hess, and Nathan Sivin, eds. 1988. *Science and Medicine in Twentieth-Century China: Research and Education*. Ann Arbor: University of Michigan Center for Chinese Studies.
Boym, Michał Piotr. 1656. *Flora sinensis, fructus floresque* [. . .]. Vienna: Matthæi Rictii. Digitized by Royal Botanic Gardens, Kew, Library, Art & Archives. https://doi.org/10.5962/bhl.title.123322.
Bray, Francesca. 1988. "Essence et utilité: la classification des plantes cultivées en Chine." *Extrême-Orient Extrême-Occident* 10: 13–26.
———. 2007. "Introduction." In *Graphics and Texts in the Production of Technical Knowledge in China: The Warp and the Weft*, edited by

Francesca Bray, Vera Dorofeeva-Lichtmann, and Georges Métaillé, 1–78. Leiden, Netherlands: Brill.

Bray, Francesca, Vera Dorofeeva-Lichtmann, and Georges Métailié, eds. 2007. *Graphics and Texts in the Production of Technical Knowledge in China: The Warp and the Weft.* Leiden, Netherlands: Brill.

Brenier, Joël, Colette Diény, Jean-Claude Martzloff, and Wladyslaw de Wieclawik. 1989. "Shen Gua (1023–1095) et les sciences." *Revue d'histoire des sciences: problèmes d'histoire des sciences en Chine (I): méthodes, contacts et transmissions* 42, no. 4 (October–December): 333–51.

Bretschneider, Emil. 1871. *On the Study and Value of Chinese Botanical Works, with Notes on the History of Plants and Geographical Botany from Chinese Sources.* Foochow: Rozario, Marcal & Co.

———. 1895. *Botanicon Sinicum: Notes on Chinese Botany from Native and Western Sources.* Shanghai, Hong Kong, Yokohama, and Singapore: Kelly & Walsh, Limited.

———. 1898. *History of European Botanical Discoveries in China.* London: Sampson Lows, Marston and Company.

Bridson, Gavin D. R. 2010. "From Xylography to Holography: Five Centuries of Natural History Illustration." *Archives of Natural History* 16, no. 2: 121–41.

Briggs, Barbara G. 1991. "One Hundred Years of Plant Taxonomy, 1880–1989." *Annals of the Missouri Botanical Garden* 78, no. 1: 19–32.

Briggs, Roy. 1993. *"Chinese" Wilson: A Life of Ernest H. Wilson, 1876–1930.* London: Her Majesty's Stationery Office.

Brockway, Lucile H. (1979) 2002. *Science and Colonial Expansion: The Role of the British Royal Botanic Gardens.* New Haven, CT; and London: Yale University Press.

Brokaw, Cynthia J., and Kai-wing Chow, eds. 2005. *Printing and Book Culture in Late Imperial China.* Berkeley, Los Angeles, and London: University of California Press.

Brooks, F. T., and Thomas Ford Chipp, eds. 1931. *Report of Proceedings, Fifth International Botanical Congress, Cambridge, 16–23 August 1930.* Cambridge: Cambridge University Press.

Bucchi, Massimiano. 1996. "When Scientists Turn to the Public: Alternative Routes in Science Communication." *Public Understanding of Science* 5: 375–94.

Buck, Peter. 1980. *American Science and Modern China, 1876–1936.* Cambridge and New York: Cambridge University Press.

Buckman, Thomas R., ed. 1966. *Bibliography and Natural History: Essays Presented at a Conference Convened in June 1964.* Library Series 27. Lawrence: University of Kansas Publications.

Bullock, Mary Brown. 1980. *An American Transplant: The Rockefeller Foundation and Peking Union Medical College.* Berkeley, Los Angeles, and London: University of California Press.

Bussotti, Michela. 2007. "Woodcut Illustration: A General Outline." In *Graphics and Texts in the Production of Technical Knowledge in China: The Warp and the Weft,* edited by Francesca Bray, Vera Dorofeeva-Lichtmann, and Georges Métailié, 461–86. Leiden, Netherlands: Brill.

Cahill, James. 2010. *Pictures for Use and Pleasure: Vernacular Painting in High Qing China.* Berkeley: University of California Press.

Cai Xitao 蔡希陶. 1937. "Pu gongying 蒲公英" (The dandelion). *Wenxue* 文學 8, no. 4: 597–607.

———. 1980. *Wode xingqu shi shenme?* 我的興趣是什麼 (What is it that interests me?). Xishuangbanna Tropical Botanical Garden website, special section celebrating the 100th anniversary of Cai Xitao's birth. www.xtbg.ac.cn/cai100/CaiWorks/201103/P020110310589292377158.pdf.

Callicott, J. Baird, and Roger T. Ames, eds. 1989. *Nature in Asian Traditions of Thought: Essays in Environmental Philosophy.* Albany: State University of New York Press.

Camus, Yves. 2007. *Jesuits' Journeys in Chinese Studies.* Macau: Macau Ricci Institute.

Cannon, Garland, Reginald Victor Jones, and William Drummond MacDonald Paton. 1975. "Sir William Jones, Sir Joseph Banks and the Royal Society." *Notes and Records of the Royal Society of London* 29, no. 2: 205–30.

Cao, Danhong, and Xu Jun. 2016. "A la recherche des germes de la modernité chinoise: traduction scientifique à la fin de la dynastie Ming et au début de la dynastie Qing." *Babel* 62, no. 4: 602–22.

Cao Hui. 2018. "Polychrome Illustrations in the Ming Bencao Literature." In *Imagining Chinese Medicine,* edited by Vivienne Lo, Penelope Barrett, David Dear, Lu Di, Lois Reynolds, and Dolly Yang, 197–208. Leiden, Netherlands; and Boston: Brill.

Carpenter, Amber D. 2010. "Embodied Intelligent (?) Souls: Plants in Plato's Timaeus." *Phronesis* 55: 281–303.

Carr, Michael Edward. 1979. "A Linguistic Study of the Flora and Fauna Sections of the Erh-Ya." PhD diss., University of Arizona.

Casamajor, Robert, letter to Ralph Peer, July 31, 1949. In the Peer Archive at the Huntington Library, Art Museum and Botanical Gardens, San Marino, CA.

Chambers, Neil, ed. 2000. *The Letters of Sir Joseph Banks: A Selection, 1786–1820.* London: Imperial College Press.

———. 2007. *Joseph Banks and the British Museum: The World of Collecting, 1770–1830.* London and New York: Routledge.

Chan, Wing-tsit. 1963. *A Source Book in Chinese Philosophy.* Princeton, NJ: Princeton University Press.

Chan, Yuen Lai Winnie. 2015. "Nineteenth-Century Canton Gardens and the East-West Plant Trade." In *Qing Encounters: Artistic Exchanges between China and the West,* edited by Petra ten-Doesschate Chu and Ning Ding, 111–23. Los Angeles: Getty Research Institute and the Technological and Higher Education Institute of Hong Kong.

Chaney, Ralph W. 1948. "As Remarkable as Discovering a Living Dinosaur: Redwoods in China." *Natural History Magazine* 47 (December): 440–44.

Chang Hung Ta (Zhang Hongda) and Bruce Bartholomew. 1984. *Camellia.* Portland, OR: Timber Press.

Chao, Yuen Ren. 1953. "Popular Chinese Plant Words: A Descriptive Lexico-Grammatical Study." *Language* 29 (July–September): 379–414.

Chapman, G. P., and Y. Z. Wang. 2002. *The Plant Life of China: Diversity and Distribution.* Berlin: Springer Verlag.

Chen Chongming 陳重明. 1980. "Wu Qijun he *Zhiwu mingshi tukao* 吳其濬和植物名實圖考" (Wu Qijun and his *Research on the Illustrations, Realities, and Names of Plants*). *Zhonghua yishi zazhi* 中華意識雜誌 10, no. 2: 65–70.

Chen Chuan 陳椽. 1991. "Ping lun jiu jing cha tu zai kao 評論九經茶荼再考" (A critique of further studies on the characters *cha* and *tu* in the nine classics). *Nongye kaogu* 農業考, no. 4: 256–60, 278.

Chen Fenghuai 陳封懷, Li Kangshou 李康壽, and Huang Chengjiu 黃成就. 1984. "Jinian wo guo jiechu zhiwuxuejia Chen Huanyong xiansheng 紀念我國接觸植物學家陳煥庸先生" (Remembering Chen Huanyong, a great Chinese botanist). *Guangdong sheng Zhiwuxue Hui hui kan* 廣東省植物學會會刊, no. 2: n.p.

Chen Haozi 陳淏子. 1688. *Hua Jing* 花鏡 (The mirror of flowers). Translated by Jules Halphen (1900) with a commentary by Georges Métailié as *Miroir des fleurs: guide pratique du jardinier amateur en Chine au XVIIe siècle* (2006). Paris: Actes Sud.

Chen, Huaiyu. 2009. "A Buddhist Classification of Animals and Plants in Early Tang Times." *Journal of Asian History* 43, no. 1: 31–51.

Chen Huanyong 陳煥鏞. 1933a. "Nonglin Zhiwu Yanjiusuo zui jinzhi gongzuo baogao 農林植物研究所最近之工作報告" (A report on recent work of the Institute for Agriculture, Forestry, and Botany). *Nong sheng yuekan* 農聲月刊 no.166–67: 118–19.

———. 1933b. *Zhiwu Yanjiusuo baogao, Guoli Zhongshan Daxue Nongxueyuan ershi yi nian baogao* 植物研究所報告, 國立中山大學農學院二十一年報告

(National Sun Yat-sen University, College of Agriculture, Institute of Botany annual report for 1932). Guangzhou: Guoli Zhongshan Daxue Nongxueyuan: 159–90.

Chen Jiarui 陈家瑞. 1978. "Dui wo guo gudai zhiwufenleixue ji qi sixiangde tantao 對我國古代植物分類學及其思想的探討" (A study of China's ancient botanical taxonomic thinking). *Zhiwufenlei xuebao* 植物分類學報 16, no. 8: 101–12.

Chen Jinzheng 陳錦正 and Zhong Renjian 鐘任建. 2008. *Zhongguo jindai zhiwuxuede kaituozhe—Zhong Guanguang* 中國近代植物學的開拓者— 鐘觀光 (1868–1940) (Zhong Guanguang (1868–1940): an innovator in modern Chinese botany). CAS Institute of Botany website, www.ibcas.ac.cn/80/expertsCharisma/zhongguanguang.html.

Chen, Linhan, and Danyan Huang. 2013. "Internationalization of Chinese Higher Education." *Higher Education Studies* 3, no. 1: 92–105.

Chen, Shisan C. (Chen Zhen). 1925. "Variation in External Characters of Goldfish, *Carassius auratus*." *Contributions from the Biological Laboratory of the Science Society of China* 1, no. 1: 1–64.

Chen, Shiwei. 1997. "Legitimizing the State: Politics and the Founding of *Academia Sinica* in 1927." *Papers on Chinese History* 6 (Spring): 23–41.

Chen Shiwen 陳世文, Yang Wenjin 楊文金, and Gu Zhixiong 古智雄. 2011. "Cong gezhi dao kexue—xifang kexue chuanru Zhongguo dui kexuejiaoyude qishi 從各致到科學— 西方科學傳入中國對科學教育的啟示" (From natural studies to science: reflections on the transmission of Western science to China). *Kexue jiaoyu yuekan* 科學教育月刊 (Science Education Monthly) 343 (October): 2–17.

Chen, Shu-fen. 2000. "A Study of Sanskrit Loanwords in Chinese." *Tsing Hua Journal of Chinese Studies*, n.s., 30, no. 3: 375–426.

Chen Siqing 陳思清. 1987. "Manghua Cai Xitao de xiaoshuo 漫話蔡希陶的小說" (A discussion of Cai Xitao's novellas). *Yunnan Shifan Daxue xuebao (zhexue ban)* 雲南師範大學學報(哲學版), no. 5: 62–68.

Chen Yiwen 陳鐿文, Kang Xiaoyu 亢小玉, and Yao Yuan 姚遠. 2008. "Du Yaquan xiansheng nianpu 杜亞泉先生年譜 1912–1933" (A biography of Mr. Du Yaquan, 1912–1933). *Xibei Daxue xuebao (ziran kexue ban)* 西北大學學報 (自然科學版) 38, no. 6: 1044–50.

Chen Yiwen 陳鐿文 and Yao Yuan 姚遠. 2008. "Du Yaquan xiansheng nianpu 杜亞泉先生年譜 (1873–1912)" (A biography of Mr. Du Yaquan (1873–1912)). *Xibei Daxue xuebao (ziran kexue ban)* 西北大學學報 (自然科學版) 38, no. 5: 845–50.

Chen Zhen. *See* Chen, Shisan C.

Cheng Meibao 程美寶. 2006. "Wan Qing guoxue dachaozhongde bowuxue zhishi—lun *Guocui xuebao* zhongde bowutuhua 晚清國學大潮中的博物學知識—論"國粹學報"中的博物圖畫" (The height of natural history in the late Qing: a study of the natural history drawings in the *Journal of National Essence*). *Shehui kexue* 社會科學, no. 8: 18–31.

———. (Ching May-bo). 2007. "Picturing Knowledge: Chinese Brushwork Illustrations of Western Natural History in a Late Qing Periodical, 1907–1911." *Journal of Modern Chinese History* 1, no. 1: 31–51.

———. 2009. "Fuzhi zhishi—*Guocui xuebao* bowutude ziliao laiyuan ji qi caiyongzhi yinshua jishu 複製知識—"國粹學報"博物圖的資料來源及其採用之印刷技術" (Knowledge reproduced: the sources of the natural history drawings in the *Journal of National Essence* and the printing technology used to publish them." *Zhongshan Daxue xuebao (shehui kexue ban)* 中山大學學報 (社會科學版) (Journal of Sun Yat-sen University (social science edition)), no. 3: 95–109.

Chinese Academy of Sciences, Kunming Institute of Botany (KIB) Editorial Committee 中國科學院昆明植物研究所簡史編輯委員會. 2008. *Zhongguo Kexueyuan Kunming Zhiwu Yanjiusuo jianshi* 中國科學院昆明植物研究所簡史 (A brief history of the Chinese Academy of Sciences Kunming Intitute of Botany). Kunming: KIB internal publication.

Ching, May-bo. *See* Cheng Meibao.

Cholet, Céline. 2019. "De l'objet du monde naturel à sa connaissance par l'image: le cas des dessins de découverte en botanique dans les publications scientifiques du Muséum National d'Histoire Naturelle de Paris." *Signata Annales des Sémiotiques/Annals of Semiotics* 10 (19): 1–14.

Chow, Kai-wing. 2004. *Publishing, Culture, and Power in Early Modern China*. Stanford, CA: Stanford University Press.

Chow, Kai-wing, Tze-ki Hon, Hung-yok Ip, and Don C. Price, eds. 2008. *Beyond the May Fourth Paradigm: In Search of Chinese Modernity*. Latham, MD: Lexington Books.

Chu, Kwei-ling, and William S. Cooper. 1950. "An Ecological Reconnaissance in the Native Home of *Metasequoia glyptostroboides*." *Ecology* 31, no. 2: 260–78.

Chu, Petra ten-Doesschate, and Ning Ding, eds. 2015. *Qing Encounters: Artistic Exchanges between China and the West*. Los Angeles: Getty Research Institute and the Technological and Higher Education Institute of Hong Kong.

Claypool, Lisa. 2005. "Zhang Jian and China's First Museum." *Journal of Asian Studies* 64, no. 3: 576–604.

Cleyer, Andreas. 1689. *Miscellanea Curiosa, sive Ephemiridum Medico-Physicarum Germanicarum Academiae Naturae Curiosorum. Decuria II, Annus VII.* Nuremberg: Wolfgang Maurith. Digital copy at the Biodiversity Heritage Library, www.biodiversitylibrary.org/item/163340#page/11/mode/1up.

Clunas, Craig. 1984. *Chinese Export Watercolours.* London: Victoria and Albert Museum.

———. 1996. *Fruitful Sites: Garden Culture in Ming Dynasty China.* Durham, NC: Duke University Press.

———. 1997. *Art in China.* Oxford and New York: Oxford University Press.

Coates, A. M. 1969. *The Quest for Plants: A History of the Horticultural Explorers.* London: Studio Vista.

Cohn, Bernard. 1996. *Colonialism and Its Forms of Knowledge: The British in India.* Princeton, NJ: Princeton University Press.

Cook, Alexandra. 2010. "Linnaeus and Chinese Plants: A Test of the Linguistic Imperialism Thesis." *Notes and Records of the Royal Society* no. 64: 121–38.

Cotton, A. D. 1944. "Introductory Remarks. Papers on the Exploration of China." *Proceedings of the Linnean Society of London*, 156th Session (1943–44): 5.

Cowan, John MacQueen, ed. 1952. *The Journeys and Plant Introductions of George Forrest, V. M H.* London: Oxford University Press for the Royal Horticultural Society.

Cowdry, E. V. 1927. "The China Foundation for the Promotion of Education and Culture." *Science* 65, no. 1676: 150–51.

Cox, E. H. M. 1945. *Plant Hunting in China: A History of Botanical Exploration in China and the Tibetan Marches.* London: Collins.

Crane, Peter. 2013. *Ginkgo: The Tree That Time Forgot.* New Haven, CT: Yale University Press.

———, ed. 2017. "John Bradby Blake." *Curtis's Botanical Magazine* 34, no. 4 (December).

Crane, Peter, and Zack Loehle. 2017. "Introduction." *Curtis's Botanical Magazine* 34, no. 4 (December): 215–30.

Creel, Herrlee. (1970) 1982. *What Is Taoism? And Other Studies in Chinese History.* Chicago: University of Chicago Press.

Dai Lijuan 戴麗娟. 2013. "Cong Xujiahui Bowuyuan dao Zhendan Bowuyuan—Faguo Yesu Huishi zai jindai Zhongguode ziranshiyanjiu huodong 從徐家匯博物院到震旦博物院— 法國耶穌會士在近代中國的自然史研究活動" (From Zikawei Museum to Heude Museum: the natural history research of French Jesuits in modern China). *Zhongyang Yanjiuyuan*

Lishi Yuyan Yanjiu Suo jikan 中央研究院歷史語言研究所集刊 (Journal of the Institute of History and Philology, Academia Sinica) 84, no. 2: 329–85.

——— (Tai, Li-chuan). 2017. "Shanghai's Zikawei Museum (1868–1952): Jesuit Contributions to the Study of Natural History in China." *Asia Major, Institute of History and Philology, Academia Sinica Taiwan. Third Series* 30, no. 1: 109–41.

D'Ambrosio, Paul. 2013. "Rethinking Environmental Issues in a Daoist Context: Why Daoism Is and Is Not Environmentalism." *Environmental Ethics* 35, no. 4: 407–17.

Darwin, Erasmus. 1791. *The Botanic Garden; A Poem, in Two Parts. Part I. Containing the Economy of Vegetation. Part II. The Loves of the Plants. With Philosophical Notes.* London: J. Johnson.

Dasgupta, Subrata. 2014. "Science Studies *sans* Science: Two Cautionary Postcolonial Tales." *Social Scientist* 42, no. 5–6: 43–61.

Daston, Lorraine. 2008. "On Scientific Observation." *Isis* 99, no. 1: 97–110.

———. 2016. "When Science Went Modern." In "Cultural Contradictions of Modern Science," *Hedgehog Review: Critical Reflections on Contemporary Culture* 18, no. 3 (Fall).

Daston, Lorraine, and Peter Galison. 1992. "The Image of Objectivity." *Representations* 40: 81–128.

Daston, Lorraine, and Elizabeth Lunbeck. 2011. *Histories of Scientific Observation.* Chicago: University of Chicago Press.

Daston, Lorraine, and Fernando Vidal. 2004. *The Moral Authority of Nature.* Chicago: University of Chicago Press.

Deleuze, Jean-Philippe-François, and Williams, Roger L., transl. 2011. "On the establishment of the principal gardens of botany: A bibliographical essay by Jean-Philippe-François Deleuze." Translated and edited by Roger L. Williams. *Huntia* 14 (2): 147–76.

Desmond, Ray. 1977. "Victorian Gardening Magazines." *Garden History* 5, no. 3: 47–66.

Ding Bo 丁波 and Chen Yan 陳燕. 2013. "Ershi shiji sishi niandai Jingsheng Shengwu Diaochasuo zai Yunnan 20世紀40年代 靜生生物調查所 在雲南" (The Fan Memorial Institute in Yunnan during the 1940s). *Yunnan Minzu Daxue xuebao (zhexue shehui kexue ban)* 雲南民族大學學報 (哲學社會科學版) (Journal of the Yunnan Nationalities University (Social Sciences)) 30, no. 5: 96–99.

Ding Lei 丁蕾. 2017. *Cong sizang dao gonggong zhanlan—Minguo shiqi Guangzhoude bowuguan he zhanlanhui* 從私藏到公共展覽— 民國時期廣州的博物館和展覽會 (From private collections to public exhibition: museums and exhibitions in Guangzhou (1910s–1940s)). Beijing: Shehui Kexue Wenxian Chubanshe.

Dingle, Edwin J. 1912. *China's Revolution, 1911–1912: A Historical and Political Record of the Civil War*. Shanghai: Commercial Press.

Djamouri, Redouane. 1993. "Théorie de la 'rectification des dénominations' et réflexion linguistique chez Xunzi." In "Le Juste Nom," *Extrême-Orient Extrême-Occident*, no. 15: 55–74.

Doležolová-Velingerová, Milena. 2014. "Modern Chinese Encyclopaedic Dictionaries: Novel Concepts and New Terminology (1903–1911)." In *Chinese Encyclopedias of New Global Knowledge (1870–1930): Changing Ways of Thought*, edited by Milena Doležolová-Velingerová and Rudolf G. Wagner, 289–328. Heidelberg and New York: Springer-Verlag.

Doležolová-Velingerová, Milena, and Rudolf G. Wagner, eds. 2014. *Chinese Encyclopedias of New Global Knowledge (1870–1930): Changing Ways of Thought*. Heidelberg and New York: Springer-Verlag.

Dongfang zazhi 東方雜誌 (Eastern miscellany). 1904. "Xinding xuewu gangyao 新定學務綱要" (Outline of the new educational curricula). 1, no. 3: 56–70. Accessed at the Chinese Text Project (CTP). https://ctext.org/library.pl?if=en&file=96754&page=120.

———. 1917. "Zhiwu da cidian fanli 植物大辭典凡例" (Guidelines for the dictionary of biology). 14, no. 10: 16. Accessed at the Chinese Text Project (CTP). https://ctext.org/library.pl?if=en&file=96908&page=19.

Drège, Jean-Pierre. 2007. "Des ouvrages classés par catégories: les encyclopédies chinoises." In "Qu'était-ce qu'écrire une encyclopédie en Chine?," *hors série* (special issue), *Extrême-Orient Extrême-Occident*, 19–38.

Du Yaquan 杜亞泉. 1902. "Bowuxue zong yi 博物學總義" (On the meaning of natural science). *Zhengyi tongbao* 政藝通報, no. 9: 6–7.

———. 1903. "Putong zhiwuxue jiaokeshu xu 普通植物學教科書序" (Preface to a standard textbook in botany). *Kexue shijie* 科學世界, no. 2: 85–90.

———, ed. 1918. *Zhiwuxue da cidian* 植物學大辞典 (Dictionary of botany). Shanghai: Shangwu Yinshuguan.

Durrant, T. 1982. *The Camellia Story*. Auckland: Heinemann.

Easterby-Smith, Sarah. 2018. *Cultivating Commerce: Cultures of Botany in Britain and France, 1760–1815*. Cambridge: Cambridge University Press.

Educational Association of China. 1904. *Technical Terms, English and Chinese, Prepared by the Committee of the Educational Association of China*. Shanghai: Presbyterian Missionary Press.

Egmond, Florike. 2010. *The World of Carolus Clusius: Natural History in the Making, 1550–1610*. London and Brookfield, VT: Pickering and Chatto.

Elliott, Jane E. 2002. *Some Did It for Civilisation, Some Did It for Their Country: A Revised View of the Boxer War*. Hong Kong: Chinese University Press.

Ellis, Markman, Richard Coulton, and Mathew Mauger. 2015. *Empire of Tea: The Asian Leaf That Conquered the World*. London: Reaktion Books.

Elman, Benjamin A. 2002. "Jesuit *Scientia* and Natural Studies in Late Imperial China, 1600–1800." *Journal of Early Modern History* 6, no. 3: 209–32.

———. 2004. "From Pre-Modern Chinese Natural Studies 格致學 to Modern Science 科學 in China." In *Mapping Meanings: The Field of New Learning in Late Qing China*, edited by Michael Lackner and Natascha Vittinghoff, 25–73. Leiden, Netherlands; and Boston: Brill.

———. 2005. *On Their Own Terms: Science in China, 1550–1900*. Cambridge, MA: Harvard University Press.

———. 2007. "Collecting and Classifying: Ming Dynasty Compendia and Encyclopedias (Leishu)." In "Qu'était-ce qu'écrire une encyclopédie en Chine?," *hors série* (special issue), *Extrême-Orient Extrême-Occident*, 131–57.

———. 2010. "The Investigation of Things (*gewu* 格物), Natural Studies (*gezhixue* 各致學), and Evidential Studies (*kaozhengxue* 考證學) in Late Imperial China, 1600–1800." In *Concepts of Nature: A Chinese-European Cross-Cultural Perspective*, edited by Hans Ulrich Vogel, Günter Dux, and Mark Elvin: 368–99. Leiden, Netherlands: Brill.

———. 2014. "Toward a History of Modern Science in Republican China." In *Science and Technology in Modern China, 1880s–1940s*, edited by Jing Tsu and Benjamin A. Elman, 15–38. Leiden, Netherlands: Brill.

Elvin, Mark. 2004. *The Retreat of the Elephants: An Environmental History of China*. New Haven, CT; and London: Yale University Press.

———. 2010. "Introduction." In *Concepts of Nature: A Chinese-European Cross-Cultural Perspective*, edited by Hans Ulrich Vogel, Günter Dux, and Mark Elvin, 56–101. Leiden, Netherlands: Brill.

———. 2014. "Scientific Curiosity in China and Europe: Natural History in the Late Ming and the Eighteenth Century." In *Environmental History in East Asia: Interdisciplinary Perspectives*, edited by Liu Ts'ui-jung, 11–39. Academia Sinica on East Asia Series. London and New York: Routledge.

Endersby, Jim. 2008. *Imperial Nature: Joseph Hooker and the Practices of Victorian Science*. Chicago: University of Chicago Press.

Ereshefsky, Marc. 2007. "Foundational Issues concerning Taxa and Taxon Names." *Systematic Biology* 56, no. 2 (April): 295–301.

———. 2010. "Darwin's Solution to the Species Problem." *Synthèse* 175, no. 3: 405–25.

Esherick, Joseph W. 1976. *Reform and Revolution in China: The 1911 Revolution in Hunan and Hubei*. Berkeley, Los Angeles, and London: University of California Press.

———. 1987. *The Origins of the Boxer Uprising*. Berkeley, Los Angeles, and London: University of California Press.

Etō, Shinkichi, and Harold Z. Schiffrin, eds. 2008. *China's Republican Revolution*. Tokyo: University of Tokyo Press.

Fan Chengxun 范承勳, ed. 1692. *Jizu Shan zhi* 雞足山志 (The gazetteer of Jizu Mountain). Scanned by the Temple Gazetteer Project at the Dharma Drum Institute of Liberal Arts, Taibei, http://buddhistinformatics.dila.edu.tw/fosizhi/.

Fan, Fa-ti. 2003a. "Science in a Chinese Entrepôt: British Naturalists and Their Chinese Associates in Old Canton." In "Science and the City," special issue edited by Sven Dierig, Jens Lachmund, and Andrew Mendelsohn. *Osiris*, 2nd series, 18: 60–78.

———. 2003b. "Victorian Naturalists in China: Science and Informal Empire." *British Journal for the History of Science* 36, no. 1: 1–26.

———. 2004a. *British Naturalists in Qing China*. Cambridge, MA: Harvard University Press.

———. 2004b. "Nature and Nation in Chinese Political Thought: The National Essence Circle in Early Twentieth-Century China." In *The Moral Authority of Nature*, edited by Lorraine Daston and Fernando Vidal, 409–37. Chicago: University of Chicago Press.

———. 2007. "Redrawing the Map: Science in Twentieth-Century China." *Isis* 98: 524–38.

———. 2008. "How Did the Chinese Become Native? Science and the Search for National Origins in the May Fourth Era." In *Beyond the May Fourth Paradigm: In Search of Chinese Modernity*, edited by Kai-wing Chow, Tze-ki Hon, Hung-yok Ip, and Don C. Price, 183–208. Latham, MD: Lexington Books.

Fan Hongye 樊洪業. 1989. "Sai xiansheng yu xin wenhua yundong—kexue shehuishide kaocha 賽先生與新文化運動—科學社會史的考察" (Mr. Science and the New Culture Movement: a study in the social history of science). *Lishi yanjiu* 歷史研究, no. 3: 39–49.

———. 1999. *Zhongguo Kexueyuan bian nian shi* 中國科學院編年史 (1949–1999) (A chronology of the Chinese Academy of Sciences (1949–1999)). Shanghai: Shanghai Kexue Jiaoyu Chubanshe.

———. 2015. "'Kexue' chuangkan licheng kaoshu "科學"創刊歷程考述" (An account of the founding of *Kexue* and its history). *Kexue* 科學 (Science) 67, no. 1: 5–7.

———. 2016. "Cong chuantong kexue dao jindai kexue 從傳統科學到近代科學" (From traditional science to modern science). *Kexue wenhua pinglun* 科學文化評論 13, no. 3: 5–14.

Fan Tiequan 範鐵權 and Han Jianjiao 韓建嬌. 2012. "Zhonghua Zirankexue She yu Minguo kexue tizhihuade yanjin 中華自然科學社與民國科學體制化的演進" (The Natural Science Society of China and science systematization in the Republic of China). *Ziran bianzhengfa yanjiu* 自然辯證法研究 (Studies in the dialectics of nature) 28, no. 8: 90–113.

Fan Wentao 范文濤 and Chen Yichan 陳義產. 1990. "Zhong Guanguang jiaoshou yu wo xiao zhiwuyuan 鐘觀光教授與我校植物園" (Zhong Guanguang and the university's botanical garden). *Zhejiang Nongye Daxue xuebao* 浙江農業大學學報 16, no. 4: 450–53.

Fan, Xiangtao. 2007. "Scientific Translation and Its Social Functions: A Descriptive-Functional Approach to Scientific Textbook Translation in China." *Journal of Specialized Translation* 7. Online journal at www.jostrans.org/issue07/art_fan.php.

Fan Yuan-lien. 1927. "Eastern Trends in the New Sciences: The China Foundation for the Promotion of Education and Culture." *News Bulletin (Institute of Pacific Relations)* (December): 17–20.

Fan Zhaoming 樊兆鳴, ed. 2011. *Jiangnan Zhizaoju Fanyiguan tuzhi* 江南製造局翻譯館圖志(An illustrated account of the Translation Department of the Jiangnan Arsenal). Shanghai Tushuguan lishi wenxian yanjiu congkan 上海圖書館歷史文獻研究叢刊 (Shanghai Library Studies on Historical Materials Series). Shanghai: Shanghai Jishu Wenxian Chubanshe.

Fang Shumei 方樹梅. 1930. *Dian nan cha hua xiao zhi* 滇南茶花小志 (A brief account of the camellias of southern Yunnan). Kunming: private printing.

Fang Wenpei 方文培, ed. 1946. *Emei zhiwu tuzhi* 峨眉植物圖志 (Icones plantarum Omeiensium). Chengdu: National Sichuan University.

Fang Wenpei 方文培 and Zhang Shufeng 章樹楓. 1928. "Chuan Kang zhiwu biaoben caiji ji 川康植物標本採集記" (An account of a collecting expedition to Sichuan and Tibet). *Kexue* 科學 (Science) 13, no. 11: 1509–13.

Farber, Paul Lawrence. 2000. *Finding Order in Nature: The Naturalist Tradition from Linnaeus to E. O. Wilson.* Baltimore: Johns Hopkins Press.

Farrer, Reginald J. 1926. *On the Eaves of the World.* 2 vols. London: Edward Arnold & Co.

Farrington, B. 1964. *The Philosophy of Francis Bacon.* Liverpool: Liverpool University Press.

Fay, Peter Ward. 1975. *The Opium War, 1840–1842.* Chapel Hill: University of North Carolina Press.

Feathers, David L., and Milton Brown, eds. 1978. *The Camellia: Its History, Culture, Genetics and a Look into Its Future Development.* Columbia, SC: American Camellia Society.

Feng Chengru 馮澄如. 1959. *Shengwu huitu fa* 生物繪圖法 (Methods of biological illustration). Beijing: Beijing Kexue Chubanshe.

Feng Shike 馮時可. 1934 . "Dian zhong chahua ji 滇中茶花記" (An account of the camellias of central Yunnan). In *Gu jin tu shu ji cheng* 古今圖書集成 (The imperially commissioned compendium of literature and illustrations, ancient and modern), edited by Jiang Tingxi 蔣廷錫. 1726. Juan 1456, part 2 of *Yunnan zong bu yiwen* 雲南總部藝文 (Literary works on Yunnan). Citations refer to the 1934 ed. Shanghai: Zhonghua Shuju. Digitized by the Chinese Text Project, http://ctext.org/library.pl?if=gb&file=91562&page=6.

Feng Tianyu 馮天瑜. 2008. "*Kexue* mingci tanyuan "科學" 名詞探源" (An exploration of the term *kexue*). *Zhongguo keji shuyu* 中國科技術語, no. 3: 78–80.

Feng Youlan. *See* Fung, Yu-Lan.

Fernsebner, Susan. 2006. "Objects, Spectacle, and a Nation on Display at the Nanyang Exposition of 1910." *Late Imperial China* 27, no. 2: 99–124.

Fèvre, Francine, and Georges Métailié. 2005. *Dictionnaire Ricci des plantes de Chine.* Paris: Collection Ricci.

Fortune, Robert. 1847. *Three Years' Wanderings in the Northern Provinces of China.* London: John Murray.

——. 1852. *A Journey to the Tea Countries of China.* London: John Murray.

——. 1857. *A Residence among the Chinese.* London: John Murray.

——. 1863. *Yedo and Peking.* London: John Murray.

Foucault, Michel. 1994. *The Order of Things: An Archaeology of the Human Sciences.* New York: Vintage Books.

Franchet, M. 1882. "Les Plantes du Père d'Incarville dans l'herbier du Muséum d'histoire naturelle de Paris." *Bulletin de la Société Botanique de France* 29, no. 1: 2–13.

Fryer, John. 1880. *An Account of the Department for the Translation of Foreign Books at the Kiangnan Arsenal, Shanghai. With Various Lists of Publications in the Chinese Language.* Shanghai: American Presbyterian Press.

——. 傅蘭雅. 1891. *Zhiwu tushuo* 植物圖說 (An illustrated botany) (a translation of Balfour 1851). Shanghai: Educational Association of China.

Fu Lei 付雷. 2014. "Zhongguo jindai shengwuxue mingcide shending yu tongyi 中國近代生物學名詞的審定與統一" (The determination and standardization of modern biological vocabulary in China). *Zhongguo keji shuyu* 中國科技術語 no. 3: 52–57.

Fu, Lei. 2017. "Nathaniel Gist Gee's Contribution to Biology in Modern China." *Protein & Cell* 8, no. 4: 237–39.

Fu, Liangyu. 2013. "Indigenizing Visualized Knowledge: Translating Western Science Illustrations in China, 1870–1910." *Translation Studies* 6, no. 1: 78–102.

Fulling, Edmund H. 1970. "*Metasequoia*: Fossil and Living: An Initial Thirty-Year (1941–1970) Annotated and Indexed Bibliography with an Historical Introduction." *Botanical Review* 42, no. 3: 215–315.

Fung, A. 1996. "Testing the Self-Strengthening: The Chinese Army in the Sino-Japanese War of 1894–1895." *Modern Asian Studies* 30, no. 4: 1007–32.

Fung, Yu-Lan (Feng Youlan). 1922. "Why China Has No Science: An Interpretation of the History and Consequences of Chinese Philosophy." *International Journal of Ethics* 32: 237–63.

Furth Charlotte. 1970. *Ting Wen-chiang: Science and China's New Culture*. Cambridge, MA: Harvard University Press.

Galison, Peter. 1998. "Judgement against Objectivity." In *Picturing Science, Producing Art*, edited by Caroline A. Jones and Peter Galison, 327–59. London and New York: Routledge.

Gao Song 高松. 1550. *Gao Song ju pu* 高松菊譜 (Gao Song's monograph on the chrysanthemum), 1959 facsimile. Beijing: Zhongguo Shudian.

Gao, Yi. 2017. "Les origines chinoises des Lumières et de la Révolution française." *Annales historiques de la Révolution française* 387, no. 1: 103–22.

Gee, Nathaniel Gist. 1918. "Preface." In *Zhiwuxue da cidian* 植物學大辭典 (Dictionary of botany), edited by Du Yaquan 杜亞泉. Shanghai: Shangwu Yinshuguan.

Gentleman's Magazine and Historical Review. 1776. "Memoir (on the occasion of his death) of John Blake." (August): 348–51.

Gezhi huibian 各致彙編 (The Chinese scientific and industrial magazine). 1876. "Gezhi lüelun: lun zhiwuxue 各致略論. 論植物學" (Topics in science: on botany). 1, no. 10: 1a–4b.

———. 1877. "Li Renshu xiansheng xu 李壬叔先生序" (Reply to Mr. Li Renshu). 2, no. 3: 3b.

Gittlen, William. 1988. *Discovered Alive: The Story of the Chinese Redwood*. Berkeley and Frankfurt: Perside Publications.

Glover, Denise, Stevan Harrell, Charles F. McKhann, and Margaret Byrne Swain, eds. 2011. *Explorers and Scientists in China's Borderlands, 1880–1950*. Seattle: University of Washington Press.

Goldin, Paul R. 2018. "Xunzi." In *The Stanford Encyclopedia of Philosophy Archive (Fall 2018 Edition)*, edited by Edward N. Zalta. Accessed at the Stanford Encyclopedia of Philosophy Archive. https://plato.stanford.edu/archives/fall2018/entries/xunzi/.

Golinski, Jan. 1990. "The Theory of Practice and the Practice of Theory: Sociological Approaches in the History of Science." *Isis* 81, no. 3: 492–505.

Gong Lixian 龔禮賢. 1934. "Piping yu taolun: yu Zhongyang Yanjiuyuan Zhong Guanguang jiaoshou lun zhiwu ke ming shu 批評與討論: 與中央研究院鍾觀光教授論植物科名書" (Critique and discussion: a letter to Professor Zhong Guanguang of Academia Sinica on the naming of families in botany). *Xue yi* 學藝 13, no. 5: 131–34.

Goodman, Jordan, and Peter Crane. 2017. "The Life and Work of John Bradby Blake." *Curtis's Botanical Magazine* 34, no. 4: 231–50.

Gordin, Michael D. 2017. "Introduction: Hegemonic Languages and Science." In "Linguistic Hegemony in the History of Science," special focus issue, *Isis* 108, no. 3: 606–11.

Gordin, Michael, Walter Grunden, Mark Walker, and Zuoyue Wang. 2002. "'Ideologically Correct' Science." In *Science and Ideology: A Comparative History*, edited by Mark Walker, 35–65. London and New York: Routledge.

Gourlay, Patrick. 2004. *Regards sur la Croisière jaune*. Morlaix, France: Skol Vreizh.

Gressitt, J. Linley. 1936. "Notes on Collecting in Hainan Island with Data on Localities." *Lingnan Science Journal* 15, no. 3.

———. 1953. "The California Academy: Lingnan Dawn Redwood Expedition." *Proceedings of the California Academy of Sciences* 28: 25–58.

Greve, Andrew Q., and Jack S. Levy. 2018. "Power Transitions, Status Dissatisfaction, and War: The Sino-Japanese War of 1894–1895." *Security Studies* 27, no. 1: 148–78.

Griffin, Anne. 2016. *Michał Piotr Boym's Flora sinensis, fructus floresque humillime [Flora of China, fruits and flowers]*. Biodiversity Heritage Library blog, http://blog.biodiversitylibrary.org/2016/12/micha-piotr-boyms-flora-sinensis.html.

Griffin, Frank, ed. 1964. *Camellian: A Compilation of Authoritative Information on Camellia Culture*. Columbia, SC: Vogue Press, Inc.

Griffiths, D. A. 1988. "A Garden on the Edge of China: Hong Kong, 1848." *Garden History* 16, no. 2 (Autumn): 189–98.

Griffiths, D. A., and S. P. Lau. 1986. "The Hong Kong Botanical Gardens: A Historical Overview." *Journal of the Hong Kong Branch of the Royal Asiatic Society* 26: 55–77.

Guo Jianyou 郭建佑. 2012. "Cong wan Qing yi shu shumude fenlei tixi lun wan Qing shi ren dui xixue leimude jiedu yu yingdui 從晚清譯書書目的分類體系論晚清士人對西學類目的解讀與應對" ([Late Qing scholars' understanding and response to the classification systems of Western learning as seen in the classification systems used for translated works]). *Tushuguan zixunxue yanjiu* 圖書館資訊學研究 (Journal of library and information science research) 6, no. 2: 139–81.

Guo Yanbing 郭燕冰. 2013. "Yu jie dong xi yang liang wenmingde bing dizhi hua Guo Cui Pai chengyuan Cai Zhefu 欲結東西洋兩文明的並蒂之花國粹派成員蔡哲夫" (Cai Zhefu, a member of the National Essence Group, who wanted to join Eastern and Western cultures like a stem and a flower). *Lingnan wen shi* 嶺南文史, no. 2: 59–64.

Haas, William J. 1988a. "Botany in Republican China: The Leading Role of Taxonomy." In *Science and Medicine in Twentieth Century China: Research and Education*, edited by J. Z. Bowers, J. W. Hess, and N. Sivin, 31–64. Ann Arbor: University of Michigan Center for Chinese Studies.

———. 1988b. "Transplanting Botany to China: The Cross-Cultural Experience of Chen Huanyong." *Arnoldia* 48, no. 2: 9–25.

———. 1996. *China Voyager: Gist Gee's Life in Science.* Armonk, NY; and London: M. E. Sharpe.

Hai Ruo 海若. 2017. *Xunzhao shiqude zongji* 尋找逝去的踪迹 (Looking for a trace of the departed). Article posted to Jiangsu Zuojia Wang 江蘇作家網 (Jiangsu Writers' Network) on August 4, 2017, www.jszjw.com/author/works/201708/t20170804_16180388.shtml.

Halphen, Jules.1900. *Miroir des fleurs: Guide pratique du jardinier amateur en Chine au XVIIe siècle*. Translation of Chen Haozi 陳淏子 1688, reprinted in 2006 with a commentary by Georges Métailié. Paris: Actes Sud.

Handel-Mazzetti, Heinrich von. 1927. *Naturbilder aus Südwest-China: Erlebnisse und Eindrücke eines Österreichischen Forschers während des Weltkrieges.* Vienna and Leipzig: Österreichischer Bundesverlag für Unterricht, Wissenschaft und Kunst.

Hao Bingjian 郝秉鍵. 2003. "Shanghai Gezhi Shuyuan ji qi jiaoyu chuangxin 上海格致書院及其教育創新" (The Shanghai Polytechnic Institution and Reading Room and its innovation in education). *Qing shi yanjiu* 清史研究 (Studies in Qing history) 3: 85–96.

Harbsmeier, Christof. 2010. "Towards a Conceptual History of Some Concepts of Nature in Classical Chinese: *zì rán* 自然 and *zì rán zhī lǐ* 自然之理." In *Concepts of Nature: A Chinese-European Cross-Cultural Perspective*, edited by Hans Ulrich Vogel, Günter Dux, and Mark Elvin, 220–54. Leiden, Netherlands: Brill.

Harrell. 2011. "Introduction." In *Explorers and Scientists in China's Borderlands, 1880–1950*, edited by Denise Glover, Stevan Harrell, Charles F. McKhann, and Margaret Byrne Swain, 3–23. Seattle: University of Washington Press.

Harrist, Robert E., Jr., 1993. "Site Names and Their Meaning in the Garden of Solitary Enjoyment." *Journal of Garden History* 13, no. 4: 199–212.

Hart, Robert, and Joseph Edkins.1886. *Zhiwuxue qimeng* 植物學啟蒙 (A primer of botany). Shanghai: Zong Shiwusi Shuju.

Haudricourt, André Georges, and Georges Métailié. 1994. "De l'illustration botanique en Chine." *Etudes chinoises* 13, no. 1–2: 381–416.

Hedberg, Inga, ed. 2005. *Species Plantarum 250 Years: Proceedings of the Species Plantarum Symposium Held in Uppsala, August 22–24, 2003.* Uppsala, Sweden: Uppsala University.

Hird, James Dennis. 1906. *A Picture Book of Evolution.* London: Watts & Co.

Hodacs, Hanna, 2016. *Silk and Tea in the North: Scandinavian Trade and the Market for Asian Goods in Eighteenth-Century Europe.* London: Palgrave Macmillan.

Holway, Tatiana. 2018. "History or Romance? Ernest H. Wilson and Plant Collecting in China." *Garden History* 46, no. 1: 3–26.

Hong Deyuan 洪德元, Chen Zhiduan陳之端, and Qiu Yinlong 仇寅龍. 2008. "Lishi cui ren fenjin, weilai ling ren chongjing—jinian Zhongguo Zhiwuxuehui chengli qishiwu zhounian 歷史催人奮進，未來令人憧憬— 紀念中國植物學會成立75週年" (History urges people to forge onwards, people will take note of this in the future: in commemoration of the 75th anniversary of the founding of the Botanical Society of China). *Zhiwu fenlei xuebao* 植物分類學報 (Journal of systematics and evolution) 46, no. 4: 439–40.

Hong, Deyuan, and Stephen Blackmore. 2015. *Plants of China: A Companion to the Flora of China.* Beijing: Science Press; and Cambridge: Cambridge University Press.

Honour, Hugh. 1962. *Chinoiserie: The Vision of Cathay.* New York: Dutton.

Hsia, Florence C. 2009. *Sojourners in a Strange Land: Jesuits and Their Scientific Missions in Late Imperial China.* Chicago: University of Chicago Press.

Hsu, Elisabeth, ed. 2001. *Innovation in Chinese Medicine.* Needham Research Institute Studies 3. Cambridge: Cambridge University Press.

Hsueh Chi-ju (Xue Jiru). 1985. "Reminiscences of Collecting the Type Specimens of *Metasequoia glyptostroboides*." *Arnoldia* 45, no. 4: 10–18.

Hu, Danian. 2005. *China and Albert Einstein: The Reception of the Physicist and His Theory in China, 1917–1979.* Cambridge, MA: Harvard University Press.

Hu Dekun 胡德焜. 2014. "Hu Xiansu yu Lushan Zhiwuyuan neide "San Lao Mu" 胡先驌與廬山植物園内的"三老墓"" (Hu Xiansu and the "Tombs of the Three Elders" in the Lushan Botanical Garden). In *Lushan Zhiwuyuan bashi chun qiu jinianji* 廬山植物園八十春秋紀念集 (A commemorative compilation on the 80th anniversary of the Lushan Botanical Garden), by Hu Zonggang 胡宗岡, 341–61. Shanghai: Jiaotong Daxue Chubanshe.

Hu, H. H. *See* Hu Xiansu.

Hu, Hsen-Hsu. *See* Hu Xiansu.

Hu Xiansu 胡先驌. 1915 and 1916. "Shuowen zhiwu gu ming jin zheng 說文植物古名今證" (Present-day names for plants in *The Analytical Dictionary of Characters (Shuo wen)*). 2 pts. *Kexue* 科學 (Science) 1, no. 7: 789–91; 2, no. 3: 311–17.

——. 1922. "Zhejiang caiji zhiwu you ji 浙江採集植物遊記" (Notes on a plant-collecting expedition to Zhejiang). 7 installments. *Xueheng* 學衡 (Critical review), nos. 1–7.

——. 1934. "Fakan ci 發刊辭" (Editorial on the first issue of this journal). *Zhongguo zhiwuxue zazhi* 中國植物學雜誌 1, no. 1: 1–2.

——. 1936. "Ruhe chongfen liyong Zhongguo zhiwuzhi fuyuan 如何充分利用中國植物之富源" (How to make full use of China's wealth of plant resources). *Kexue* 科學 (Science) 20, no. 10: 850–58.

—— (Hu, H. H.). 1938. "Recent Botanical Exploration in China." *Journal of the Royal Horticultural Society of London* 63, no. 8: 381–89.

——. 1941. "Fakan ci 發刊詞" (Editorial on the first issue of this journal). *Yunnan Nonglin Zhiwu Yanjiusuo cong kan* 雲南農林植物研究所叢刊 1, no. 1: 1–2.

—— (Hu Hsen-Hsu). 1946. "Notes on a Palaeogene Species of *Metasequoia* in China." *Bulletin of the Geological Society of China* 26 (December): 103–7.

—— (Hu, H. H.). 1948. "How *Metasequoia*, the "Living Fossil," Was Discovered in China." *Journal of the New York Botanical Garden* 49 (585): 201–7.

——. 1950. "Shuisha ji qi lishi 水杉及其歷史" (*Metasequoia* and its history). *Zhongguo zhiwuxue zazhi* 中國植物學雜誌 5, no. 1: 9–13.

—— (Hu, H. H.). 1966. "Metasequoia Poem." *Eastern Horizon* 5, no. 4: 26–28.

Hu Xiansu (Hu H. H) and Zheng Wanjun (W. C. Cheng). 1948. "On the new family Metasequoiaceae and on *Metasequoia glyptostroboides*, a living species of the genus *Metasequoia* found in Szechuan and Hupeh." *Bulletin of the Fan Memorial Institute of Biology*. n.s., 1, no. 2: 153–63.

Hu Zonggang 胡宗崗. 2000. "Beiping Jingsheng Shengwu Diaochasuode fuyuan 北平靜生生物調查所的復員" (Restoration of the Fan Memorial Institute of Biology). *Zhongguo keji shiliao* 中國科技史料 21, no. 1: 52–60.

———. 2001. "Yunnan Nonglin Zhiwu Yanjiusuo chuangban yuanqi 雲南農林植物研究所創辦緣起" (The Establishment of the Yunnan Institute of Agriculture and Forestry). *Zhongguo keji shiliao* 中國科技史料 22, no. 3: 238–48.

———. 2005a. *Bu gai yiwangde Hu Xiansu* 不該遺忘的湖先驌 (Hu Xiansu who should never be forgotten). Wuhan: Changjiang Wenyi Chubanshe.

———. 2005b. *Jingsheng Shengwu Diaochasuo shigao* 靜生生物調查所史稿 (Historical sketch of the Fan Memorial Institute of Biology). Jinan: Shandong Jiaoyu Chubanshe.

———. 2006. "Liu Shen'e xibei kexue kaocha kao 劉慎諤西北科學考察考" (A study of Liu Shen'e and his scientific survey of the northwest). *Zhongguo kexueshi zazhi* 中國科學史雜誌 27, no. 4: 348–52.

———. 2009. *Lushan Zhiwuyuan zui chu sanshi nian* 廬山植物園最初三十年, *1934–1964* (The first thirty years of the Lushan Botanical Garden, 1934–1964). Shanghai: Jiaotong Daxue Chubanshe.

———. 2013a. *Huanan Zhiwuxue Yanjiusuo zao qi shi* 華南植物學研究所早期史— 中山大學農林植物研究所史事 *(1928–1954)* (A history of the early years of the South China Institute of Botany: the history of the Sun Yatsen University Institute of Agriculture, Forestry and Botany). Shanghai: Jiaotong Daxue Chubanshe.

———. 2013b. *Jian cao shi mu liushi nian—Wang Wencai zhuan* 箋草釋木六十年— 王文採傳 (Sixty years of writing about grasses and expounding on trees: a biography of Wang Wencai). Shanghai: Jiaotong Daxue Chubanshe and Zhongguo Kexue Jishu Chubanshe.

———. 2013c. "Zhongyangg Yanjiuyuan kaichu Chen Huanyong pingyiyuan an 中央研究院開除陳煥庸評議員案" (The case of the expulsion of Chen Huanyong from Academia Sinica). *Kexue wenhua pinglun* 科學文化評論 10, no. 3: 115–21.

———. 2014. *Lushan Zhiwuyuan bashi chun qiu jinianji* 廬山植物園八十春秋紀念集 (A collection in celebration of the 80th anniversary of the Lushan Botanical Garden). Shanghai: Jiaotong Daxue Chubanshe.

———. 2018. *Yunnan zhiwuyanjiu shilüe* 雲南植物研究史略 (A historical outline of botany in Yunnan). Shanghai: Jiaotong Daxue Chubanshe.

Hua shi 花史 (The history of flowers) [Ming Dynasty, n.d.], digitized copy from Beijing University Library, available from the Chinese Text Project. Accessed April 25, 2016. http://ctext.org/library.pl?if=en&res=82208.

Huang Kewu 黄克武 (Huang, Max K.). 2008. "Xin mingcizhi zhan: Qing mo Yan Fu yiyu yu hezhi Hanyude jingsai 新名詞之戰: 清末嚴復譯語與和製漢語的競賽" (The war of neologisms: the competition between the newly translated terms coined by Yan Fu and by the Japanese in the late

Qing). *Zhongyang Yanjiuyuan jindaishi yanjiu jikan* 中央研究院近代史研究所集刊 62: 1–42.

Huang Libo 黃立波 and Zhu Zhiyu 朱志瑜. 2012. "Wan Qing shiqi guanyu fanyi zhengcede taolun 晚清時期關於翻譯政策的討論" (Debates on translation policies during the late Qing period). *Zhongguo fanyi* 中國翻譯 (Chinese journal of translators), no. 3: 26–33.

Huang, Max K. *See* Huang Kewu.

Huang, Rui-lan. 2016. "Prof. Huan-Yong Chen: A Leading Botanist and Taxonomist, One of the Pioneers and Founders of Modern Plant Taxonomy in China." *Protein & Cell* 7, no. 11: 773–76.

Hummel, Arthur W. 1970. *Eminent Chinese of the Ch'ing Period (1644–1912)*. Washington, DC: Library of Congress, Orientalia Division.

Hutton, Eric L. 2014. *Xunzi: The Complete Text*. Princeton, NJ; and Oxford: Princeton University Press.

Jacobson, Dawn. 1993. *Chinoiserie*. London: Phaidon Press.

Jalais, Annu. 2018. "Reworlding the Ancient Chinese Tiger in the Realm of the Asian Anthropocene." *International Communication of Chinese Culture* 5, no. 2: 121–44. https://doi.org/10.1007/s40636-018-0123-8.

Jami, Catherine. 1999. "'European Science in China' or 'Western Learning'? Representations of Cross-Cultural Transmission, 1600–1800." *Science in Context* 12, no. 3: 413–34.

Jardine, Nicholas, James Secord, and Emma Spary, eds. 1996. *Cultures of Natural History*. Cambridge: Cambridge University Press.

Jarvis, C. E., and P. H. Oswald. 2015. "The Collecting Activities of James Cuninghame FRS on the Voyage of the *Tuscan* to China (Amoy) between 1697 and 1699." *Notes and Records: The Royal Society Journal of the History of Science* 69, no. 2: 135–53.

Jiang Jiafa 江家發 and Wang Gang 王剛. 2011. "*Yaquan zazhi* chuanbo jindai huaxuede lishi chengjiu 亞泉雜誌傳播近代化學的歷史成就" (The historical achievement of the *Yaquan Journal* in popularizing chemistry). *Huaxue jiaoyu* 化學教育, no. 1: 78–80.

Jiang Lijing. 2016. "Retouching the Past with Living Things: Indigenous Species, Tradition and Biological Research in Republican China, 1918–1937." *Historical Studies in the Natural Sciences* 46, no. 2: 154–206.

Jiang, Shuyong. 2007. "Into the Source and History of Chinese Culture: Knowledge Classification in Ancient China." *Libraries and the Cultural Record* 42, no. 1: 1–20.

Jiang Tingxi 蔣廷錫, ed. 1726. *Gu jin tu shu ji cheng* 古今圖書集成 (The imperially commissioned compendium of literature and illustrations, ancient and modern). Citations refer to the 1934 ed. Shanghai: Zhonghua Shuju.

Digitized by the Chinese Text Project, http://ctext.org/library.pl?if=gb&file=91940&page=82.

Jiang Weiqiao 蔣維喬. 1941. "Zhong Xianchang Xiansheng zhuan 鐘憲鬯先生傳" (Biography of Mr. Zhong Xianchang). *Shijie wenhua* 世界文化 2, no. 5: 57–60.

Jiang Yuping 姜玉平. 2003. "Beiping Yanjiuyuan Zhiwu Yanjiusuode ershi nian 北平研究院植物研究所的二十年" (Twenty years of the Institute of Botany of the National Academy of Beiping). *Zhongguo keji shiliao* 中國科技史料 24, no. 1: 34–46.

———. 2005. "Jingsheng Shengwu Diaochasuo xueshu jianzhi yanjiu 靜生生物調查所學術建制研究" (A study on the institution of the Fan Memorial Institute of Biology). *Zhongguo keji shiliao* 中國科技史料 26, no. 4: 291–311.

Jiang Yuping 姜玉平 and Zhang Binglun 張秉倫. 2002. "Cong Ziran Lishi Bowuguan dao Dongwu Yanjiusuo he Zhiwu Yanjiusuo 從自然歷史博物館到動物研究所和植物研究所" (From the Natural History Museum to the Institute of Zoology and the Institute of Botany). *Zhongguo keji shiliao* 中國科技史料 23, no. 1: 18–30.

Jiang Zong 江總. n.d. "Jiang Zong quan ji 江總全集" (Complete works of Jiang Zong). *Shi ci ming ju* 詩詞名句 (Famous poetry and verse). Accessed July 18, 2017. www.shicimingju.com/chaxun/zuozhe/386.html.

Jin Guantao 金觀濤 and Liu Qingfeng 劉青峰. 2005. "*Keju* he *kexue* zhongda shehui shijian he guannian zhuanhuade anli yanjiu "科舉" 和 "科學" 重大社會事件和觀念轉化的案例研究" (A case study of the social reality and changing perceptions of *keju* [examination system] and *kexue* [science]). *Kexue wenhua pinglun* 科學文化評論 2, no. 3: 5–15.

Jin Tao 金濤. 1999. "Wu Zhonglunde "Yunnan zhiwu kaocha riji" 吳中倫的 "雲南植物考察日記"" (The field diaries of Wu Zhonglun's *Botanical Expedition in Yunnan*). *Zhongguo keji shiliao* 中國科技史料, no. 2: 179–89.

Jing Libin 經利彬, Wu Zhengyi 吳徵鎰, Kuang Keren 匡可任, and Cai Dehui 蔡德惠. 1945. *Diannan bencao tupu* 滇南本草圖譜 (Illustrated materia medica of Southern Yunnan). Kunming: Guoli Guoyiyao Yanjiusuo, Yunnan Shengli Yaowu Gaijinsuo 國立國醫藥研究所, 雲南省立藥物改進所 (National Institute of Chinese Medicine).

Johnson, Brian. 2007. "The Changing Face of the Botanical Garden." In *Botanic Gardens: A Living History*, edited by Nadine Käthe Monem and Blanche Craig, 62–81. London: Black Dog.

Johnson, Nuala C. 2011. *Nature Displaced, Nature Displayed: Order and Beauty in Botanical Gardens*. Tauris Historical Geography Series, Book 7. London and New York: I. B. Tauris.

Jones, Caroline A., and Peter Galison. 1998. *Picturing Science, Producing Art.* London and New York: Routledge.

Journal of the China Branch of the Royal Asiatic Society. 1886. "The Advisability, or the Reverse, of Endeavouring to Convey Western Knowledge to the Chinese through the Medium of Their Own Language," n.s., 2, no. 1: 1–21.

Jullien, François. 1990. "Présentation: l'art de la liste." *Extrême-Orient Extrême-Occident*, no. 12: 7–12.

Kaempfer, Engelbert. 1712. *Amoenitatum exoticarum politico-physico medicarum.* Lemgoviae (Lemgo).

Kang Kaiyuan 康凱原. 2011. "Chaye shide ling yi jiaodu: lun Yingguo zhiwu lieren Fuqiongde kexue diaocha yi ji chamiao yu jishude yizhi 茶業史的另一角度：論英國植物獵人福瓊的科學調查以及茶苗與技術的移植" (Another angle on the history of tea: the British plant hunter Robert Fortune, scientific research, tea seedlings, and technology transfer). *Huilan chun qiu* 洄瀾春秋, no. 8: 151–70.

Karl, Rebecca E., and Peter Zarrow, eds. 2002. *Rethinking the 1898 Reform Period: Political and Cultural Change in Late Qing China.* Harvard East Asian Monographs 214. Cambridge, MA; and London: Harvard University Press.

Ke Zunke 柯尊科 and Li Bin 李斌. 2016. "Zhongguo Kexueshede xingwang—yi *Kexue* zazhi xiansuode kaocha 中國科學社的興亡-以"科學"雜誌線索的考察" (The rise and fall of the Science Society of China: an investigation based on the study of its journal *Kexue*). *Ziran bianzhengfa tongxun* 自然辯證法通訊 (Journal of the dialectics of nature) 38, no. 3: 21–33.

Keng, Hsuan. 1974. "Economic Plants of Ancient North China as Mentioned in *Shih Ching* (Book of Poetry)." *Economic Botany* 28, no. 4 (October–December): 391–410.

Kexue 科學 (Science). 1915a. "Fakan ci 發刊詞" (Editorial on the launch of this journal). 1, no. 1: 3–7.

———. 1915b. "Liyan 例言" (Introduction). 1, no. 1: 1–2.

———. 1922. "Ben she shengwu yanjiusuo kaimu ji 本社生物研究所開幕記" (The opening ceremony of the society's biology laboratory). 7, no. 8: 846–48.

Kilpatrick, Jane. 2007. *Gifts from the Gardens of China.* London: Frances Lincoln Ltd.

———. 2014. *Fathers of Botany: The Discovery of Chinese Plants by European Missionaries.* Royal Botanic Gardens, Kew: Kew Publishing.

Kim, Yung Sik. 2004. "The 'Why Not' Question of Chinese Science: The Scientific Revolution and Traditional Chinese Science." *East Asian Science, Technology, and Medicine* 22: 96–112.

Kindall, Elizabeth. 2012. "Experiential Readings and the Grand View: Mount Jizu by Huang Xiangjian (1609–1673)." *Art Bulletin* 94, no. 3: 412–36.

Kirkland, Russell. 2004. *Taoism: The Enduring Tradition*. London and New York: Routledge.

Kohler, Robert E. 2002. "Place and Practice in Field Biology." *History of Science* 40: 189–210.

———. 2007. "Finders, Keepers: Collecting Sciences and Collecting Practice." *History of Science* 45: 428–54.

Koon, Yeewan. 2014. *A Defiant Brush: Su Renshan and the Politics of Painting in Early 19th-Century Guangdong*. Hong Kong: Hong Kong University Press; Honolulu: University of Hawaii Press.

Kwok, D. W. Y. 1965. *Scientism in Chinese Thought, 1900–1950*. New Haven, CT; and London: Yale University Press.

Kwong, Luke S. K. 2000. "Chinese Politics at the Crossroads: Reflections on the Hundred Days Reform of 1898." *Modern Asian Studies* 34, no. 3: 663–95.

Kyna, Rubin. 2016. "The Metasequoia Mystery." *Landscape Architecture Magazine* (January 19): 120–27.

Lackner, Michael, Iwo Amelung, and Joachim Kurtz, eds. 2001. *New Terms for New Ideas: Western Knowledge and Lexical Change in Late Imperial China*. Leiden, Netherlands; Boston; and Cologne: Brill.

Lackner, Michael, and Natascha Vittinghoff, eds. 2004. *Mapping Meanings: The Field of New Learning in Late Qing China*. Leiden, Netherlands; and Boston: Brill.

Lamy, Denis. 2008. "Le dessin botanique dans la transmission des connaissances." In *Passions botaniques: naturalistes voyageurs au temps des grandes découvertes*, edited by Yves-Marie Allain, Lucile Allorge, and Cécile Aupic, 139–54. Rennes, France: Editions Ouest- France.

Lary, Diana. 2010. *Warlord Soldiers: Chinese Common Soldiers, 1911–1937*. Cambridge: Cambridge University Press.

Latour, Bruno. 1986. "Visualization and Cognition: Thinking with Eyes and Hands." *Knowledge and Society: Studies in the Sociology of Culture Past and Present* 6: 1–40.

———. 1999. *Politiques de la nature: comment faire entrer les sciences en démocratie*. Paris: La Découverte.

Laufer, Berthold. 1934. "The Lemon in China and Elsewhere." *Journal of the American Oriental Society* 54, no. 2: 143–60.

Le Rougetel, Hazel. 1982. "The Fa Tee Nurseries of South China." *Garden History* 10, no. 1 (Spring): 70–73.

Le Tianyu 樂天宇 and Xu Weiying 徐維英. 1957. *Shaan Gan Ning Pendi zhiwuzhi* 陝甘寧盆地植物志 (The flora of the Shaan[xi], Gan[su], Ning[xia] Basin). Beijing: Linye Chubanshe.

Legge, James. 1861. *The Chinese Classics. Vol. 1. Confucian Analects, The Great Learning, and the Doctrine of the Mean.* London: Trübner.

Lei, Fu. 2017. "Nathaniel Gist Gee's contribution to biology in modern China." *Protein & Cell* 8, no. 4: 237–39.

Lennox, J., and Bolton, R., eds. 2010. *Being, Nature, and Life in Aristotle.* Cambridge: Cambridge University Press.

Leys, Simon, trans. 1997. *The Analects of Confucius.* New York and London: Norton.

Li Ang 李昂 and Chen Yue 陳悅. 2015. "Zhongwen guji zhong zhiwu tuxiang biaodadian chuyi 中文古籍中植物圖像表達點芻議" (A preliminary study on the characteristic form of expression of plant illustrations in ancient Chinese books). *Ziran kexueshi yanjiu* 自然科學史研究 34, no. 2: 164–81.

Li Haimin 李海珉. 2013. "Yitan san jue Tan Yuese 藝壇三絕談月色" (Tan Yuese, incomparable in three fields of the arts). *Shouzang jie* 收藏界, no. 1: 100–103.

Li, Hongqi. 1991. *China and Europe: Images and Influences in Sixteenth to Eighteenth Centuries.* Monograph series, Chinese University of Hong Kong, Institute of Chinese Studies, 12. Hong Kong: Chinese University Press.

Li, Hui-lin. 1944. "Botanical Exploration in China during the Last Twenty-Five Years." *Proceedings of the Linnean Society of London*, 156th Session (1943–44): 25–44.

Li, Ming, and Le Kang. 2010. "Bing Zhi: Pioneer of Modern Biology in China." *Protein & Cell* 1, no. 7: 613–15.

Li Nan 李楠 and Yao Yuan 姚遠. 2011. "*Bowuxue zazhi* bankan sixiang tanyuan 博物學雜誌辦刊思想探源" (An investigation into the thinking behind the founding of *Bowuxue zazhi* [Journal of natural history]). *Bianji xuebao* 編輯學報 (Acta editologica) 23, no. 5: 398–400.

Li Nanqiu 黎難秋, ed. 1996. *Zhongguo kexue fanyi shiliao* 中國科學翻譯史料 (Materials on the history of science translation in China). Hefei: Zhongguo Kexue Jishu Daxue Chubanshe.

———. 1999. "Minguo shiqi Zhongguo kexue fanyi huodong gaikuang 民國時期中國科學翻譯活動概況" (A review of activities in scientific translation during the republican period). *Zhongguo keji fanyi* 中國科技翻譯 12, no. 4: 42–49.

Li Nanqiu 黎難秋, Xu Ping 徐萍, and Zhang Fan 張帆. 1999. "Zhongguo kexuefanyishi ge shiqide tedian, chengguo, ji jianping 中國科學翻譯史各時期的特點, 成果及簡評" (The various stages of scientific translation in China:

special characteristics, achievements, and a critique). *Zhongguo fanyi* 中國翻譯 (Chinese journal of translators), no. 4: 43–46.

Li Shanlan 李善蘭, Alexander Williamson (Wei Lianchen 威廉臣), and Joseph Edkins (Ai Yuese 艾約瑟). 1858. *Zhiwuxue* 植物學 (Botany). Shanghai: Mohai Shuguan [Inkstone Press].

Li Shizhen 李時珍. 1596. *Bencao gangmu* 本草綱目 (Classification of materia medica). Shunzhi edition (1645) digitized by the Chinese Text Project at http://ctext.org/library.pl?if=gb&file=52746&page=82.

Li, T. June, and Suzanne E. Wright. 2016. *Gardens, Art, and Commerce in Chinese Woodblock Prints*. San Marino, CA: Huntington Library, Art Collections and Botanical Gardens.

Li Xueqin 李學勤 and Lü Wenyu 呂文鬱. 1996. *Siku da cidian* 四庫大辭典 (Dictionary of the Siku [quan shu] [Complete collection of the four treasuries]). Changchun: Jilin Daxue Chubanshe.

Li Yuanyang 李元陽, ed. 1563. *Jiajing Dali Fu zhi* 嘉靖大理府志 (The Jiajing-era gazetteer of Dali Prefecture). 1983 reprint by the Dali Bai Autonomous Prefecture Cultural Bureau.

Li Zhongli 李中立. 1612. *Bencao yuanshi* 本草原始 (Sources for *Materia Medica*). Using Ming, Wanli edition (1612) digitized by Beijing University Library at the China-America Digital Academic Library (CADAL), https://archive.org/details/02092990.cn/mode/2up.

Liang Congguo 梁從國. 2013. "Wan Qing Dao Xian shiqi xifang shengwuxue zhishi zai hua chuanbo kaocha 晚清道咸時期西方生物學知識在華傳播考察 (An investigation into the spread of biology in China during the Dao[guang] and Xian[feng] periods of the Qing." *Guangxi Minzu Daxue xuebao (ziran kexue ban)* 廣西民族大學學報 (自然科學版) 19, no. 2: 21–25.

Lindley, John. 1832. *An Introduction to Botany*. London: Longman, Rees, Orme, Brown, Green and Longman.

———. 1847. *Elements of Botany, Structural and Physiological*. 5th ed. London: Bradbury and Evans.

Linnaeus, Carolus. *See* Linné, Carl von.

Linné, Carl von (Carolus Linnaeus). 1735. *Systema naturae, sive, regna tria naturae systematice proposita per classes, ordines, genera, et species*. Lugduni Batavorum [Leiden, Netherlands]: Theodorum Haak.

———. 1762–63. *Species plantarum, exhibentes plantas rite cognitas, ad genera relatas, cum differentiis specificis, nominibus trivialibus, synonymis selectis, locis natalibus, secundum systema sexuale digestas*. Editio secunda (2nd ed.). 2 vols. Holmiae [Stockholm]: Impensis Direct. Laurentii Salvii.

Linnean Society of London. 1943–44. "Papers on the Exploration of China." *Proceedings of the Linnean Society of London*, 156th Session.

Lippert, Wolfgang. 2001. "Language in the Modernization Process: The Integration of Western Concepts and Terms into Chinese and Japanese in the Nineteenth Century." In *New Terms for New Ideas: Western Knowledge and Lexical Change in Late Imperial China*, edited by Michael Lackner, Iwo Amelung, and Joachim Kurtz, 57–66. Leiden, Netherlands; Boston; and Cologne: Brill.

Liu Dayou 劉大猷. 1904. *Zhiwuxue jiaoke shu* 植物學教科書 (Textbook of botany). Shanghai: Nongxuehui, Nongxue Congshu. Translation of Matsumura's *Shokubutsugaku kyōkasho* 植物學教科書 (A textbook of botany), Tokyo: Keigyōsha, Meiji 23 [1890].

Liu Qizhen 劉啟振 and Wang Siming 王思明. 2014. "Zai lun Hu Xiansu xuanzhi Lushan chuangban zhiwuyuande dongyin 再論胡先驌選址廬山創辦植物園的動因" (Reconsidering Hu Xiansu's motivation for selecting Lushan as the location for a botanical garden). *Zhongguo kexueshi zazhi* 中國科學史雜誌, no. 2: 204–6.

Liu Shoudan 劉壽聃. 1919, 1920, and 1922. "Zhiwu yiwu tongming kao 植物異物同名考" (On plants that are different but have the same name). 3 pts. *Guoli Wuchang Gaodeng Shifan Xuexiao bowuxuehui zazhi* 國立武昌高等師範學校博物學會雜誌 2, no. 4: 46–48; 3, no. 3: 49–50; 4, no. 3, 56–57.

Liu, Ts'ui-jung. 2014. *Environmental History in East Asia: Interdisciplinary Perspectives*. Academia Sinica on East Asia Series. London and New York: Routledge.

Liu Weimin 劉為民. 2012. "Wentan ming jia Hu Xiansu 文壇名家胡先驌" (Hu Xiansu, a famous literary figure). *Kexue yuanliu* 科學源流 64, no. 6: 50–54.

Liu Xian 劉咸. 1936. "Ben she di ershi yi ci nian hui jishi 本社第二十一次年會記事" (Report on the 21st annual meeting of the society). *Kexue* 科學 (Science) 20, no. 10: 871–83.

Liu Xiao 劉曉. 2013. "Zhongguo Kexueyuan jian yuan chu qide keyan jigou tiaozheng gongzuo 中國科學院建院初期的科研機構調整工作" (The reorganization of research institutions during the early phase of the establishment of the Chinese Academy of Sciences). *Zhongguo kejishi zazhi* 中國科技史雜誌 34, no. 3: 301–15.

Liu Xiaogan 劉笑敢. 2004. "Renwen ziran yu tiandi ziran 人文自然與天地自然" (The *ziran* of human culture and the *ziran* of heaven and earth). *Nanjing Shifan Daxue Wenxueyuan xuebao* 南京師範大學文學院學報 3: 1–12.

Liu Yongli 劉永利. 2017. "Zuo wei fanyi jiade Du Yaquan 作為翻譯家的杜亞泉" (Du Yaquan as a translator). *Shanghai fanyi* 上海翻譯 (Shanghai journal of translators), no. 1: 62–67.

Liu Youlin 劉有林. 2013. "Jindai di yi nü yin ren—Tan Yuese 近代第一女印人— 談月色" (Tan Yuese, the first woman in modern seal carving). *Qi cai yuwen (Xiezi yu shufa)* 七彩語文 (寫字與書法), no. 9: 7–9.

Liu Zhenyu 劉振宇 and Wei Wei 維微. 2005. *Zhongguo Li Zhuang: kangzhan liuwang xuezhede renwen dang'an* 中國李莊: 抗戰流亡學者的人文檔案 (Li Zhuang, China: archives in the humanities from academic refugees during the War of Resistance). Chengdu: Sichuan Renmin Chubanshe.

Lo, Vivienne, Penelope Barrett, David Dear, Lu Di, Lois Reynolds, and Dolly Yang. 2018. *Imagining Chinese Medicine*. Leiden, Netherlands; and Boston: Brill.

Lone, Stewart. 1994. *Japan's First Modern War: Army and Society in the Conflict with China, 1894–95*. New York: St. Martin's Press.

Long Bojian 龍伯堅. 1957. *Xian cun bencao shulu* 現存本草書錄 (A catalogue of extant materia medica). Beijing: Renmin Weisheng Chubanshe.

Long Chengpeng 龍成鵬. 2016. *Zou dao shijie jintoude zhiwuxue jia—zhiwu lieren Weiersun zai zhongguo xibude tanxian* 走到世界盡頭的植物學家— 植物獵人威爾遜在中國西部的探險 (The exploration of western China by [E.H.] Wilson, a dedicated botanist and plant hunter). *Jinri minzu* 今日民族, no. 8: 33–37.

López, Antonio Mezcua. 2013. "Gazing at the Mountains, Tasting Tea: The Relation between Landscape Culture and Tea Culture in Song China." *Studies in the History of Gardens and Designed Landscapes* 33, no. 3: 139–47.

Lovell, Julia, 2011. *The Opium War: Drugs, Dreams, and the Making of China*. Oxford: Picador.

Lu Di 蘆笛. 2014. "Yingguo Qiu Yuan he waiguoren zai Zhongguode zhiwu caiji huodong 英國邱園和外國人在中國的植物採集活動" (England's Kew Gardens and foreign plant hunters' collecting activities in China). *Zhongguo yesheng zhiwu ziyuan* 中國野生植物資源 (Chinese wild plant resources) 33, no. 1: 55–62.

Lu Houyuan, Jianping Zhang, Yimin Yang, Xiaoyan Yang, Baiqing Xu, Wuzhan Yang, Tao Tong, Shubo Jin, Caiming Shen, Huiyun Rao, Xingguo Li, Hongliang Lu, Dorian Q. Fuller, Luo Wang, Can Wang, Deke Xu, and Naiqin Wu. 2016. "Earliest Tea as Evidence for One Branch of the Silk Road across the Tibetan Plateau." *Scientific Reports* 6, article 18955.

Lu, Tracey L-D. 2014. *Museums in China: Power, Politics, and Identities*. London and New York: Routledge.

Luesink, David. 2012. "Dissecting Modernity: Anatomy and Power in the Language of Science in China." PhD diss., University of British Columbia.

Lung, Rachel. 2016. "The Jiangnan Arsenal: A Microcosm of Translation and Ideological Transformation in 19th-Century China." *Meta* 61: 37–52.

Luo Guihuan 羅桂環. 1987. "Wo guo zao qide liang ben zhiwuxue yizhu *Zhiwuxue* he *Zhiwu tushuo* ji qi shuyu 我國早期的兩本植物學譯著—"植物學" 和 "植物圖說" 及其術語" (The two earliest botanical books to be translated in China, *Zhiwuxue* [Botany] and *Zhiwu tushuo* [An illustrated botany] and their terminology). *Ziran kexueshi yanjiu* 自然科學史研究 6, no. 4: 383–87.

———. 2002. "Jindai xifang ren dui Wuyishande shengwuxue kaocha 近代西方人對武夷山的生物學考察" (Biological exploration in the Wuyi Mountains by modern Westerners). *Zhongguo keji shiliao* 中國科技史料 23, no. 1: 23–37.

———. 2006. "Shilun 20 shiji qianqi 'Zhongyang Guwu Baoguan Weiyuanhui' de chengli ji yiyi 試論 20 世紀前期 '中央古物保管委員會' 的成立及意義" (On the founding and the significance of the National Committee for the Preservation of Antiquities in the early twentieth century). *Zhongguo keji shi zazhi* 中國科技史雜誌 27, no. 2: 138–39.

———. 2011. "Minguo shiqi dui xifangren zai hua shengwu caijide xianzhi 民國時期對西方人在華生物採集的限制" (The restrictions on collecting botanical specimens by Westerners in republican China). *Ziran kexueshi yanjiu* 自然科學史研究 30, no. 4: 450–59.

Luo Zimei 羅自梅. 1981. "Hu Xiansu 胡先驌" (Hu Xiansu). *Jiangxi shehui kexue* 江西社會科學: 132–34.

Lydekker, Richard. 1893–96. *The Royal Natural History*. Vols. 1–6. New York: Frederick Warne & Co.; and Edinburgh: Morrison and Gibb.

Ma, Jinshuang. 2003. "The Chronology of the 'Living Fossil' *Metasequoia glyptostroboides*." *Harvard Papers in Botany* 8, no. 1: 9–18.

Ma, Jinshuang, and Kerry Barringer. 2005. "Dr. Hsen-Hsu Hu (1894–1968): A Founder of Modern Plant Taxonomy in China." *Taxon* 54, no. 2: 559–66.

Ma, Jinshuang, and Steve Clemants. 2006. "A History and Overview of the *Flora Reipublicae Popularis Sinicae* (FRPS, Flora of China, Chinese edition, 1959–2004)." *Taxon* 55, no. 2: 451–60.

Ma, Jinshuang (Ma J. S.) and Guofan Shao. 2003. "Rediscovery of the First Collection of the 'Living Fossil,' *Metasequoia glyptostroboides*." *Taxon* 52, no. 3: 585–88.

Ma Zuyi 馬祖毅. 1998. *Zhongguo fanyi jianshi: "wusi" yiqian bufen* 中國翻譯簡史: "五四" 以前部分 (A brief history of translation in China: before the May 4th [Movement]). Beijing: Zhongguo Dui Wai Fanyi Chuban Gongsi.

MacKinnon, Stephen R. (1992). *Power and Politics in Late Imperial China: Yuan Shikai in Beijing and Tianjin, 1901–1908*. Berkeley, Los Angeles, and London: University of California Press.

Macoboy, Stirling. 1981. *The Colour Dictionary of Camellias*. Sydney, Auckland, London, New York: Lansdowne Press.

Magee, Judith. 2011. *Images of Nature: Chinese Art and the Reeves Collection*. London: Natural History Museum.

Magnin-Gonze, Joëlle. 2009. *Histoire de la botanique*. Paris: Delachaux et Niestlé.

Marcon, Federico. 2015. *The Knowledge of Nature and the Nature of Knowledge in Early Modern Japan*. Chicago: Chicago University Press.

Marder, Michael. 2014. *The Philosopher's Plant: An Intellectual Herbarium*. New York: Columbia University Press.

Martzloff, J. C. 1997. *A History of Chinese Mathematics*. Berlin, Heidelberg: Springer.

Masini, Federico. 1993. "The Formation of the Modern Chinese Lexicon and Its Evolution Toward a National Language: The Period from 1840 to 1898." *Journal of Chinese Linguistics Monograph*, series 6. Hong Kong: Chinese University Press on behalf of the Project on Linguistic Analysis.

Matsumura Jinzō 松村任三. 1890 [Meiji 23]. *Shokubutsugaku kyōkasho* 植物學教科書 (A textbook of botany). Tokyo: Keigyōsha.

McCord, Edward A. 1993. *The Power of the Gun: The Emergence of Modern Chinese Warlordism*. Berkeley, Los Angeles, and London: University of California Press.

McCracken, Donald P. 1997. *Gardens of Empire: Botanical Institutions of the Victorian British Empire*. London and Washington, DC: Leicester University Press.

McLean, B. 2004. *George Forrest, Plant Hunter*. Edinburgh: Antique Collectors' Club and Royal Botanic Garden.

McOuat, Gordon. 2001. "Cataloguing Power: Delineating 'Competent Naturalists' and the Meaning of Species in the British Museum." *British Journal for the History of Science* 34, no. 1: 1–28.

Meng Shiyong孟世勇, Liu Huiyuan劉慧圓, and Yu Mengting余夢婷. 2018. "Zhongguo zhiwu caiji xian xingzhe Zhong Guanguangde caiji kaozheng 中國植物採集先行者鐘觀光的採集考證" (A study of the collecting practices of Zhong Guanguang, China's pioneering plant collector). *Shengwu duo yang xing* 生物多樣性 (Biodiversity) 26, no. 1: 79–88.

Mengxue bao 蒙學報 (Journal of pedagogy). 1897. "Zhiwu wen da 植物問答" (Questions and answers about plants). 8: 30–38.

Merrill, E. D. 1948. "*Metasequoia*, Another 'Living Fossil.'" *Arnoldia* 8, no. 1: 1–8.

Métailié, Georges. 1981. "La création lexicale dans le premier traité de botanique occidentale publié en chinois (1858)." *GRECO: documents pour l'histoire du vocabulaire scientifique* 2: 65–73.

———. 1988. "Des mots et des plantes dans le *Bencao gangmu* de Li Shizhen." In "Effets d'ordre dans la civilisation chinoise. Rangements à l'oeuvre, classifications implicites," *Extrême-Orient Extrême-Occident*, no. 10: 27–43.

———. 1989. "Histoire naturelle et humanisme en Chine et en Europe au XVIe siècle: Li Shizhen et Jacques Dalechamp." *Revue d'histoire des sciences* 42, no. 4: 353–74.

———. 1994. "A propos du sexe des fleurs: le cas des 'rui.'" *Cahiers de linguistique: Asie orientale* 23, no. 1: 223–30.

———. 1998. "A propos de quatre manuscrits chinois de dessins de plantes." *Arts asiatiques* 53: 32–38.

———. 2001a. "The *Bencao Gangmu* of Li Shizhen: An Innovation in Natural History?" In *Innovation in Chinese Medicine*, edited by Elisabeth Hsu, 221–61. Needham Research Institute Studies 3. Cambridge: Cambridge University Press.

———. 2001b. "The Formation of Botanical Terminology: A Model or a Case Study?" In *New Terms for New Ideas: Western Knowledge and Lexical Change in Late Imperial China*, edited by Michael Lackner, Iwo Amelung, and Joachim Kurtz, 327–28. Leiden, Netherlands; Boston; and Cologne: Brill.

———. 2002. "Comparative Study of the Introduction of Modern Botany in Japan and China." *Historia Scientarium: International Journal of the History of Science Society of Japan* 11, no. 3: 205–17.

———. 2007. "The Representation of Plants: Engravings and Paintings." In *Graphics and Texts in the Production of Technical Knowledge in China: The Warp and the Weft*, edited by Francesca Bray, Vera Dorofeeva-Lichtmann, and Georges Métailié, 487–520. Leiden, Netherlands: Brill.

———. 2012. "The Botany of Cheng Yaotian (1725–1814): Multiple Perspectives on Plants." In *Antiquarianism and Intellectual Life in Europe and China, 1500–1800*, edited by Peter N. Miller and François Louis, 250–62. Bard Graduate Center Cultural Histories of the Material World. Ann Arbor: University of Michigan Press.

———. 2015. *Traditional Botany: An Ethnobotanical Approach*, translated by Janet Lloyd. Vol. 6, pt. 4 of *Science and Civilisation in China*, edited by Joseph Needham. Cambridge: Cambridge University Press.

Miki, S. 1941. "On the Change of Flora in Eastern Asia since the Tertiary Period (I): The Clay or Lignite Beds Flora in Japan with Special Reference to the *Pinus trifolia* beds in Central Hondo." *Japanese Journal of Botany* 11: 237–304.

Miller, Peter N., and François Louis, eds. 2012. *Antiquarianism and Intellectual Life in Europe and China, 1500–1800*. Bard Graduate Center Cultural Histories of the Material World. Ann Arbor: University of Michigan Press.

Mittag, Achim. 2010. "Becoming Acquainted with Nature from the *Odes*: Sidelights on the Study of the Flora and Fauna in the Song Dynasty's *Shijing* 詩經 (Classic of Odes) Scholarship." In *Concepts of Nature: A Chinese-European Cross-Cultural Perspective*, edited by Hans Ulrich Vogel, Günter Dux, and Mark Elvin, 310–44. Leiden, Netherlands: Brill.

Monem, Nadine Käthe, and Blanche Craig, eds. 2007. *Botanic Gardens: A Living History*. London: Black Dog.

Mossman, Samuel. 1893. *The Great Taiping Rebellion: A Story of General Gordon in China*. London: Griffith, Farran, Browne & Co.

Mu Fengliang 穆風良. 2004. "Siyi Guan yu Tongwen Guan mingcheng kao 四夷館與同文館名稱考" (A survey of the name change from Siyi Guan to Tongwen Guan). *Qinghua Daxue xuebao (zhexue shehuikexue ban)* 清華大學學報 (哲學社會科學版) (Journal of Tsinghua University (Philosophy and Social Sciences)) 19, no. 1: 64–67.

Mu Yu 穆宇 and Wang Zhao 王釗. 2017. "From Materia Medica to Art: Dialog Between Eastern and Western Botanical Illustration." In *Walking the Path to Eternal Fragrance*, compiled by the Organizing Committee of the XIX International Botanical Congress, 282–322. Nanjing: Jiangsu Phoenix Science Press.

Mueggler, Erik. 2005. "The Lapponicum Sea: Matter, Sense, and Affect in the Botanical Exploration of Southwest China and Tibet." *Comparative Studies in Society and History* 47, no. 3: 442–79.

———. 2011. *The Paper Road: Archive and Experience in the Botanical Exploration of West China and Tibet*. Berkeley, Los Angeles, and London: University of California Press.

Mun, Seung-Hwan. 2013. "Printing Press without Copyright: A Historical Analysis of Printing and Publishing in Song China." *Chinese Journal of Communication* 6, no. 1: 1–23.

Musillo, Marco. 2017. *The Shining Inheritance: Italian Painters at the Qing Court, 1699–1812*. Los Angeles: Getty Research Institute.

Nakayama, Shigeru, and Nathan Sivin, eds. 1973. *Chinese Science: Explorations of an Ancient Tradition, Compiled in Honor of the Seventieth Birthday of Joseph Needham, F. R. S.* Cambridge, MA: MIT Press.

Nappi, Carla. 2009a. "Bolatu's Pharmacy Theriac in Early Modern China." *Early Science and Medicine* 14, no. 6: 737–64.

———. 2009b. *The Monkey and the Inkpot*. Cambridge, MA: Harvard University Press.
Needham, Joseph. 1943. "Science in Chungking." *Nature* 152 (July 17): 64–66.
———. 1943–46. "Organizations Visited and People Met in China, 1943–1946. Accessed through the Needham Research Institute, Cambridge (U.K.) at http://www.nri.cam.ac.uk/JN_wartime.htm.
———. 1954. *Science and Civilisation in China*. Cambridge: Cambridge University Press.
———. 1967. "The Roles of Europe and China in the Evolution of Oecumenical Science." *Journal of Asian History* 1, no. 1: 3–32.
Needham, Joseph, Lu Gwei-Djen, and Huang Hsing-Tsung. 1986. *Biology and Biological Technology*. Vol. 6, pt. 1 of *Science and Civilisation in China*, edited by Joseph Needham. Cambridge: Cambridge University Press.
Nelmes, E. 1944. "Robert Fortune, Pioneer Collector, the Centenary of Whose Departure for China Has Fallen in This Year." *Proceedings of the Linnean Society of London*, 156th Session (1943–44): 8–15.
Nicolson, Dan H. 1991. "A History of Botanical Nomenclature." *Annals of the Missouri Botanical Garden* 78, no. 1: 33–56.
Nie Genjin 倪根金 and Zhou Miya 周米亞. 2014. "Chuantong jupuzhongde yiju jishu tanxi 傳統菊譜中的藝菊技術探析" (An investigation of ornamental chrysanthemums in traditional monographs on the chrysanthemum). *Nongye kaogu* 農業考古, no. 1: 285–96.
Nivison, D., and Bryan van Norden. 1996. *The Ways of Confucianism*. Chicago: Open Court.
Novotná, Zdenka. 1967. "Linguistic Factors of the Low Adaptability of Loan-Words to the Lexical System of Modern Chinese." *Monumenta Serica* 26: 103–18.
Nyberg, Kenneth. 2009. "Linnaeus' Apostles, Scientific Travel and the East India Trade." *Zoologica Scripta* 38, suppl. 1 (February): 7–16.
O'Brien, Seamus. 2011. *In the Footsteps of Augustine Henry and His Chinese Plant Collectors*. Woodbridge, Suffolk, UK: Garden Art Press.
Ogilvie, Brian W. 2006. *The Science of Describing: Natural History in Renaissance Europe*. Chicago: University of Chicago Press.
Olohan, Maeve. 2007. "The Status of Scientific Translation." *Journal of Translation Studies* 10, no. 1: 131–44.
Organizing Committee of the XIX International Botanical Congress, eds. 2017. *Walking the Path to Eternal Fragrance*. Nanjing: Jiangsu Phoenix Science Press.
Osbeck, Peter. 1771. *A Voyage to China and the East Indies. Together with a Voyage to Suratte by Olof Toreen and An Account of Chinese Husbandry by*

Captain Charles Gustavus Eckeberg, translated from the German by John Reinhold Forster, F. A. S. London. London: Benjamin White at Horace's Head, in Fleet Street.

Outram, Dorinda. 1996. "New Spaces in Natural History." In *Cultures of Natural History*, edited by Nicholas Jardine, James Secord, and Emma Spary, 249–65. Cambridge: Cambridge University Press.

Paine, S. C. M. 2003. *The Sino-Japanese War of 1894–1895: Perceptions, Power, and Primacy.* New York: Cambridge University Press.

Pan Jixing. 潘吉星. 1984. "Tan *zhiwuxue* yi ci zai Zhongguo he Riben de youlai 談"植物學"一詞在中國和日本的由來" (The origin of the word *zhiwuxue* [botany] between China and Japan). *Da ziran tansuo* 大自然探索, no. 3: 167–72.

Pan Xun 潘洵 and Peng Xinglin 彭星霖. 2007. "Kangzhan shiqi da houfang keji shiyede "Nuoya fang zhou" Zhongguo Xibu Kexueyuan yu da houfang Beibei keji wenhua zhongxinde xingcheng 抗戰時期大後方科技事業的'諾亞方舟' 中國西部科學院與大後方北碚科技文化中心的形成" (The West China Academy as the "Noah's Ark" of the technology sector in the rear areas during the Anti-Japanese War period and the emergence of the Beibei Science and Technology Center). *Xinan Daxue xuebao (shehui kexue ban)* 西南大學學報 (社會科學版) 33, no. 6: n.p.

Pang, Alex Soojung-Kim. 1997. "Visual Representation and Post-Constructivist History of Science." *Historical Studies in the Physical and Biological Sciences* 28, no. 1: 139–71.

Pavord, Anna. 2005. *The Naming of Names: The Search for Order in the World of Plants.* London: Bloomsbury Publishing.

Peake, Cyrus H. 1934. "Some Aspects of the Introduction of Modern Science into China." *Isis* 22, no. 1: 173–219.

Peer, Ralph S. 1949. "Newly Discovered Chinese *Reticulata* for the Southern California Camellia Garden." *Southern California Camellia Society Bulletin* 11, no. 2: 9–11.

———. 1950. "New Varieties of *Camellia reticulata*." *The Camellian* 1, no. 1: 12–13.

———. 1951. "New Varieties of *Camellia reticulata* Established in California." *Journal of the Royal Horticultural Society* 76, no. 8: 301–7.

Perkins, Franklin. 2005. "Following Nature with Mengzi or Zhuangzi." *International Philosophical Quarterly* 45, no. 3: 327–40.

Perry, Elizabeth J. 1980. *Rebels and Revolutionaries in North China, 1845–1945.* Stanford, CA: Stanford University Press.

Petiver, James. 1702a. "A Description of Some Coralls, and Other Curious Submarines Lately Sent to James Petiver, Apothecary and Fellow of the

Royal Society, from the Philippine Isles by the Reverend George Joseph Camel; As Also an Account of Some Plants from Chusan an Island on the Coast of China. Collected by Mr James Cuninghame, Chyrugeon & F. R. S." *Philosophical Transactions of the Royal Society* 23: 1419–29.

———. 1702b. "Part of two Letters to the Publisher from Mr. James Cunningham, F. R. S. and Physician to the English at Chusan in China, giving an account of his Voyage thither to the Island of Chusan, of the several sorts of Tea, of the Fishing, Agriculture of the Chinese, &c. with several Observations not hitherto taken notice of." *Philosophical Transactions of the Royal Society* 23: 1201–9.

Phillips, Andrew. 2011. "Saving Civilization from Empire: Belligerency, Pacifism and the Two Faces of Civilization during the Second Opium War." *European Journal of International Relations* 18, no. 1: 5–27.

Piementel, Juan. 2009. "Baroque Natures: Juan E. Nieremberg, American Wonders, and Preterimperial Natural History." In *Science in the Spanish and Portuguese Empires, 1500–1800*, edited by Daniela Bleichmar, Paula de Dos, Kristin Huffine, and Kevin Sheehan, 93–114. Stanford, CA: Stanford University Press.

Pim, S. 1966. *The Wood and the Trees: A Biography of Augustine Henry*. London: Macdonald.

Platt, Stephen R. 2012. *Autumn in the Heavenly Kingdom: China, the West, and the Epic Story of the Taiping Civil War*. New York: Alfred A. Knopf.

Pollini, Jacques. 2013. "Bruno Latour and the Ontological Dissolution of Nature in the Social Sciences: A Critical Review." *Environmental Values* 22, no. 1: 25–42.

Porter, David. 1996. "Writing China: Legitimacy and Representation, 1606–1773." *Comparative Literature Studies* 33, no. 1 (East-West Issue): 98–122.

———. 2001. *Ideographia: The Chinese Cipher in Early Modern Europe*. Stanford, CA: Stanford University Press.

Prince, Linda M. 2007. "A Brief Nomenclatural Review of Genera and Tribes in Theaceae." *Aliso: A Journal of Systematic and Evolutionary Botany* 24, no. 1: 8.

Pu Jingzi 撲静子. 1719. *Cha hua pu* 茶花譜 (The compendium of camellias). In *Xu si ku quan shu* 續修四庫全書 (The expanded complete library in four sections), vol. 1116, 603–18. 1995 reprint, Shanghai: Guji Chubanshe.

Puga, Rogério Miguel. 2012. "The First Museum in China: The British Museum of Macao (1829–1834) and Its Contribution to Nineteenth-Century British Natural Science." *Journal of the Royal Asiatic Society*, 3rd series, 22, no. 3/4 (July & October): 575–86.

Qian Chongshu 錢崇澍. 1915. "Ping *Bowuxue zazhi* 評博物學雜誌" (An assessment of the *Journal of Natural History*). *Kexue* 科學 (Science) 1, no. 5: 605–6.

Qiao Feng 喬峯. 1931. "Xin jin qushide er zhiwuxue jia 新近去世的二植物學家" (On the recent passing of two botanists). *Ziran jie* 自然界 6, no. 2: 16–17.

Qin Renchang 秦仁昌. 1931. "Di Wu Ci Shijie Zhiwuxue Hui ji shi lu 第五次世界植物學會紀事錄" (A report on the Fifth International Botanical Congress). *Kexue* 科學 (Science) 15, no. 3: 448–50.

———. 1939. "Yunnan san da minghua 雲南三大名花" (Three famous flowers of Yunnan). *Xinan bianjiang* 西南邊疆, no. 6: 25–29.

———. 1940. "Qiaozhi Fulaisi (George Forrest) shi yu Yunnan xibu zhiwuzhi fuyuan 乔治福莱斯 (George Forrest) 氏與雲南西部植物之富源" (George Forrest and Yunnan's rich botanical resources). *Xinan bianjiang* 西南邊疆, no. 9: 1–24.

Qin Renchang and Hu Xiansu 胡先驌. 1930–37. *Zhongguo juelei zhiwu tupu* 中國蕨類植物圖譜 (Icones filicum Sinicarum) [Illustrated manual of the cryptogams of China]. 4 vols. Beiping: Jingsheng Shengwu Diaochasuo (Fan Memorial Institute of Biology).

Qin Shao. 2004. "Exhibiting the Modern: The Creation of the First Chinese Museum, 1905–1930." *China Quarterly* 179 (September): 684–702.

Raj, Kapil. 2007. *Relocating Modern Science: Circulation and the Construction of Knowledge in South Asia and Europe, 1650–1900*. Basingstoke, UK: Palgrave Macmillan.

Rakow, Donald A., and Sharon A. Lee. 2015. "Western Botanical Gardens: History and Evolution." *Horticultural Review* 43: 269–310.

Raven, Charles Earle. 2009. *John Ray, Naturalist: His Life and Works*. Cambridge: Cambridge University Press.

Raven, Peter, and Zack Loehle, eds. 2017. "John Bradby Blake Special Part." *Curtis's Botanical Magazine* 34, no. 4.

Ray, John. 1686–1704. *Historia plantarum; species hactenus editas insuper multas noviter inventas & descriptas complectens*. 3 vols. London: M. Clark, H. Fairthorne.

Reardon-Anderson, James. 1991. *The Study of Change: Chemistry in China, 1840–1949*. Cambridge and New York: Cambridge University Press.

Reilly, Thomas H. 2014. *The Taiping Heavenly Kingdom: Rebellion and the Blasphemy of Empire*. Seattle and London: University of Washington Press.

Ren Hongjun 任鴻雋. 1915. "Shuo Zhongguo wu kexuezhi yuanyin 說中國無科學之原因" (On the reasons that China does not have science). *Kexue* 科學 (Science) 1, no. 1: 8–13.

———. 1983. "Zhongguo Kexueshe she shi jianshu 中國科學社社史簡述" (Brief history of the Science Society of China). *Zhongguo keji shiliao* 中國科技史料 4, no. 1: 2–13.

Reynolds, David C. 1991. "Redrawing China's Intellectual Map: Images of Science in Nineteenth-Century China." *Late Imperial China* 12, no. 1: 27–61.

Rhoads, Edward J. M. 2005. "In the Shadow of Yung Wing." *Pacific Historical Review* 74, no. 1: 19–58.

Rieu, Alain-Marc, ed. 2009. *Knowledge and Society Today (Multiple Modernity Project)*. Lyon: HAL Archives Ouvertes.

Rinaldi, Bianca Maria. 2016. *The "Chinese Garden in Good Taste": Jesuits and Europe's Knowledge of Chinese Flora and Art of the Garden in the 17th and 18th Centuries*. Munich: Martin Meidenbauer.

Ritvo, Harriet. 1992. "At the Edge of the Garden: Nature and Domestication in Eighteenth- and Nineteenth-Century Britain." *Huntington Library Quarterly* 55, no. 3: 363–78.

———. 1997. *The Platypus and the Mermaid and Other Figments of the Classifying Imagination*. Cambridge, MA; and London: Harvard University Press.

Roberts, J. A. G. 1997. "L'image de la Chine dans l'*Encyclopédie*." *Recherches sur Diderot et sur l'Encyclopédie* 22: 87–108.

Robinet, Isabelle. 1997. *Taoism: Growth of a Religion*. Stanford, CA: Stanford University Press.

Roetz, Heiner. 2010. "On Nature and Culture in Zhou China." In *Concepts of Nature: A Chinese-European Cross-Cultural Perspective*, edited by Hans Ulrich Vogel, Günter Dux, and Mark Elvin, 198–219. Leiden, Netherlands: Brill.

Ryle, G. 1938. "Categories." *Proceedings of the Aristotelian Society*, n.s., 38 (1937–38): 189–206.

Salama-Carr, Myriam, ed. 2007. *Translating and Interpreting Conflict*. Approaches to Translation Studies series, vol. 28. Amsterdam and New York: Rodopi.

San, Duanmu. 2007. *The Phonology of Standard Chinese*. Oxford: Oxford University Press.

Saunders, Gill. 1995. *Picturing Plants: An Analytical History of Botanical Illustration*. London: Victoria and Albert Museum; and Berkeley: University of California Press.

Save the Redwood League. n.d. "Dawn Redwoods." Accessed July 5, 2019. www.savetheredwoods.org/redwoods/dawn-redwoods/.

Schäfer, Dagmar. 2011. *The Crafting of the 10,000 Things.* Chicago: University of Chicago Press.

———. 2017. "Thinking in Many Tongues: Language(s) and Late Imperial China's Science." *Isis* 108, no. 3: 621–28.

Schafer, Edward H. 1965. "The Idea of Created Nature in T'ang Literature." *Philosophy East and West* 15, no. 2 (April): 153–60.

Schiebinger, Londa. 2004. *Plants and Empire: Colonial Bioprospecting in the Atlantic World.* Cambridge, MA: Harvard University Press.

Schiebinger, Londa, and Claudia Swan, eds. 2007. *Colonial Botany: Science, Commerce, and Politics in the Early Modern World.* Philadelphia: University of Pennsylvania Press.

Schlegel, Gustave. 1894. "Scientific Confectionary." *T'oung Pao* 5, no. 2: 147–51.

Schmalzer, Sigrid. 2007. "On the Appropriate Use of Rose-Colored Glasses: Reflections on Science in Socialist China." *Isis* 98, no. 3: 571–83.

Schneider, Laurence A. 1982. "The Rockefeller Foundation, the China Foundation, and the Development of Modern Science in China." *Social Science and Medicine* 16: 1217–21.

———. 2003. *Biology and Revolution in Twentieth-Century China.* Lanham, MD: Rowman and Littlefield.

———. 2012. "Michurinist Biology in the People's Republic of China, 1948–1956." *Journal of the History of Biology* 45, no. 3: 525–56.

Schwartz, Benjamin I. 1985. *The World of Thought in Ancient China.* Cambridge, MA: Belknap Press of Harvard University Press.

Science. 1943. "Joint Annual Meeting of Scientific Societies in China." 98 (2548): 380–81.

Scott, James C. 1998. *Seeing like a State: How Certain Schemes to Improve the Human Condition Have Failed.* New Haven, CT: Yale University Press.

Secord, James A. 2004. "Knowledge in Transit." *Isis* 95, no. 4: 654–72.

Seemann, Bertold. 1859. "Synopsis of the Genera *Camellia* and *Thea.*" *Transactions of the Linnean Society of London* 22, no. 4 (November): 337–52.

Shan, Patrick Fuliang. 2018. *Yuan Shikai: A Reappraisal.* Vancouver: University of British Columbia Press.

Shen Dongmei 沈冬梅. 2007. *Cha yu Song dai shehui shenghuo* 茶與宋代社會生活 (Tea and social life in the Song Dynasty). Beijing: China Social Sciences Press.

Shen, Grace Yen. 2009. "Taking to the Field: Geological Fieldwork and National Identity in Republican China." *Osiris* 24, no. 1: 231–52.

Shen Gua 沈括. [1088?] *Mengxi bitan* 夢溪筆談 (Dream pool essays). Citations refer to the 2015 edition, Shanghai: Guji Chubanshe.

Shen, Guowei. 2001. "The Creation of Technical Terms in English-Chinese Dictionaries from the Nineteenth Century." In *New Terms for New Ideas: Western Knowledge and Lexical Change in Late Imperial China*, edited by Michael Lackner, Iwo Amelung, and Joachim Kurtz, 287–304. Leiden, Netherlands; Boston; and Cologne: Brill.

———. 2009. "Yan Fu yu *kexue* 嚴復與"科学"" (Yan Fu and the term *kexue*). *Journal of East Asian Cultural Interaction Studies* 東アジア文化交渉研究 4: 143–62.

———. 2014. "Science in Translation: Yan Fu's Role." In *Science and Technology in Modern China, 1880s–1940s*, edited by Jing Tsu and Benjamin A. Elman, 93–113. Leiden, Netherlands; and Boston: Brill.

Shen Huanzhang 沈煥章. 1917. "Guoxue yu kexue qi xingzhi you wu maodun 國學與科學其性質有無矛盾" (Is there a contradiction between Chinese studies and science?). *Yunnan xueshu piping chu zhoukan* 雲南學術批評處週刊 (The Yunnan scholarly and critical weekly) 7 (January): 7.

Shennong bencao jing 神農本草經 (The classical pharmacopoeia of the heavenly husbandman). 2nd to 1st centuries BCE.

Shi, Aijie. 2017. "Nationalizing Science in Republican China: Academia Sinica's Policy on International Biological Expeditions." Master's thesis, Dietrich School of Arts and Sciences, University of Pittsburgh.

Shi Yang 施楊. 2014. "Guoli Bianyiguan yu Minguo kexue jishuyu tongyi 國立編譯館與民國科學術語統一" (The National Bureau of Translation and the standardization of scientific terminology during the early republican period). *Zhonghua kejishi xuehui xuekan* 中華科技史學會學刊 19, no. 12: 9–14.

Silbey, David. 2012. *The Boxer Rebellion and the Great Game in China*. New York: Hill and Wang.

Silverman, Milton. 1948. "Science Makes a Spectacular Discovery." *San Francisco Chronicle*, March 25, 1948.

———. 1990. *The Search for the Dawn Redwoods*. San Francisco: published by the author.

Simpson, Niki, and Peter G. Barnes. 2008. "Photography and Contemporary Botanical Illustration." *Curtis's Botanical Magazine* 25, no. 3: 258–80.

Sivin, Nathan. 1973. "Preface." In *Chinese Science: Explorations of an Ancient Tradition. Compiled in Honor of the Seventieth Birthday of Joseph Needham, F. R. S.*, edited by Shigeru Nakayama and Nathan Sivin, xi–xxxvi. Cambridge, MA: MIT Press.

Skott, Christina. 2014. "Expanding Flora's Empire: Linnaean Science and the Swedish East India Company." In *The Routledge History of Western Empires*, edited by Robert Aldrich and Kirsten McKenzie, 238–53. Abingdon, UK; and New York: Routledge.

Spary, E. C. 2000. *Utopia's Garden: French Natural History from Old Regime to Revolution.* Chicago and London: University of Chicago Press.

Spence, Jonathan D. 1980. *To Change China: Western Advisers in China.* New York and London: Penguin Books.

———. 1984. *The Memory Palace of Matteo Ricci.* New York: Elisabeth Sifton Books/Viking.

———. 1996. *God's Chinese Son: The Taiping Heavenly Kingdom of Hong Xiuquan.* New York and London: W. W. Norton & Company.

Stemerding, Dirk. 1993. "How to Make Oneself Nature's Spokesman? A Latourian account of Classification in Eighteenth- and Early Nineteenth-Century Natural History." *Biology and Philosophy* 8, no. 2: 193–223.

Sterckx, Roel. 2008. "The Limits of Illustration: Animalia and Pharmacopeia from Guo Pu to *Bencao Gangmu*." *Asian Medicine* 4; 357–94.

Stevens, P. F. 1984. "Metaphors and Typology in the Development of Botanical Systematics, 1690–1960, or the Art of Putting New Wine in Old Bottles." *Taxon* 33, no. 2: 169–211.

Strasser, Bruno J. 2012. "Collecting Nature: Practices, Styles, and Narratives." *Osiris* 27: 303–40.

Stross, Randall. 1986. *The Stubborn Earth: American Agriculturists on Chinese Soil.* Berkeley: University of California Press.

Sun Yingbao 孫英寶, Hu Zonggang 胡宗崗, Ma Lüyi 馬履一, and Fu Dezhi 傅德志. 2010. "Feng Chengru yu shengwu huitu 馮澄如與生物繪圖" (Feng Chengru and biological illustration). *Guangxi zhiwu* 廣西植物 (Guihuaia) 30, no. 2: 152–54.

Sun Yingbao 孫英寶, Ma Lüyi 馬履一, and Qin Haining 覃海寧. 2008. "Zhongguo zhiwuxuehua xiao shi 中國植物學畫小史" (A brief history of botanical scientific illustration in China). *Zhiwu fenlei xuebao* 植物分類學報 (Journal of systematics and evolution) 46, no. 5: 772–84.

Sun Yingbing 孫應冰 and Ma Haoyuan 馬浩原. 2017. "Shengwu xueshuyu chuangyi shijuexiade Qing dai lai Hua chuanjiaoshi shengwuxue yizhu ji qi kexue jiazhi 生物學術語創譯視角下的清代來華傳教士生物學譯著及其科學價值" (Research on biological translations by missionaries coming to China during the Qing dynasty and their scientific value from the perspective of the creation of translated biological terminology). *Jiangsu Kejidaxue xuebao (shehui kexue ban)* 江蘇科技大學學報 (社會科學版) 17, no. 2: 57–62.

Synge, P. M., ed. 1950. *Camellias and Magnolias: Report of the Conference Held by the Royal Horticultural Society, April 4–5, 1950.* London: Royal Horticultural Society.

Szczesniak, Boleslaw. 1955. "The Writings of Michael Boym." *Monumenta Serica* 14 (1949–1955): 481–538.

Tai, Li-chuan. *See* Dai Lijuan.

Tan Jiazhen 談價偵, ed. 1985. "Zhongguo xiandai kexuejia zhuan 中國現代科學家傳" (Biographies of modern Chinese scientists). Changsha: Hunan Kexue Chubanshe.

Tang Jin 唐進. 1931. "Shanxi sheng zhiwu caiji ji 山西省植物採集記" (Account of a botanical tour in Shanxi Province). *Jingsheng Shengwu Diaochasuo huibao* 靜生生物調查所彙報 (Bulletin of the Fan Memorial Institute of Biology) 2, no. 4 (April 20).

Tao Hongjing 陶弘景. 500 CE? *Shennong bencao jing jizhu* 神農本草經集注 (Collected commentaries on the materia medica).

Teng, Ssu-yü. 1971. *The Taiping Rebellion and the Western Powers: A Comprehensive Survey.* Oxford: Clarendon Press.

Theophrastus. 1916. *Enquiry into Plants, Volume I*, translated by Arthur F. Hort. Loeb Classical Library 70. Cambridge, MA: Harvard University Press.

Thiselton-Dyer, W. T. 1880. *The Botanical Enterprise of the Empire.* London: Her Majesty's Stationery Office.

Thomas, Adrian P. 2006. "The Establishment of Calcutta Botanic Garden: Plant Transfer, Science and the East India Company, 1786–1806." *Journal of the Royal Asiatic Society*, 3rd series, 16, no. 2: 165–77.

Thopa Xin 拓跋欣. 480–535 CE. *Wei Wang hua mu zhi* 魏王花木志 (A book of flowers and trees by the Prince of Wei), using the reconstructed text in *Shuo Fu* 說郛 (Environs of fiction) ca. 1368 CE, 66–71. Accessed July 11, 2017, at the Chinese Text Project. http://ctext.org/library.pl?if=gb&file=66723&page=67.

Tie Zheng 鐵錚 and Wang Xiqun 王希群. 2007. "Jiemi shuisha faxian guochengde mimi 解密水杉發現過程的秘密" (Solving the mystery of the discovery of *Metasequoia glyptostroboides*). *Kexue shibao* 科學時報 (blog), http://tech.sina.com.cn/d/2007-07-06/10531602233.shtml.

Trehane, Jennifer. 1998. *Camellias: The Complete Guide to Their Cultivation and Use.* Portland, OR: Timber Press.

Tsoong, K. K. *See* Zhong Guanguang.

T'sou, Benjamin K. 2001. "Language Contact and Lexical Innovation." In *New Terms for New Ideas: Western Knowledge and Lexical Change in Late Imperial China*, edited by Michael Lackner, Iwo Amelung, and Joachim Kurtz, 35–56. Leiden, Netherlands; and Boston: Brill.

Tsu, Jing, and Benjamin A. Elman, eds. 2014. *Science and Technology in Modern China, 1880s–1940s.* Leiden, Netherlands; Boston; and Cologne: Brill.

Tsukagoshi, Minoru, Arata Momohara, and Mutsuhiko Minaki. 2011. "*Metasequoia* and the Life and Work of Dr. Shigeru Miki." *Japanese Journal of the History of Botany* 19, no. 1–2: 1–14.

Tucker, Mary Evelyn, and John Berthrong, eds. 1998. *Confucianism and Ecology: The Interrelation of Heaven, Earth, and Humans*. Cambridge, MA: Harvard University Press.

Twitchett, D. 1983. *Printing and Publishing in Medieval China*. New York: Frederic C. Beil.

United States Department of Agriculture. 1955. *Plant Inventory No. 156: January 1 to December 1, 1948*. Washington, DC: United States Department of Agriculture.

Unschuld, Paul, ed. 1989. *Approaches to Traditional Chinese Medical Literature*. Dordrecht, Netherlands: Kluwer Academic Publishers.

Valder, Peter. 1999. *The Garden Plants of China*. Portland, OR: Timber Press.

Van Dyke, Paul A. 2007. *The Canton Trade: Life and Enterprise on the China Coast, 1700–1845*. Hong Kong: Hong Kong University Press.

Van Houten, Richard. 1988. "Nature and *Tzu-jan* in Early Chinese Philosophical Literature." *Journal of Chinese Philosophy* 15: 33–49.

Vittinghoff, Natasha. 2004. "Social Actors in the Field of New Learning in Nineteenth Century China." In *Mapping Meanings: The Field of New Learning in Late Qing China*, edited by Michael Lackner and Natascha Vittinghoff, 75–118. Leiden, Netherlands; and Boston: Brill.

Vogel, Hans Ulrich, Günter Dux, and Mark Elvin. 2010. *Concepts of Nature: A Chinese-European Cross-Cultural Perspective*. Papers from a conference on "Understanding Nature in China and Europe until the Eighteenth Century: A Cross-Cultural Project," held in Rheine, Westphalia, Germany, March 22–25, 2000. Leiden, Netherlands: Brill.

Wakeman, Frederic, Jr. 1978. "The Canton Trade and the Opium War." *The Cambridge History of China, Vol. 10: Late Ch'ing, 1800–1911*, 163–212. Cambridge: Cambridge University Press.

———. 1997. "A Revisionist View of the Nanjing Decade: Confucian Fascism." *China Quarterly*, 150 (June): 395–432.

Waley, Arthur. (1958) 2005. *The Opium War through Chinese Eyes*. London and New York: Routledge.

Walker, Mark, ed. 2002. *Science and Ideology: A Comparative History*. London and New York: Routledge.

Wan Kaiyuan 萬開元. 2020. *Zhong Guanguang hua zhuan* 鐘觀光畫傳 (A biography of Tsoong Kuankuang). Hangzhou: Zhejiang Renmin Meishu Chubanshe.

Wang, Frédéric. 2009. "The Relationship between Chinese Learning and Western Learning according to Yan Fu (1854–1921)." In *Knowledge and Society Today (Multiple Modernity Project)*, edited by Alain-Marc Rieu, 47–56. Lyon: HAL Archives Ouvertes.

Wang Guoquan 汪國權. 1985. "Zhongguo zhiwuyuande youlai he fazhan 中國植物園的由來和發展" (The origin and development of botanical gardens in China). *Zhongguo keji shiliao* 中國科技史料 6, no. 4: 10–17.

———. 1986. "Yuantao shenchude jiyi—ji Lushan Zhiwuyuande dansheng 綠濤深處的記憶— 記廬山植物園的誕生" (Memories welling up from deep within: remembering the birth of the Lushan Botanical Garden). *Zhongguo yuanlin* 中國園林, no. 3: 31–36.

Wang Hongkai 王宏凱. 1998. "Qingmo 'Xuexiao Jiaokeshu Weiyuanhui' shilüe 清末 "學校教科書委員會"史略" (A brief history of the late Qing "Textbook Commission"). *Shoudu Shifandaxue xuebao (shehui kexue ban)* 首都師範大學學報 (社會科學版), no. 3: 75–80.

Wang Hongxia 王紅霞. 2009. "Wan Qing kexueshuyu fanyi—yi Fulanya wei shidian 晚清科學術語翻譯— 以傅蘭雅為視點" (Translation of technical terminology during the late Qing, viewed from the perspective of John Fryer). *Fujian luntan (sheke jiaoyu)* 福建論壇 (社科教育), no. 2: 104–6.

Wang, Hui. 2011. "The Concept of 'Science' in Modern Chinese Thought." *Journal of Modern Chinese History* 5, no. 1: 45–67.

Wang Jie 王介. 1220. *Lü chanyan bencao* 履巉嚴本草 (The cliff walker's herbal). Digital copy in black and white. Accessed June 7, 2019, at the Internet Archive. https://archive.org/search.php?query=creator%3A%22%E9%A1%8C%28%E5%AE%8B%29%E7%8E%8B%E4%BB%8B%E6%92%B0%22.

Wang Qi 王圻 and Wang Siyi 王思義, eds. 1609. *San cai tu hui* 三才 圖會 (The universal encyclopedia). Accessed November 1, 2017, at the Chinese Text Project. https://ctext.org/library.pl?if=gb&res=3438.

Wang Qizhen 王啟振 and Wang Siming 王思明. 2014. "Zailun Hu Xiansu xuanzhi Lushan Zhiwuyuande dongyin 再論胡先驌選址廬山植物園的動因" (A further discussion on Hu Xiansu's motives for selecting the location of the Lushan Botanical Garden). *Zhongguo kejishi zazhi* 中國科技史雜誌 35, no. 2: 204–6.

Wang Rongbao 汪榮寶 and Ye Lan 葉瀾. 1903. *Xin er ya* 新爾雅 (The new literary expositor). Shanghai: Wenming Shujü.

Wang Xirong 王細榮. 2013. "Jindai kexue Zhongguohuade shijianzhe: Yu Heqin 近代科學中國化的實踐者: 虞和欽" (Yu Heqin, a facilitator of the sinicization of modern science). *Ziran bianzhengfa tongxun* 自然辯證法通訊 (Journal of the dialectics of nature) 35, no. 4: 99–107.

Wang Yangzong 王揚宗. 1988. "Jiangnan Zhizao Jü Fanyiguan shilüe 江南製造局翻譯館史略" (A brief history of the Translation Department of the Jiangnan Arsenal). *Zhongguo keji shiliao* 中國科技史料 3: 65–74.

———, ed. 2009. *Jindai kexue zai Zhongguode chuanbo* 近代科學在中國的傳播 (The dissemination of modern science in China). Jinan: Shandong Jiaoyu Chubanshe.

Wang Yanzu 王彥祖. 1916. "Zhiwu putong ming yu Lading kexue ming duizhao biao 植物普通名與拉丁科學名對照表" (A comparative table of plants with their vernacular and scientific Latin names). *Kexue* 科學 (Science) 2, no. 12: 1341–44.

Wang Yonghou 王永厚. 1994. "*Jiuhuang Bencao* de banben yuanliu "救荒本草" 的版本源流" (The sources of editions of *Treatise on Plants for Use in Emergency*). *Zhongguo nongshi* 中國農史, no. 3: 117–18.

Wang Zhenru 汪振儒, Liang Jiamian 梁家勉, and Wang Zongxun 王宗訓, eds. 1994. *Zhongguo zhiwuxue shi* 中國植物學史 (The history of botany in China). Beijing: Kexue Chubanshe.

Wang Zifan 王子凡, Zhang Mingmei 張明妹, and Dai Silan 戴思蘭. 2009. "Zhongguo gudai júhua pulu cunshi xianzhuang ji zhuyao neirongde kaozheng 中國古代菊花譜錄存世現狀及主要內容的考證" (The status of extant classical Chinese monographs on the chrysanthemum and a study of their main contents). *Ziran kexueshi yanjiu* 自然科學史研究, no 1: n.p.

Wang, Zuoyue. 2002. "Saving China through Science: The Science Society of China, Scientific Nationalism, and Civil Society in Republican China." In "Science and Civil Society," special issue, *Osiris*, 2nd series, 17: 291–322.

———. 2007. "Science and the State in Modern China." *Isis* 98, no. 3: 558–70.

———. 2015. "The Chinese Developmental State during the Cold War: The Making of the 1956 Twelve-Year Science and Technology Plan." *History and Technology* 31, no. 3: 180–205.

Ward, Julian. 2001. *Xu Xiake (1587–1641): The Art of Travel Writing*. Abingdon, UK; and New York: Routledge.

Watson, Philip. 2004. "Famous Gardens of Luoyang by Li Gefei." *Studies in the History of Gardens and Designed Landscapes* 24, no. 1: 38–54.

Watt, Sir George. 1907. "Tea and the Tea Plant." *Journal of the Royal Horticultural Society* 32: 64–96.

Wen Changbin 溫昌斌. 2005. "Zhongguo Kexueshe wei tongyi keji yiming er jinxingde gongzuo 中國科學社為統一科技譯名而進行的工作" (Work carried out by the Science Society of China to standardize translations of technical vocabulary). *Kexue jishu yu bianzhangfa* 科學技術與辯證法 22, no. 5: 86–92.

Weston, Timothy B. 1998. "The Formation and Positioning of the New Culture Community, 1913–1917." *Modern China* 24, no. 3: 225–84.

———. 2002. "The Founding of the Imperial University and the Emergence of Chinese Modernity." In *Rethinking the 1898 Reform Period: Political and*

Cultural Change in Late Qing China, edited by Rebecca Karl and Peter Zarrow, 99–123. Harvard East Asian Monographs 214. Cambridge, MA; and London: Harvard University Press.

Wilkinson, Anne. 2006. *The Victorian Gardener: The Growth of Gardening and the Floral World*. Stroud: History Press.

Wilson, Ernest Henry, and Charles Sprague Sargent. 1913. *A naturalist in western China, with vasculum, camera, and gun; being some account of eleven years' travel, exploration, and observation in the more remote parts of the Flowery kingdom, by Ernest Henry Wilson . . . with an introduction by Charles Sprague Sargent*. London: Methuen & Co. Ltd.

Winstanley, David, 1996. *A Botanical Pioneer in Southwest China: Experiences and Impressions of an Austrian Botanist During the First World War*. Chippenham, UK: Antony Rowe Ltd. www.paeo.de/h1/hand_maz/pioneer/ootitel.html.

Witteven, Joeri. 2016. "Suppressing Synonymy with a Homonym: The Emergence of the Nomenclatural Type Concept in Nineteenth Century Natural History." *Journal of the History of Biology* 49: 135–89.

Wong, Lawrence Wang-Chi. 2004. "Beyond Xin Da Ya: Translation Problems in the Late Qing." In *Mapping Meanings: The Field of New Learning in Late Qing China*, edited by Michael Lackner and Natascha Vittinghoff, 239–64. Leiden, Netherlands; and Boston: Brill.

———. 2007. "Translators and Interpreters During the Opium War between Britain and China (1839–1842)." In *Translating and Interpreting Conflict*, edited by Myriam Salama-Carr, 41–57. Approaches to Translation Studies series, vol. 28. Amsterdam and New York: Rodopi.

Wright, David. 1995. "Careers in Western Science in Nineteenth-Century China: Xu Shou and Xu Jianyin." *Journal of the Royal Asiatic Society*, 3rd series, 5, no. 1 (April): 49–90.

———. 1996. "John Fryer and the Shanghai Polytechnic: Making Space for Science in Nineteenth-Century China." *British Journal for the History of Science* 29, no. 1 (March): 1–16.

———. 1998. "The Translation of Modern Western Science in Nineteenth-Century China, 1840–1895." *Isis* 89, no. 4: 653–73.

———. 2000. *Translating Science: The Transmission of Western Chemistry into Late Imperial China, 1840–1900*. Leiden, Netherlands: Brill.

———. 2001. "Yan Fu and the Tasks of the Translator." In *New Terms for New Ideas: Western Knowledge and Lexical Change in Late Imperial China*, edited by Michael Lackner, Iwo Amelung, and Joachim Kurtz, 235–55. Leiden, Netherlands: Brill.

Wright, Mary Clabaugh. 1978. *China in Revolution: The First Phase, 1900–1913*. New Haven, CT: Yale University Press.

Wu Bingxin 吳冰心. *See* Wu Jiaxu 吳家煦.

Wu Deling 吳德鄰. 2008. "Jinian Chen Huanyong yuanshi 紀念陳煥鏞院士" (Remembering academician Chen Huanyong). In "Zhongguo Kexueyuan Huanan Zhiwuyuan 80 zhounian yuanqing wencui 中國科學院華南植物園 80 週年園慶文萃" (Festschrift in honor of the 80th anniversary of the Chinese Academy of Sciences Huanan Botanical Garden). Guangzhou: Chinese Academy of Sciences Huanan Botanical Garden.

Wu Heng 吳衡. 1987. "Shaan Gan Ning bianqude ziran ziyuan diaocha gongzuo 陝甘寧邊區的自然資源調查工作" (Surveys of the natural resources of the Shaan[xi] Gan[su] Ning[xia] border region). *Zhongguo keji shiliao* 中國科技史料 8, no. 5: 3–13.

Wu Jiarui 吳家睿. 1989. "Jingsheng Shengwu Diaochasuo jishi 靜生生物調查所紀事" (A chronicle of the Fan Memorial Institute of Biology). *Zhongguo keji shiliao* 中國科技史料 10, no. 1: 26–36.

Wu Jiaxu 吳家煦 [Wu Bingxin 吳冰心]. 1914a. "*Bowuxue zazhi* xuli 博物學雜誌序例" (Preface to the *Journal of Natural History*). *Bowuxue zazhi* 博物學雜誌 1, no. 1: 1–5.

———. 1914b. "Diaocha. Jiangsu zhiwu bolüe 調查. 江蘇植物博略" (Survey: an extensive study of the flora of Jiangsu). *Bowuxue zazhi* 博物學雜誌 1, no. 1: 135–44.

———. 1915a. "Zhiwu biaoben zhizuofa 植物標本製作法" (The methodology of preparing plant specimen vouchers). *Zhonghua xuesheng jie* 中華學生界 1, no. 3: 1–10.

———. 1915b. "Zhiwu caiji fa 植物采集法" (The methodology of plant collecting). *Zhonghua xuesheng jie* 中華學生界, 1, no. 1: 1–9.

———. 1915c. "Zhiwu jizai fa 植物記載法" (The methodology of recording plants). *Zhonghua xuesheng jie* 中華學生界 1, no. 4: 1–10.

———. 1915d. "Zhiwu jizai fa—xu 植物記載法(續)" (The methodology of recording plants (cont'd.)). *Zhonghua xuesheng jie* 中華學生界 1, no. 12: 11–32.

Wu Qijun 吳其濬. 1848. *Zhiwu mingshi tukao* 植物名實圖考 (Research on the illustrations, realities, and names of plants). Using Guangxu 光緒 6 (1880) reprint of the Daoguang 道光 28 (1848), Taiyuan Fu Shu edition.

Wu Yuandi 吳元滌. 1914. "Niushoushan lüxing caiji ji 牛首山旅行採集記" (An account of a botanizing trip to Ox Head Mountain). *Bowuxue zazhi* 博物學雜誌 1, no. 1: 99.

———. 1918. "Zhongshan Linggusi lüxing caiji ji 鍾山靈谷寺旅行採集記" (Account of a botanizing trip to the Linggu Temple at Zhongshan). *Kexue* 科學 (Science) 3, no. 9: 1029–32.

Wu Zhengyi 吳徵鎰. 1953. "Zhongguo zhiwuxue lishi fazhande guocheng he xiankuang 中國植物學歷史發展的過程和現況" (The process of historical development of Chinese botany and its current status). *Kexue tongbao* 科學通報, no. 1: 12–20.

Wu, Zhengyi and S. C. Chen. 1959–2004. *Zhongguo zhiwu zhi* 中國植物志 (Flora Reipublicae Popularis Sinicae). 80 volumes. Beijing: Science Press.

Wu, Zhengyi, and Peter Raven, eds. 1994–2013. *Flora of China*. Beijing: Science Press; and St. Louis: Missouri Botanical Garden. Accessed at www.efloras.org/florataxon.aspx?flora_id=2&taxon_id=105380.

Wu Zixiu 吳子修. 1924. "Yantai caiji jilüe 煙臺採集紀略" (A brief account of a botanizing trip to Yantai). *Bowuxue zazhi* 博物學雜誌 1, no. 2: 121–23.

Xiang, Lanxin. 2003. *The Origins of the Boxer War: A Multinational Study*. London: Routledge.

Xiao Lei 肖蕾. 2014. "Minguo shiqide Zhongguo Zhiwuxue Hui 民國時期的中國植物學會" (The Botanical Society of China During the Period of the Republic of China). *Hebei Beifangxueyuan xuebao (shehui kexue ban)* 河北北方學院學報 (社會科學版) 30, no. 3: 41–45.

Xiao, Xiaosui. 2004. "The 1923 Scientific Campaign and Dao discourse: A Cross-Cultural Study of the Rhetoric of Science." *Quarterly Journal of Speech* 90, no. 4: 469–92.

Xie Zhaozhe 謝肇淛. [1608?]. *Wu za zu* 五雜組 (Fivefold miscellany). Citations refer to the 2001 edition, Shenyang: Liaoning Jiaoyu Chubanshe.

Xie Zhensheng 謝振聲. 1989. "Shanghai Kexue Yiqiguan yu 'Kexue Shijie' 上海科學儀器館與'科學世界'" (The Shanghai Scientific Instruments Academy and "Science World"). *Zhongguo keji shiliao* 中國科技史料 10, no. 2: 61–66.

Xin, Tong, Jan de Riek, Huijun Guo, Devra Jarvis, Lijuan Ma, and Chunlin Long. 2015. "Impact of traditional culture on *Camellia reticulata* in Yunnan, China." *Journal of Ethnobiology and Ethnomedicine*, no. 11: 74.

Xing Shuzhi 幸樹幟. 1928. "Guangxi zhiwu caiji jilüe 廣西植物採集紀略" (An account of a botanical collecting expedition to Guangxi). *Ziran kexue* 自然科學 1, no. 1: 123–29.

Xu, Congrong, Hu Haisheng, Wu Zhangwen, Zheng Yanping, and Zhong Linsheng. 2009. "The Ideal Landscape and Its Ecological Planning Approach in Mount Lushan National Park, China." *Journal of Forestry Research* 20, no. 3: 270–84.

Xu Guangtai 徐光台. 2011. "Li Madou yu Xie Zhaozhe 利瑪竇與謝肇淛" (Matteo Ricci and Xie Zhaozhe). *Qinghua xuebao* 清華學報 41, no. 2: 259–97.

Xu Guoli 徐國利. 1998. "Guanyu 'Kangzhan shiqi gaoxiao neiqian' de jige wenti 關於'抗戰時期高校內遷'的幾個問題." (Some questions regarding the internal displacement of higher education institutions during the Anti-Japanese War). *Kangri Zhanzheng yanjiu* 抗日戰爭研究 2: 1–16.

Xu Jun 許鈞. 1998. *Fanyi sikao lu* 翻譯思考錄 (A record of reflections on translation). Hankou: Hubei Jiaoyu Chubanshe.

———. 2003. *Fanyi lun* 翻譯論 (On translation). Wuhan: Hubei Jiaoyu Chubanshe.

Xu Xiake 徐霞客. 1776. *Xu Xiake you ji* 徐霞客遊記 (The travels of Xu Xiake). Citations refer to the 1996 reprint of the 1928 edition by Ding Wenjiang 丁文江, Beijing: Shangwu Yinshuguan.

Xu Zaifu 許再富. 2011. "Cai Xitao shengping jieshao 蔡希陶生平介紹" (An introduction to the life of Cai Xitao). In "Zhiwu ziyuan kechixu liyong ji Cai Xitao xueshu sixiang yantaohui lunwen zhaiyao 植物資源可持續利用暨蔡希陶學術思想研討會論文摘要" (Selected papers from the symposium on Cai Xitao's thinking and the sustainability of botanical resources), edited by Zhongguo Kexueyuan Xishuangbanna Redai Zhiwuyuan 中國科學院西雙版納熱帶植物園 (Chinese Academy of Sciences Xishuangbanna Tropical Botanical Garden, 3–11. Menglun: Xishuangbanna Tropical Botanical Garden.

Xu Zaifu 許再富 and Pei Shengji 裴盛基. 1985. "Cai Xitao 蔡希陶" (Cai Xitao). In *Zhongguo xiandai kexuejia zhuan* 中國現代科學家傳 (Biographies of modern Chinese scientists), edited by Tan Jiazhen 談價偵, 439–52. Changsha: Hunan Kexue Chubanshe.

Xun Zi 荀子 (The book of Master Xun). ca. 241 BCE. Siku quan shu edition. Accessed December 20, 2018, at China-America Digital Academic Library (CADAL). https://archive.org/details/cadal?and%5B%5D=荀子&sin=.

Xunzi. 2014. *Xunzi: The Complete Text*. Translated by Eric L. Hutton. Princeton, NJ: Princeton University Press.

Yang, Chen-pang. 2017. "From Modernizing the Chinese Language to Information Science: Chao Yuen Ren's Route to Cybernetics." In "Linguistic Hegemony in the History of Science," special focus issue, *Isis* 108, no. 3: 553–80.

Yang, Guangjun 楊光駿. 1928. "Suzhou Nü Zhong Fu Xiao liu nian ji caiji zhiwu 蘇州女中附校六年級採集植物" (Sixth graders from the Primary School of the Suzhou Women's Middle School on a botanical collecting trip). *Tuhua shibao* 圖畫時報, no. 433: 1.

Yang Tsui-hua. 1991. *Patronage of Science: The China Foundation for the Promotion of Education and Culture*, translated by Chi-Chu Chen and Yu-wen Su. Monograph series no. 65. Taipei: Academia Sinica Institute of Modern History.

Ye Lan 葉瀾. 1898. *Zhiwuxue gelüe* 植物學歌略 (A verse primer of botany). Shanghai: publisher not known.

Yen, Hsiao-pei. 2015. "From Paleoanthropology in China to Chinese Paleoanthropology: Science, Imperialism and Nationalism in North China, 1920–1939." *History of Science* 53, no. 1: 21–56.

Yu Dejun 俞德濬. 1934. "Sichuan zhiwu caiji ji 四川植物採集記" (A record of a botanical collecting expedition to Sichuan). Pts. 1 and 2. *Zhongguo zhiwuxue zazhi* 中國植物學雜誌 1, no. 3: 325–44; 1, no. 4: 442–64.

———. (Yü, T. T.). 1947. *The Garden Camellias of Yunnan* [unpublished typescript]. Kunming: Kunming Institute of Botany Archives. Facsimile in Frank Griffin 1964, 27–108.

———. (Yü, T. T.). 1950. "*Camellia reticulata* and Its Garden Varieties." In *Camellias and Magnolias: Report of the Conference Held by the Royal Horticultural Society, April 4–5, 1950*, edited by P. M. Synge, 13–26. London: Royal Horticultural Society.

———. 1981. "Jingsheng Shengwu Diaochasuo 靜生生物調查所" (The Fan Memorial Institute of Biology). *Zhongguo keji shiliao* 中國科技史料, no. 4: 84–85, 42.

———. 1985. "Hu Xiansu 胡先驌" (Hu Xiansu). In *Zhongguo xiandai kexuejia zhuan* 中國現代科學家傳 (Biographies of modern Chinese scientists), edited by Tan Jiazhen 談價偵, 70–85. Changsha: Hunan Kexue Chubanshe.

Yu Heyin 虞和寅. 1903. "Zhiwuxue lüe shi 植物學略史" (A brief history of botany). *Kexue shijie* 科學世界 1, no. 1: 15–19.

Yü, T. T. *See* Yu Dejun.

Yu Yifei 於一飛 and Chen Jinzheng 陳錦正. 1985. "Zhong Guanguang 鐘觀光" (Zhong Guanguang). In *Zhongguo xiandai kexuejia zhuan* 中國現代科學家傳 (Biographies of modern Chinese scientists), edited by Tan Jiazhen 談價偵, 1–10. Changsha: Hunan Kexue Chubanshe.

Yuncker, T. G. (1930) 1931. "The Fifth International Botanical Congress." *Proceedings of the Indiana Academy of Sciences* 40: 61–66.

Zeller, Suzanne E. 1987. *Inventing Canada: Early Victorian Science and the Idea of a Transcontinental Nation*. Toronto and Buffalo: University of Toronto Press.

Zeng Zhaolun 曾昭掄. 1936. "Zhongguo kexue huishe gaishu 中國科學會社概述" (A survey of scientific associations in China). *Kexue* 科學 (Science) 20, no. 10: 798–843.

Zhang Dawei 張大為, Hu Dexi 胡德熙, and Hu Dekun 胡德焜. 1995. "Hu Xiansu wencun 胡先驌文存" (The writings of Hu Xiansu). Nanchang: Jiangxi Gaoxiao Chubanshe.

Zhang Fan 張帆. 2009. "Cong 'gezhi' dao 'kexue': wan Qing xueshu tixide guodu yu bieze 從 "格致" 到 "科學": 晚清學術體系的過渡與別擇 (1895–1905 年)" (From *gezhi* [natural studies] to *kexue* [science]: the transitions and distinctions made by intellectuals during the late Qing (1895–1905)). *Xueshu yanjiu* 學術研究, no. 12: 102–14.

———. 2016. "Minchu guoxue yanjiu zhong 'kexue' fanshide bianqian—yige gainianshide kaocha 民初國學研究中 "科學" 範式的變遷— 一個概念史的考察" (The transformation of the "science" paradigm in Chinese studies in the early period of republican China: an exploration of the concept's history). *Jindai shi yanjiu* 近代史研究, no. 5: 125–42.

Zhang Fang 張鈁2015. "*Bencao yuanshi* de shengwu tuxiang liubian ji qishi 本草原始的生物圖像流變及啟示" (Changes in and influence of the images in the *Sources for Materia Medica*). *Ziran kexue shi yanjiu*自然科學史研究 34, no. 3: 279–93.

Zhang Hanliang 張漢良. 2011. "Ruidian zhiwuxue jia Lin'nai yu cha ye de xi chuan 瑞典植物學家林奈與茶葉的西傳" (The Swedish botanist Linnaeus and the arrival of tea in the West). *Min shang wenhua yanjiu* 閩商文化研究, no. 2: 59–64.

Zhang, He. 2018. "The Symbol of the Spread of Modern Western Botany into China: Chih-wu Hsüeh, an Unconventional Translation in the Late Qing Dynasty." *Protein & Cell* 9, no. 6: 511–15.

Zhang Hongda. *See* Chang Hung Ta.

Zhang Jian 張劍. 2002. "Chuantong yu xiandai zhi jian—Zhongguo Kexueshe lingdao qun ti fenxi 傳統與現代之間— 中國科學社領導群體分析" (Between past and present: an analysis of the leadership of the Science Society of China). *Shilin* 史林, no. 1: 83–93.

———. 2003. "Minguo kexue shetuan yu shehui bianqian 民國科學社團與社會變遷" (Scientific associations and a changing society during the republican period). *Shilin* 史林, no. 5: 64–74.

Zhang Jianhong 張建紅 and Zhao Yulong 趙玉龍. 2004. "Minguo chuniande *Zhiwuxue da cidian* 民國初年的植物學大辭典" (*The Dictionary of Botany* from the early years of the republic). *Huaxia wenhua* 華夏文化, no. 4: 55.

Zhang Junli 張君勱 and Ding Wenjiang 丁文江. 1924. *Kexue yu rensheng guan* 科學與人生觀 (Science and an outlook on life). Shanghai: Taidong Shujü.

Zhang Longxi. 1988. "The Myth of the Other: China in the Eyes of the West." *Critical Inquiry* 15, no. 1: 108–31.

Zhang Mengwen 張孟聞. 1934a. "Guonei kexue: Zhongguo Kexueshe Shengwu Yanjiusuo zhanlanhui ji 國內科學: 中國科學社生物研究所展覽會記" (Science in China: a report on the exhibition at the Science Society of China's Biological Laboratory). *Kexue* 科學 (Science) 18, no. 4: 550–70.

———. 1934b. "Zhongguo Kexueshe Shengwu Yanjiusuo zhanlanhui ji 中國科學社生物研究所展覽會記" (A report on the exhibition at the Science Society of

China's Biological Laboratory). *Kexue huabao* 科學畫報 (La science populaire/Popular science) 1, no. 19: 722–31.

Zhang Shufeng 章樹楓. 1928. "Chuan Kang zhiwu biaoben caiji ji 川康植物標本採集記 (2)" (An account of a botanical collecting expedition to Sichuan and Tibet, part 2.) *Kexue* 科學 (Science) 13, no. 11: 1513–21.

Zhang Tingmao 張廷茂. 2004. *Ming Qing shiqi Aomen haishang maoyi shi* 明清時期澳門海上貿易史 (History of Macao's maritime trade during the Ming-Qing period). Macao: Ao Ya Kan Publishing Co.

Zhang Wei 張衛 and Zhang Ruijian 張瑞堅. 2010. "*Bencao yuanshi* banben kaocha 本草原始版本考察" (A study of the publishing history of the *Sources for Materia Medica*). *Zhongyi wenxian zazhi* 中醫文獻雜誌, no. 1: 2–5.

Zhang Yaqun 張亞群. 2005. "Fei keju yu xueshu zhuanxing—lun Qing mo kexuejiaoyude fazhan 廢科舉與學術轉型— 論清末科學教育的發展" (The abolition of the civil service exams and the transformation of scholarship: on the development of science education during the late Qing.) *Dongnan xueshu* 東南學術, no. 4: 47–52.

Zhao Jian 趙堅. 2006. "Xiandai Hanyude kouyu wailai yu 現代漢語的口語外來語" (Japanese loanwords in modern Chinese). *Journal of Chinese Linguistics* 43, no. 2: 306–27.

Zheng Jinsheng 鄭金生. 1989. "The Collation and Annotation of the Rare Book *Lü Chanyan Bencao*: A Medical Literature Research Project." In *Approaches to Traditional Chinese Medical Literature*, edited by Paul Unschuld, 29–39. Dordrecht: Kluwer Academic Publishers.

———. 2003. "Mingdai huajia caise bencao chatu yanjiu 明代畫家彩色本草插圖研究" (A study of polychrome materia medica illustrations by Ming artists). *Xin shixue* 新史學 14, no. 4: 65–120.

———. 2018. "Observational Drawing and Fine Art in Chinese Materia Medica Illustration." In *Imagining Chinese Medicine*, edited by Vivienne Lo, Penelope Barrett, David Dear, Lu Di, Lois Reynolds, and Dolly Yang, 151–60. Leiden, Netherlands; and Boston: Brill.

Zheng Qiao 鄭樵. 1150. *Tupu lüe* 圖譜略 (A brief account of the graphic arts). Juan 72 in Tong Zhi 通志 (Historical collections). Siku quan shu edition. Accessed June 7, 2019, at the Chinese Text Project. https://ctext.org/library.pl?if=gb&file=4248&page=16.

Zhengyi tongbao 正藝通報. 1902. "Zhong xi gezhi yitongkao 中西格致異同考" (An investigation into the differences between Chinese and Western science). No. 10: 7–8.

Zhong Guanguang 鐘觀光. 1921–22. "Lüxing caiji ji 旅行採集記" (Report of a botanical collecting expedition). *Dixue zazhi* 地學雜誌 (Earth sciences journal), 7 installments from vol. 11, no. 7 to vol. 12, no. 5.

———. (K. K. Tsoong). 1932a. "Lun zhiwuxue bang mingzhi zhongyao ji qi zhengli fa 論植物邦名之重要及其整理法" (On the importance of Chinese names for plants, with suggestions for a proper system of nomenclature). *Sinensia* 3, no. 1 (July): 1–8.

———. 1932b. "Kexue Mingci Shencha Hui zhiwuxue mingci shenchaben zhiwu shumingzhi xiaoding 科學名詞審查會植物學名詞審查本植物屬名之校訂" (Criticisms and corrections of the botanic terms as they appear in the third report of the General Committee on Scientific Terminology). *Sinensia* 3, no. 1: 9–52.

———. 1933. "Zhong Ri liang guo zhiwuxuejiazhi yiqu 中日兩國植物學家之異趣" (The different interests of Chinese and Japanese botanists). *Kexuede Zhongguo* 科學的中國 1, no. 4: 5–7.

Zhongguo Kexueyuan Huanan Zhiwu Yanjiusuo 中國科學院華南植物研究所 (South China Institute of Botany editorial committee). 1996. *Chen Huanyong jinian wenji* 陳煥鏞紀念文集 (Collected papers in memory of Chen Huanyong). Guangzhou: Huanan Zhiwuyanjiusuo.

Zhongguo Kexueyuan Xishuangbanna Redai Zhiwuyuan 中國科學院西雙版納熱帶植物園 (Chinese Academy of Sciences Xishuangbanna Tropical Botanical Garden), ed. 2011. "Zhiwu ziyuan kechixu liyong ji Cai Xitao xueshu sixiang yantaohui lunwen zhaiyao 植物資源可持續利用暨蔡希陶學術思想研討會論文摘要" (Selected papers from the symposium on Cai Xitao's thinking and the sustainability of botanical resources). Menglun: Xishuangbanna Tropical Botanical Garden.

Zhou Cheng 周程. 2009. "*Kexue* yi ci bing fei cong riben yinjin "科學"一詞並非從日本引進" (The term *kexue* was certainly not an import from Japan). *Zhongguo wenhua yanjiu* 中國文化研究 (Summer): 182–87.

Zhou Cheng 周程 and Ji Xiufang 紀秀芳. 2009. "Jiujing shei zai Zhongguo zui xian shiyongle *kexue* yi ci? 究竟誰在中國最先使用了 "科學" 一詞? (Who really was the first to use the term *kexue* in China?). *Ziran bianzhengfa tongxun* 自然辯證法通訊 (Journal of the dialectics of nature) 31, no. 4: 93–98.

Zhou Wu 周武. 2016. "Du Yaquan yu *Dongfang zazhi* 杜亞泉與 "東方雜誌"" (Du Yaquan and *Eastern Miscellany*). *Kexue* 科學 (Science), no. 2: 41–42.

Zhu Fajian 朱發建. 2003. "Qing mo guoren kexueguande yanhua: cong 'gezhi' dao 'kexue' de ciyi kaobian 清末國人科學觀的演化：從 "格致" 到 "科學" 的詞義考辨" (The evolution of the Chinese view of science in the late Qing dynasty: a semantic study on Chinese expression of modern science and technology). *Hunan Shifan Daxue shehui kexue xuebao* 湖南師範大學社會科學學報 32, no. 4: 79–82.

Zhu Jiachun 朱嘉春. 2016. "任鴻雋的科學翻譯活動" (Ren Hongjun's scientific translations). *Zhongguo keji fanyi* 中國科技翻譯 29, no. 3: 54–57.

Zhu Jifa 竺濟法. 2016. "Yi 'shu tu' queren 6000 nian rengong zaipei chashugen ke xin ma? 以 "熟土" 確認 6000年人工栽培茶樹根可信嗎?" (Is it credible to use evidence of "worked soils" to claim that tea has been cultivated for 6,000 years?). *Nongye kaogu* 農業考古, no. 2: 214–16.

Zhu Xiao 朱橚. 1406. *Jiuhuang bencao* 救荒本草 (Treatise on plants for use in emergency). References are to the 1988 facsimile of the Ming Jiajing 4 (1525) edition in *Zhongguo gudai banhua congkan*中國古代版畫叢刊 (Ancient Chinese woodblock prints series), vol. 2: 372–73. Shanghai: Shanghai Guji Chubanshe.

Zhu Zongyuan 朱宗元 and Liang Cunzhu 梁存柱. 2005. "Zhong Guanguang Xianshengde zhiwu caiji gongzuo jian ji woguo di yige zhiwu biaobenshide jianli 鐘觀光先生的植物採集工作兼記我國第一個植物標本室的建立" (The story of Mr. Zhong Guanguang's botanical collecting expeditions is also the story of setting up China's first herbarium). *Beijing Daxue xuebao (ziran kexue ban)* 北京大學學報 (自然科學版) (Acta scientarum naturalium Universitatis Pekinensis) 41, no. 6: 825–32.

Zi, Zhongyun. 1995. "The Rockefeller Foundation and China." *American Studies Quarterly* 2: n.p.

Zong Zeya 宗澤亞. 2012. *Qing Ri Zhanzheng* 清日戰爭, *1894–1895* (The Sino-Japanese War, 1894–1895). Beijing: Shijie Tushu Chubanshe.

Zou Bingwen 鄒秉文. 1916. "Wan Guo zhiwuxue ming ding mingli 萬國植物學名定名例" (Regulations from the International Code of Plant Nomenclature). *Kexue* 科學 (Science) 2, no. 9: 1015–29.

Zou Xiangui 鄒賢桂. 1998. "Zhiwu kexuehua zai zhiwuxue yanjiu zhongde yiyi 植物科學畫在植物學研究中的意義" (The importance of botanical illustration in botanical research). *Guangxi zhiwu* 廣西植物 8, no. 3: 29–32.

Zou Yigui 鄒一桂. 1756. *Xiao Shan hua pu* 小山畫譜 (Xiao Shan's treatise on painting). Using *Si ku quan shu* 四庫全書 (The complete library in four sections), 1773 edition.

Zuo Jinglie 左景烈. 1934. "Hainan Dao 海南島" (Hainan Island). 2 pts. *Zhongguo zhiwuxue zazhi* 中國植物學雜誌 1, no. 1: 59–89; 1, no. 2: 215–35.

Zuo, Ya. 2018. *Shen Gua's Empiricism*. Cambridge, MA: Harvard-Yenching Institute Monograph series 113.

Zürcher, E. 2007. *The Buddhist Conquest of China*. 3rd ed. Leiden, Netherlands: Brill.

Zurndorfer, Harriet. 2009. "China and Science on the Eve of the 'Great Divergence,' 1600–1800: A Review of Recent Revisionist Scholarship in Western Languages." *History of Technology* 29 (January): 81–101.

INDEX

A

Academia Sinica (Zhongyang Kexueyuan), 40, 115–16, 154, 156; National Natural History Museum, 160; wartime relocation of, 154, 156; Zoological and Botanical Institute, 161

Account of the Camellias of Central Yunnan, An (Dian zhong cha hua ji), 18

Agricultural Association (Nongxue Hui), 37, 142

agriculture, 37, 108, 161, 169; *Agriculture News* (Nong xue bao), 142; in higher education, 36

algae and fungi (*jun zao*), 130, 143; as a category of living being, 45, 108, 110

Amoy (Xiamen), 39, 127, 128*fig.*

An Ji, 113

Analytical Dictionary of Characters (Shuo wen jie zi), 112

anatomy (plant), 30*fig.*, 55, 69–71, 116, 129; reproductive, 35, 57, 70*fig.*, 102, 106; pistils, 69, 73, 102; stamens, 10, 21, 73; wood anatomy, 151, 156

Anhui, 85, 92

animals, 29, 100, 111, 114, 147; a class of living beings, 35, 42, 44–45, 54, 106–8; in the Confucian classics, 105–6; in scientific illustration 130, 133

anthologies, 63; in works on plants, 10, 15–19, 119, 122–23

apricot (*Prunus mume*), 10, 209*n*15

Arabic, 56, 67

Arnold Arboretum, 27, 176; Chen Huanyong at, 86; Hu Xiansu at, 85, 166, 168

Assembly of Perfumes, The (Qun fang pu), 53, 108

associations, 29, 130, 140, 144–47, 149

azalea (*Azalea* spp.), 98, 156

B

bamboo (*zhu*), 46, 124

bandits and banditry, 27, 91, 158

Banks, Sir Joseph, 77

Batavia (Jakarta), 31, 63

Beibei, 146, 153–54

Beijing, 39, 168, 171; Beijing Teachers' College (Jing Shi Yixue Guan), 37; School of Combined Learning (Tongwen Guan), 57, 65; as seat of

Beijing (*continued*)
    government, 36, 83, 141. *See also* Beiping
Beiping, 149–50, 178; Beiping Natural History Museum (Beiping Tianran Bowuguan), 166; Japanese Occupation of, 85, 89, 136; National Academy of Beiping (Beiping Academy), 149, 152, 157. *See also* Beijing; Fan Memorial Institute of Biology; Peking University
Bing Zhi, 62, 113, 134, 178; co-director of the Fan Memorial Institute 41, 148, 150; founding member of the Science Society of China 7, 42, 147
biodiversity, 6, 28, 96, 137, 150
Biological Laboratory of the Science Society of China (Zhongguo Kexue She Shengwu Yanjiusuo), 8, 40, 147–48, 154; natural history exhibition (1933), 159, 162–63
biology, 42, 148, 150; biological illustration, 119, 133; in colleges and universities, 38–39, 84; in the popular press, 34
birds; a class of living things, 106–7; naming of, 100, 105–6
Blake, John Bradby, 30*fig.*, 127–28
Book of Poems (Shijing), 105
botanical artists: Bun-Ko, Doctor, 127, 128*fig.*; Cai Shou, 129–33, 132*fig.*, 165; Kuang Keren, 137–39, 138*fig.*; Mai Xiu (Māāk or Mauk Sow-u), 128. *See also* Feng Chengru
botanical gardens, 28, 158–59, 163–71; Buitzenborg (Bogor), 31; Calcutta, 164; of the College of Agriculture (Hangzhou), 84, 166; Hong Kong, 88, 164–66, 165*fig.*; *hortus botanicus* (Leiden), 164; Jardin des Plantes / Muséum d'Histoire Naturelle, 22, 31, 79, 164; Jiangsu Agricultural Academy arboretum (Nanjing), 166; Kew Gardens, 31, 79, 80–81, 129, 164; Lushan Arboretum and Botanical Garden, 85, 151, 167–72, 170*fig.*; Peradeniya (Ceylon), 164; Real Jardín Botánico (Madrid), 164; Royal Botanic Garden (Edinburgh), 128, 169; Specimen Garden at Sun Yat-sen University (Guangzhou), 149, 158, 166–67; Xishuangbanna Tropical Botanical Garden, 91. *See also* gardens
botanical illustration, 73, 104, 117; characteristics of, 117–19, 131, 134; commissioned by East India Company, 30*fig.*, 31, 127–29, 128*fig.*; drawn from life, 23, 120, 124, 131, 132*fig.*; drawn from vouchers in herbaria, 117; in premodern texts, 19, 119–27, 123*fig.*, 125*fig.*, 126*fig.*; technical innovations in, 118–19, 134–36; in translated texts, 35, 70, 70*fig.*, 129; using woodblock prints, 15, 120–21. *See also* botanical artists
botanical names, 84, 98–99, 131, 162; assigned by Westerners, 7, 22, 39, 80, 104; binomial system of, 73, 102, 108, 112–14; derived from Chinese names, 71–72, 104–5; standardization of, 72–74. *See also* naming; nomenclature; translation
Botanical Society of China (Zhongguo Zhiwuxuehui), 41, 105, 146–47, 172
botanists, 3–4, 75, 101, 140–41; biographies of, 82–91; global community of, 8, 53, 74, 88, 176; and the legacy of the past, 46–48, 53–54, 74, 77–79, 179–80; relations with

Japanese and Western botanists, 39–40, 81, 84; as reporters, 93–94, 97, 180; trained abroad, 10, 40, 116, 166; Western botanists in China, 3, 22–23, 75, 79–81, 94–98

botany, 7–8, 10, 68–69; a break with traditional knowledge or transition, 5, 21, 29–31, 42, 52–54, 179–80; economic botany, 80, 154–55, 164; as a field science, 75–77, 101, 180; and imperialism, 31–36, 51, 98, 104; and nationalism, 51–52, 75–76, 180–81; in schools and higher education, 28–29, 34–37, 109–10, 148–49, 170; translated as *zhiwuxue*, 46, 159; tropical botany, 91. See also ethnobotany; terminology; translation

Botany (*Zhiwuxue*) (translation of John Lindley's *Elements of Botany*), 7, 32, 35, 54, 70*fig.*, 178; introducing Linnaean classification, 104, 109; a source of new terminology, 46, 69–71. See also Li Shanlan; terminology; Williamson, Alexander

bowuxue (the broad learning of things), 29, 47–50, 129, 143

Boxer Rebellion (1899–1901), 36, 145, 148

Boym, Michał Piotr, 29, 79

Bretschneider, Emil, 22, 26, 59

Brewster, Christopher, 127–28, 128*fig.*

Brief Account of the Camellias of Southern Yunnan, A (Dian nan cha hua xiao zhi), 9–10, 180

Brief Account of the Graphic Arts, A (Tupu lüe), 119

Buddhism, 57–58, 95; classification of living beings in monasteries, 108; Jiayepo (Kāśyapa), 95; Mile (Maitreya), 95; temples as habitat for plants, 14, 23, 96–98. See also sutras

Bulletin of the Botanical Society of China (Zhongguo Zhiwuxue Huibao), 146

Bulletin of the Fan Memorial Institute of Biology, 137, 173–74, 178

Bulletin of the Geological Society of China, 176

C

Caesalpino, Andrea, 102

Cai Shou (Cai Xun), 129–33, 132*fig.*, 165

Cai Xitao, 25, 43–44, 82, 156, 179; biography of, 89–91, 90*fig.*

California: Descanso Gardens, 25; University of California (Berkeley), 66, 81, 85, 167, 176

Camellia (*shan cha*): *Camellia japonica*, 13, 21; *Camellia oleifera* (tea oil), 11; *Camellia reticulata*, 9, 10, 18, 22–26, 24*fig.*, 57; *Camellia sinensis*, (tea bush), 11–13, 21, 122, 123*fig.*; *mantuoluo*, 52; mountain tea flower (*shan cha hua*), 9–21; in premodern texts, 14–20, 16*fig.*, 17*fig.*, 52–53, 98, 121; *Swatea fl. rubro*, 13; taxonomic classification in Chinese, 56–57, 68; taxonomic relationship with tea, 11–14, 21; tsubakki, 11, 12*fig.*, 13

Canton, 11, 31, 79, 127–29. See also Guangzhou

cells, 35, 55; translated as *xibao* ("fine bladder" or "womb"), 70

Chaney, Ralph, 176–77, 179

change: cosmology of change 6, 42–43, 119, 179; *li* (order within change), 43, 101
Chen Fenghuai, 169–71
Chen Haozi, 122
Chen Huanyong (W. Y. Chun, Chun Woon-Young), 40, 87*fig.*, 134, 165; biography of, 86–88; director of Sun Yat-sen University's Institute of Agriculture, Forestry, and Botany, 149, 158, 161, 166–67; international networks, 74, 82, 149
Chen Rong, 166
Cheng Yaotian: *Notes on All the Arts* (Tong yi lu), 78
Chiang Kai-shek (Jiang Jieshi), 24, 40, 168, 169
Chickenfoot Mountain. *See* Jizu Shan
China Foundation for the Promotion of Education and Culture (Zhonghua Jiaoyu Wenhua Jijinhui), 41, 148–51, 153, 169
Chinese Academy of Sciences (CAS), 5, 28, 41, 105; absorbs existing research institutions, 6, 157, 171. *See also names of individual scientific institutions*
Chinese Journal of Botany (Zhongguo Zhiwuxue Zazhi), 146, 175
Chinese Scientific and Industrial Magazine (Gezhi huibian), 34, 65, 109, 129, 142
Chinese students abroad, 35, 71, 212n12; in Japan, 49, 149; in the United States, 38, 144–46
Chinese Students' World (Zhonghua xuesheng jie), 76
Chōei, Takano, 48
Chongqing, 27, 41, 91, 175; wartime seat of government, 41, 75, 153–54

chrysanthemum (*Chrysanthemum* spp.), 10; *Gao Song's Compendium on the Chrysanthemum* (Gao Song Ju Pu), 124, 125*fig.*
Chun Woon-Young (W. Y. Chun). *See* Chen Huanyong
Chusan (Zhoushan), 11
civil service examinations (*keju*), 34, 48–49, 51, 82; abolition of, 8, 36, 61
civil war, 24, 29, 89, 156, 171
Cixi (Empress Dowager), 36
Classical Pharmacopoeia of the Heavenly Husbandman, The (Shennong bencao jing), 107
classics, 8, 48, 54, 124, 130, 193; as a source for new terminology, 46–47, 58, 69–72. *See also* language; naming; nomenclature
classification, 4–6, 42, 44–47, 169, 178; grouping and splitting in, 100–103; Linnaean system of, 13, 104–5, 109, 114; in monasteries, 108; by morphology, 102, 106–7; premodern systems of, 54, 105–8; scientific systems of, 50, 71–72, 110, 163; by toxicity, 107. *See also* ordering; taxonomy
Classification of Materia Medica (Bencao gangmu), 5–6, 19, 55, 74, 107–8, 121
Cleyer, Andreas, 11, 12*fig.*
Cliff Walker's Herbal (Lü chanyan bencao), 120–21
collaboration, 141–44; with international peers, 39, 146, 178; in translations, 34, 58, 60, 62–63
collecting: by anonymous "native collectors," 80, 82, 178; botanical, 75–77, 110, 180; expeditions, 27, 80–81, 91–94, 149, 150–51; for

herbaria, 28, 147, 168; for Western gardens, 22–23, 32, 79–80. *See also* Cai Xitao; Chen Huanyong; Hu Xiansu; plant hunters; Zhong Guanguang

Communist Party of China, 28, 41, 154, 171

compendia, 10, 107, 120–21; *Gao Song's Compendium on the Chrysanthemum* (Gao Song Ju Pu), 124, 125*fig.*; *The Compendium of Camellias* (Cha hua pu), 15; *The Imperially Commissioned Compendium of Literature and Illustrations, Ancient and Modern* (Gu jin tu shu ji cheng), 18, 53

Compiled and Translated Encyclopedia for General Education, The (Bianyi putong jiaoyu baike quan shu), 71

Confucianism, 44, 47, 100, 163; Confucius, 105; neo-Confucianism, 59

Contributions from the Institute of Botany (Beiping yanjiuyuan zhiwuxue yanjiu congkan), 152

cosmology, 42–43, 103, 106

Critical Review (Xueheng), 85, 92

cryptogams (ferns), 110, 165; *Icones filicum Sinicarum* (Illustrated manual of the cryptogams of China), 134, 151

Cultural Revolution, 85, 88, 137, 166, 171

culture, 130–31, 167, 176; plants as cultural objects, 14–16, 45; of science, 38, 53, 181

Cuninghame, James, 11, 13, 79; botanical drawings commissioned by, 127

curricula, 36, 49, 54, 109–10, 170

cycles, 43–44, 45, 54, 179

D

d'Incarville, Père Pierre, 29, 79

Daléchamps, Jacques, 6

Dali (Bai Autonomous Prefecture), 93, 94, 97

Daoism, 43, 44

Darwin, Charles, 69

David, Père Armand, 3, 33; *Davidia involucrata* (handkerchief tree), 22

dawn redwood. *See Metasequoia glyptostroboides*

de Jussieu, Antoine Laurent, 5, 109

de Jussieu, Bernard, 79

deforestation, 166, 167–68

description: of Chinese flora by Europeans, 11–13, 22, 79; premodern description of plants, 9–10, 16–20, 53, 121–122, 123*fig.*; scientific, 20–21, 75, 50, 117, 147

Dictionary of Botany (Zhiwuxue da cidian), 7, 20, 55, 66, 73–74

Ding Wenjiang, 76

dissemination, 7, 65, 140, 147–48

DNA, 5, 13, 103

Dream Pool Essays (Mengxi bitan), 77

Du Yaquan, 45, 49, 54, 109; editor-in-chief of the *Dictionary of Botany*, 68, 73; editor of *Science World*, 83, 146

Dutch East India Company (Vereenigde Oostindische Compagnie), 13, 31, 164

E

Earth Sciences Journal (Dixue zazhi), 39, 142

East India Company (British), 23, 31, 79, 127–28; and the tea trade, 11, 80

East India Trading Company (Swedish), 79
Eastern Miscellany (Dongfang zazhi), 37, 61, 65, 74, 142
ecology, 28, 69, 156, 176
Edinburgh, 24, 166; Royal Botanic Garden, 129, 169
education, 142, 148, 151; Ministry of, 36, 66, 83, 85; reform of, 28, 34, 61
Educational Association of China (Yizhi Shuhui), 61, 70, 129
Elements of Botany. See *Botany (Zhiwuxue)*
Elman, Benjamin A., 179
Elvin, Mark, 78
encyclopedias, 5, 29, 46, 105
English, 33–34, 68; a language of science, 56, 68, 74, 113; publications in, 41, 88, 146
epistemology, 4–5, 39, 53, 143, 179
ethics, 44, 47, 59
ethnobotany, 103. See also traditional knowledge
Europe: botanical illustration in, 30*fig.*, 31, 118, 127–28, 128*fig.*; Chinese plants in, 11, 29, 79–80, 112, 151; origins of scientific botany in, 6, 50, 76–77
evolution, 69, 130, 162
exhibitions, 159; Biological Laboratory exhibition (Nanjing, 1934), 162–63; Guangzhou City Exposition (1933), 149, 161; Nanyang Exposition (1910), 161

F

family (botanical, *ke*), 57, 74, 110–12, 116; concept introduced in *Botany*, 70–72, 104–5

Fan Chengda, 18
Fan Memorial Institute of Biology (Jingsheng Shengwu Diaochasuo), 85, 149–51, 157, 173; Bing Zhi, co-director, 41, 150; Hu Xiansu, co-director, 3, 41, 86*fig.*; wartime relocation to Kunming, 41, 154–156. See also Yunnan Institute of Agriculture, Forestry, and Botany (Yunnan Nonglin Zhiwu Yanjiusuo)
Fan Yuanlian (Fan Jingsheng), 149–50
Fang Shumei, 9–10, 23, 180
Feng Chengru, 119, 133–36, 135*fig.*, 151; death by suicide, 171; first drawing of *Metasequoia glyptostroboides*, 137, 174*fig.*, 178
Feng Shike, 18
Feng Youlan (Fung Yu-Lan), 43, 52
ferns. See cryptogams
field assistants, 33, 82, 83*fig.*, 90*fig.*; 92–93, 98; "Old Yao," 88; recorded as "native collector" in herbaria, 80, 178; Wang Hanchen, 80, 82; Zhao Chengzhang, 82
field sciences, 7, 75–77, 148, 180, 205n1; reports, 93–94, 110
First World War, 37, 39, 96
Five Phases (*wu xing*), 101, 106, 119; as five elements (earth, fire, metal, water, wood), 43
Fivefold Miscellany (Wu za zu), 78
flora: botanical names of, 73, 104; of China, 3, 28, 51, 74, 178; inventories of, 33, 149, 155; reference works on, 53, 81, 118, 134, 155; symbols of the nation, 7, 115, 180; Western interest in, 22, 79–80, 127, 164
Flora of China, 5, 53, 77, 133, 139
Flora Reipublicae Popularis Sinicae, 105

Flora sinensis (1656), 29, 79
flowers: anatomy of, 35, 69–71, 70*fig.*; a class of living being, 42, 46, 106, 108; in Linnaean classification, 5, 109, 128; ornamentals, 32, 46, 80, 122. *See also* botanical illustration
Flowers and Trees of the King of Wei, The (Wei Wang hua mu zhi), 14
Ford, Charles, 164–65
forest management: in Guangdong, 149; in Yan'an, 155; in Yunnan, 156; at Lushan, 167–68
Forrest, George, 4, 23, 80
Fortune, Robert, 3, 23, 32, 80, 118
France, 33, 160; French (a language of the sciences), 56, 68, 146
Franchet, Adrien, 33
fruits (*guo*), 22, 67, 79, 93, 154; in plant classification, 42, 46, 107–8, 112; in botanical illustration, 30*fig.*, 127, 130, 131
Fryer, John, 63–66, 141–42; *Illustrated Botany* (Zhiwu tushuo), 35, 70; *Plants Illustrated and Explained*, 129
Fujian, 15, 84, 124
Furth, Charlotte, 76

G

Gan Duo, 177
Garden Camellias of Yunnan, The, 23, 24*fig.*
Garden of Solitary Pleasures (Dule Yuan), 163
gardens, 15, 22–26, 32, 93, 163; gardening manuals, 16*fig.*, 108, 122–23. *See also* botanical gardens

gazetteers, 93, 95*fig.*, 113, 131, 168; of Dali Prefecture, 94. *See also* geography
Gee, Nathaniel Gist, 39
genetics, 69, 134, 154
genus (*shu*), 57, 62, 74, 114; *Camellia*, 10–11; determining genus, 103–5, 110–12, 116, 118
geography, 5, 18, 58, 77, 99, 115. *See also* gazetteers
geology, 49, 75–76, 148
Germany, 59, 80, 146; German (a language of the sciences), 60, 68, 73, 74, 113
gewu zhizhi (the investigation of things and extension of knowledge), 29, 47–48, 101; *gezhixue* used for "natural studies," 129, 143; replaced by *kexue* (science), 48–50
Government University of Peking. *See* Peking University
grains (*gu*), 46, 78; a category of living beings, 6, 107, 109, 112
grasses (*cao*), 43, 46, 78; a category of living beings, 6, 107, 109, 112
Great Director of the Multitudes (Da Situ), 106
Great Learning, The (Da Xue), 47
Greek, 56, 60, 71
Gressitt, Linley, 81, 177
Groff, George Weidmann, 84
Guangdong, 84, 124, 130, 149, 158, 167
Guangxi, 39, 84, 93, 153, 160
Guangzhou, 11, 57, 88; City Exposition (1933), 149, 161. *See also* Canton
Guizhou, 18, 137
Guling (Kuling), 168, 170–71
Guo Moruo, 157
Guomindang (Kuomintang), 152, 170; Central Political Council of, 40
gymnosperms (*luo zi*), 110, 175

H

habitat, 68, 96, 106, 108, 167
Hainan, 79, 158, 162, 167, 181; expeditions to, 81, 86–87, 87*fig.*, 149, 151
Hangzhou, 83–84, 92, 120, 166
Harvard, 80, 168; Qian Chongshu at, 88, 110, 143. *See also* Arnold Arboretum
Henry, Augustine, 81, 88
herbaria, 38, 105, 115, 148, 162; in exchanges of plant material, 88, 149, 151; vouchers in, 76, 80, 112, 149, 159; in Western collections, 77, 85, 96, 112–13, 159
Heude, Père Pierre-Marie, 33, 159–60
History of Flowers, The (Hua Shi), 15, 16*fig.*, 123
Hong Kong, 25, 32, 63, 86–88; Hong Kong Botanical Gardens, 88, 164–66, 165*fig.*
Hooker, Sir Joseph, 77, 164
horticulture, 22, 31, 80, 103, 127; manuals of, 15, 29, 53, 118, 122
Hu Dekun, 171
Hu Xiansu (Hu Buzeng, Hu Hsen-Hsu), 23, 40, 112, 179; attacked by Red Guards, 85, 171; biography of, 84–86; co-director of the Fan Memorial Institute, 3, 86*fig.*, 89, 150–51; collecting in Zhejiang, 92–93; director of the Yunnan Institute of Agriculture, Forestry, and Botany, 41, 75, 155–56; editor of the *Chinese Journal of Botany* (Zhongguo Zhiwuxue Zazhi), 146; establishes Lushan Botanical Garden, 151, 167–69; and *Metasequoia glyptostroboides* 173–77; writing in literary Chinese, 62, 82, 92
Huang Yiren, 81
Hubei, 8, 88, 175; Hubei Self-Strengthening School (Hubei Ziqiang Xuetang), 34
humankind: and nature, 8, 44–45, 54, 103, 143
Hunan Normal College, 83
Hundred Days' Reform (1898), 36
Hydrangeaceae (Xiuqiuke), 72

I

Icones filicum Sinicarum (Illustrated manual of the cryptogams of China) (Zhongguo jue lei tupu), 134, 151
Icones plantarum Omeiensium (Illustrated plants of Emei) (Emei zhiwu tuzhi), 137
identification: of categories of living beings, 5, 44–46, 54, 169; and classification, 140, 159, 167, 180; of herbarium specimens, 77, 80, 117, 154; of newly discovered species, 22–23, 51, 72; in premodern texts, 84, 92, 112; from illustrations, 118, 121–24, 128
Illustrated Botany (Zhiwu tushuo), 35, 70
Illustrated Manual of Trees (Shumu tushuo), 134
Illustrated Materia Medica of Southern Yunnan (Diannan bencao tupu), 138, 138*fig.*, 156
Illustrated Pharmacopeia (Bencao tu jing), 120
Illustrated Plants of China (Zhongguo zhiwu tupu), 134

Imperial University of Peking. *See* Peking University
imperialism, 39; and botany, 31, 51, 98, 104, 164
Imperially Commissioned Compendium of Literature and Illustrations, Ancient and Modern, The (Gu jin tu shu ji cheng), 18, 53
India, 58, 67–68, 176; tea plantations in, 11, 80
indigenous, 139, 156; plant names, 22, 116
Inkstone Press (Mohai Shuguan), 32, 63, 129, 141
insects, 19, 133; a category of living being, 106–7
institutions, 8, 29, 40–41, 85; wartime relocation of, 144, 152–56. *See also* research; *and under names of institutions*
International Botanical Congress, 29, 40, 73, 104, 139; Fifth International Botanical Congress, 4, 74, 88, 166
international exchanges, 25, 88, 151, 164–65, 170
inventory, 10, 33, 101, 150, 180

J

Japan, 15, 36, 175; botanical works translated into Chinese, 37, 50; Chinese students in, 73, 81, 137, 146, 149; occupation of China, 41, 88, 141, 144, 153, 170; Sino-Japanese war (1894–95), 28, 34, 48, 62; source of scientific terminology, 48–49, 61, 71, 74, 112–13
Jardin des Plantes (known before 1793 as Jardin du Roi), 22, 31, 164
Jesuits, 30, 33, 80; Michał Piotr Boym, 29, 79; Giuseppe Castiglione, 125; Père Pierre d'Incarville, 29, 79; Père Jean-Marie Delavay, 33; dissolution of (1773), 30, 58; Père Pierre-Marie Heude, 33, 159–60; scientific translations, 47, 58, 63, 67; Pierre André Soulié, 33
Jizu Shan (Chickenfoot Mountain), 94–98, 95*fig.*
Jiang Menglin, 144
Jiang Tingxi, 18, 53
Jiang Zong, 14–15
Jiangnan (Lower Yangzi), 33, 62, 136
Jiangnan Arsenal (Jiangnan Zhizao Ju), 34, 64–66, 141
Jiangnan Arts Academy School of Biological Illustration (Jiangnan Meishu Zhuanmen Xuexiao Shengwuhua Zhuanxiuke), 136
Jiangsu, 81, 110, 134, 148; Jiangsu Higher Agricultural College (Jiangsu Jiazhong Nongye Xuexiao), 38, 166; Jiangsu Provincial Educational Association (Jiangsu Sheng Jiaoyu Hui), 114; Jiangsu Provincial First Normal College, 91
Jiangxi: Jiangxi College of Agriculture, 151, 167–68; Jiangxi Provincial Forestry School (Jiangxi Sheng Linxiao), 167; National Chung Cheng University, 85, 155
Jing, Tsu, 179
Jingsheng Shengwu Diaochasuo. *See* Fan Memorial Institute of Biology
Joint Scientific Terminology Commission (Kexue Mingci Shenchahui), 114
Journal of National Essence (Guocui xuebao), 129–33

Journal of Natural History (Bowuxue zazhi), 44, 110, 143–44
Journal of Pedagogy (Mengxue bao), 55
journals, 34, 66, 109, 160; in English, 41, 88, 146; professional journals, 40, 142, 158; scientific journals, 8, 37–38, 61, 143. *See also specific journal titles*

K

Kaempfer, Engelbert, 13
Kang Youwei, 49
Kew Gardens, 31, 79, 80–81, 129, 164
Kexue. See *Science*
knowledge, 4, 21, 105, 180; local, 82, 93, 116; about plants, 32, 47, 56, 115; production of, 8, 45, 54, 76–77; scientific, 131, 140, 148, 179; Western, 28, 58, 62–63. *See also* epistemology; traditional knowledge
Kuang Keren, 137–39, 138*fig.*
Kunming, 41, 75, 93, 155–56, 173; camellias in, 9, 23–24, 98; Heilong Tan district, 89, 155, 171; Kunming Institute of Botany (Kunming Zhiwu Yanjiusuo), 25
Kwangtung University, 88

L

laboratory, 149, 151, 156; laboratory sciences (in contrast to field sciences), 76–77, 158
lacquer (*Toxicodendron vernicifluum*), 156
landscape, 51, 93–99, 131, 158
language, 56–60, 67, 146, 178; *baihua* ("plain language"), 61, 203n9; literary (classical) Chinese, 82, 85, 92, 142–43; modernization and reform of, 60–61; official language (*guanhua*), 61; suitability of Chinese for scientific writing, 58–60; vernacular, 61–62, 72, 112, 181. *See also* classics; terminology; translation
Latin, 7, 29, 56, 147, 178; for botanical names, 68, 72–74, 99, 105, 112; questioning the use of, 39, 62, 90, 111, 113–14
Latour, Bruno, 5
Le Tianyu, 155
Leibniz, Gottfried Wilhelm, 59
lemon (*ningmeng*), 67, 132*fig.*
li (order within constant change), 43; *ziranzhi li* (self-generation's pattern), 47, 101
Li Hui-lin, 39
Lijiang, 96, 171; Lijiang Research Station, 156
Li Shanlan, 32, 46, 55, 63–65, 64*fig.*, 69–71
Li Shizhen, 19, 77, 113. See also *Classification of Materia Medica* (Bencao gangmu)
Li Yuying, 152
Li Zhongli, 107, 121
libraries, 76, 88, 105, 141, 157; National Library, Beijing, 148
Lindley, John. See *Botany* (Zhiwuxue)
Lingnan University (Canton Christian College), 38–39, 81, 84
Linnaeus (Linné), Carl, 13, 21–22, 179; classification system of, 5, 73, 104, 109, 114
Literary Expositor, The (Er ya), 106–7, 113, 124, 131
literature, 5, 89, 213, 180

Literature (Wenxue), 43, 89
Liu Dayou, 37
Liu Shen'e, 152
Liu Shoudan, 113
Liu Youtang (Y. T. Liu), 23–24
livestock, 106, 108
London, 31, 77, 80, 127, 151; London Missionary Society, 32, 63
Lu Di, 80
Lushan, 146, 159, 167; Lushan Arboretum and Botanical Garden, 85, 151, 157, 168–171, 170*fig.*
Lu Zuofu, 153

M

Macao, 29, 58, 63, 127
Madrid, 77, 164
Malacca, 63
Manchus, 28, 32, 57, 130; Manchuria, 175
Manual of Botany (John Balfour), 35, 70
manuals, 36, 109, 136; agricultural, 14, 53, 108, 118; horticultural, 15, 29, 53, 122–24
Mao Zedong, 62, 177; at Lushan, 171
mapping, 7, 75, 92
materia medica, 51, 53, 138, 163; illustrations in, 120–22, 123*fig.*; taxonomies in, 102, 107–8. *See also* medicinal plants; *and titles of individual works*
mathematics, 34, 47, 58, 142
Matsuda, Sadahisa, 81
Matsumura Jinzō, 37, 55, 109, 113
May 4th Movement, 39, 61, 133, 140
medicinal plants, 19–20, 46, 137–38, 154, 163; properties of, 55, 82, 122
Merrill, E. D., 84, 176–77, 179
Métailié, Georges, 46, 53, 69, 78

Metasequoia glyptostroboides (dawn redwood), 8, 137, 157, 173–78, 174*fig.*
microorganisms, 45
Miki Shigeru, 175
minerals, 19, 42, 44, 54, 107
ming. See tea
Ming dynasty, 29, 57–58, 93, 120, 163
minorities: ethnic, 14, 89, 137, 158, 177
Mirror of Flowers, The (Hua jing), 122
missionaries, 22, 33, 114, 141; as natural history collectors, 51, 80, 159–60; as translators, 28, 32, 34–35, 60, 62–65. *See also* Jesuits
modernization, 9, 38–39, 66, 99, 140, 145; language and, 60–61; and modernity, 50–52, 160
monographs, 5, 9, 14, 122–24, 159
morphology, 35, 55, 71, 134; in classification, 5, 102, 106
mountains, 93–94, 96, 106; as plant habitat, 46, 108, 168
museums, 7, 151, 158–61; Academia Sinica National Natural History Museum (Guoli Zhongyang Yanjiuyuan Ziran Lishi Bowuguan), 160; Beiping Natural History Museum (Beiping Tianran Bowuguan), 166; Musée de Zikawei (Musée Heude), 33, 159–60; Muséum National d'Histoire Naturelle, 33, 79, 160; Nanjing Natural History Museum, 147; Nantong Museum, 160; Natural History Museum (London), 13, 31; Shanghai Museum of the North China Branch of the Royal Asiatic Society, 160
myriad things, the (*wan wu*), 6, 42–43, 54, 76, 101; ordering of, 103, 105–8

N

naming, 7, 54, 73, 106, 109; correct naming 6, 100–101, 107, 163; of cultivars and varieties, 15–19, 23–24; indigenous names, 22, 92, 115–16; "nameless" (*wu ming*) plants, 133. *See also* botanical names; nomenclature; terminology

Nanjing, 64, 91, 148, 152, 161; Nanjing Higher Education Normal School (Nanjing Gaodeng Shifan Xuexiao), 134; Nanjing Natural History Museum, 147; Nanjing Teachers' College, 85; seat of the Nationalist government, 29, 40–41, 58, 114; University of, 88

National Central University, 175, 177

National Chung Cheng University, 85, 155

National Peking University. *See* Peking University

National Southeastern University (Guoli Dongnan Daxue), 39, 88, 134

nationalism, 39, 115, 180

Nationalist Party. *See* Guomindang

Natural Science Society of China (Zhonghua Ziran Kexue She), 146

nature, 4, 6, 42–45, 47, 54; natural history, 33, 50, 104, 159–60; natural history drawings, 129–33, 132*fig.*; natural sciences; 29, 140; natural theology, 69. *See also bowuxue* (the broad learning of things); *ziran*

Needham, Joseph, 53, 137, 154

networks, 4, 51, 140; research networks, 41, 77, 88, 164

new learning, 8, 34, 65, 77, 141. *See also* science; Western: sciences

New Literary Expositor (Xin er ya), 50, 55, 71

1911 Revolution, 28, 37, 38, 65, 133

nomenclature, 54, 72, 103–4, 112–13; and traditional names, 78, 84, 105, 181; International Code of Botanical Nomenclature, 68, 73–74, 104, 111–12, 147. *See also* naming; terminology; translation

North China Branch of the Royal Asiatic Society, 59, 131, 160

Northwest China Botanical Institute (Zhongguo Xibei Zhiwu Diaochasuo), 152, 154

Northwestern College of Agriculture (Wugong), 152

Note about the Plants of Central Yunnan, A (Dian zhong hua mu ji), 98

nurseries, 22, 31, 32, 80, 118

O

observation, 6, 28, 50, 101, 124; of the myriad things, 49, 52; as scientific practice, 59, 75–79

Opium Wars, 22, 57, 79; First Opium War and the Treaty of Nanking (1842), 28, 32, 80, 164; Second Opium War and the Treaty of Tientsin (Tianjin) (1858), 33

orchids, 43, 93, 149, 158, 167

ordering, 50, 100–102, 103–5; within self-generation's pattern (*ziranzhi li*), 43, 101. *See also* classification; naming; taxonomy

P

Pacific War, 153, 170

paleontology, 51, 78

Paris, 77, 80, 104, 127
Peking Magazine (Zhong xi jian wen lu), 34
Peking University: Beiping University (1927–49), 144; Government University of Peking (1911–19), 38, 83–84; Imperial University of Peking (1898–1912), 36, 84; National Peking University (1919–27), 39, 81
People's Republic of China, 8, 28–29, 160, 176
Petiver, James, 11, 127
Philippines, 6, 84
philosophy, 43–45, 60, 143; Aristotle and Plato, 102
phonetics, 67–68, 72, 116
photography, 119, 151
phylum, 62, 110–11
physiology, 6, 50, 55, 71, 109
pirates, 91
pitcher plant (*ping cao*) (*Nepenthes* sp.), 131
plant hunters, 3, 7, 22, 80–82, 177, 180. *See also* Forrest, George; Fortune, Robert; Rock, Joseph; von Handel-Mazzetti, Heinrich; Wilson, E. H.
plantations, 11, 80, 164
plants: a category of living beings, 35, 42, 44–46, 54, 106–8; as cultural objects, 14–16, 21–22; distribution of, 6, 156, 176; edible, 82, 107, 121; translated as *zhiwu*, 45–46. *See also* anatomy (plant); classification; medicinal plants; nomenclature; taxonomy
plum, 106, 124
poetry, 14–15, 84; by Hu Xiansu, 62, 177, 179; by Xu Xiake, 97
pomegranate (*shi liu*), 14, 93

poplar (*Populus* spp.), 106, 116
Presbyterian Missionary Press, 60
propagation, 15, 23, 68, 122, 154
Pu Jingzi, 15
Pu Yi (Xuantong emperor), 37
publishing, 37, 63, 66, 120, 142; publication of new species, 7, 112
pumpkin (*Cucurbita moschata* Duchesne), 93, 138*fig*.

Q

qi (primordial matter), 43, 119, 131
Qian Chongshu, 38, 40, 88, 110, 143
Qianlong emperor, 31
Qin dynasty, 52, 130
Qin Renchang, 88, 134, 151, 166, 169–71
Qing dynasty, 29, 33–37, 48, 57–59, 64–65

R

Rawes, Captain Richard, 23
Ray, John, 6
rectification of names, 54, 100, 107, 115–16, 180
Red Guards, 137, 171, 178
Reeves, John, 129
Ren Hongjun, 52, 144, 147, 157
Republic of China, 40, 58, 96
research, 23, 39, 62, 169; research institutions, 28–29, 40, 89, 147–52; as scientific practice, 50, 76–79, 179–80. *See also* institutions
Research on the Illustrations, Realities, and Names of Plants (Zhiwu mingshi tukao), 6, 19, 32, 46, 113, 178; illustrations in, 124–25, 126*fig*. *See also* Wu Qijun

Rhododendron spp., 22, 80–81, 156
Ricci, Matteo, 29, 58; *Dictionnaire Ricci des plantes de Chine*, 5
Rites of Zhou, The (Zhou li), 46, 106
Ritvo, Harriet, 101
Rock, Joseph, 4
roots, 102, 109, 118, 131
Royal Asiatic Society, 59, 131, 160
Royal Botanical Society, 10, 178
Royal Horticultural Society, 3, 23, 25, 32, 80
Royal Society, 11, 31, 77, 79
Russia, 22, 34, 127

S

san qi (*Panax notoginseng*), 156
Sanskrit, 67, 71, 95
Sargent, Charles Sprague, 86, 168
scholar-officials (literati), 6, 8, 47–48, 54; after abolition of civil service exams, 36, 51, 77; as translators, 32, 57–58, 62–66
science: in contrast to traditional knowledge, 3, 42, 52–54, 97; field sciences, 75–77, 96–99, 148, 180; and modernization, 9, 38, 61, 66, 140; Mr. Science (*Sai xiansheng*), 39, 51, 181; in school and university curricula, 34–36, 109–10, 170; theories of, 4–5, 100–101, 143–44; translated as *kexue*, 37, 48–50. *See also* botany; new learning; scientific method; Western: sciences
Science (Kexue), 38, 42, 52, 111, 143–44, 157
Science Society of China (Zhongguo Kexue She), 27, 38, 72, 145–46, 151, 157; convenor of meetings of scientific associations, 144–45, 154,
169. *See also* Biological Laboratory of the Science Society of China (Zhongguo Kexue She Shengwu Yanjiusuo)
Science World (Kexue shijie), 37, 61, 65, 83, 142, 146
scientific method, 50–52, 78; experimentation, 52, 78; hypothesis, 7, 52; rationality, 52, 59; recording of observations, 91–94, 100–101
scientists, 7, 39, 118, 144; activities of, 4, 33, 77; as professionals, 51, 140–41
Second World War, 25
seeds, 11, 20, 55, 80; exchanges of, 164, 170, 176
Shaanxi, 152; *Flora of the Shaan Gan Ning Basin* (Shaan Gan Ning pendi zhiwu zhi), 155
Shandong, 39, 91, 160
Shang dynasty, 105
Shanghai, 8, 32, 62–66, 131; Shanghai Polytechnic and Reading Room (Gezhi Shuyuan) 34, 65, 141–42; Shanghai Science Commission, 157; Shanghai Scientific Instruments Factory (Shanghai Kexue Yiqi Guan), 82–83, 142
Shang Zhi Society (Shang Zhi Xuehui), 149–50
Shen Gua (also known as Shen Kuo), 77–78
Sichuan, 27, 91, 148, 153–54, 175; Sichuan (Shu) camellia, 21
Silverman, Milton, 176
Sima Guang, 163
Sinensia (Guoli zhongyang yanjiuyuan ziran lishi bowuguan tekan), 40
Sino-Japanese War (1894–95), 28, 34, 48, 62
Sivin, Nathan, 43

Society for Preserving National Learning (Guoxue Baocun Hui), 130–31
Sources for Materia Medica (Bencao yuanshi), 107, 121–22, 123*fig.*
South China Institute of Botany, 88
southwestern China, 14, 41, 89, 143, 154; botanical diversity of, 150
species (*zhong*): of camellia 10–14, 22–24; Chinese terminology for, 57, 178; identification of, 80, 101–3, 110–11, 175; naming of, 7, 84, 104, 112–13
Stapf, Otto, 23
Su Song, 120
Sun Yat-sen University (Guangzhou), 41, 81; Institute of Agriculture, Forestry and Botany, 148–49, 151; Specimen Garden, 166–67
surveys, 91, 120, 137, 151, 155
sutras: *Four Part Vinaya*, 108; *Lotus Sutra*, 95; translation of, 57–56
Swift, Jonathan, 59

T

Taiping Rebellion, 28, 32–33, 62–64
Tang dynasty, 48, 91, 108
Tao Hongjing, 107
taxonomy, 100, 109, 115, 180; in Buddhist monasteries, 108; of camellias and tea, 10–14; principles of, 5, 101–4, 109–11; terminology for, 35, 62, 71–72, 113–14; traditional systems of, 53, 105–8
tea (*cha*), 19–21, 53; "*cha*" and "*ming*" from leaves picked at different seasons, 20, 122, 123*fig.*; tea oil (*Camellia oleifera*), 11; in Linnaeus, 13, 21; in trade with the West, 11–13, 80. *See also* Camellia (*shan cha*)
Tengyüeh (now Tengchong), 23, 80
terminology, 14, 20–21, 56, 68–69; for botanical description, 56, 178; coining new terminology, 35–36, 46, 60–62, 66–67, 116; derived from classical Chinese, 57, 69–70; derived from Japanese, 48–49, 71, 113; standardization of, 7–8, 66, 72–74. *See also* translation; nomenclature
Textbook of Botany (Zhiwuxue Jiaoke Shu), 37, 109–10
textbooks, 7, 35–37, 62, 109–10, 147
Theaceae. *See* Camellia (*shan cha*)
Theophrastus, 55, 101
Thopa Xin, 14
Tianjin (Tientsin), 141, 142; Treaty of (1858), 33
Tibet, 27, 92, 151
tobacco (*Nicotiana tabacum*), 72, 156
trade, 11–13, 164; in plants, 22–23, 79–80
traditional knowledge, 4–7, 13, 68–69, 130–31, 180; in ethnographic research, 14, 81
translation, 7, 55, 56–58, 71–72; collaboration between missionaries and Chinese scholars, 32, 62–66; Jesuit scientific translations, 47–48, 58; lexical substitution in, 46, 67; phonetic adaptation in, 67, 116; at the School of Combined Learning, 33–34, 57, 65; semantic adaptation in, 67; of textbooks, 37, 101–10, 146. *See also* language; terminology; nomenclature
travel writing, 18, 93–96, 180

Treatise on Plants for Use in Emergency (Jiuhuang bencao), 121
Treatise on the Mushrooms of the Lower Yangtze (Wu xun pu), 143
Treaty Ports, 8, 32–33, 39, 141. *See also* Opium Wars
trees, 105–6, 169; a category of living being, 6, 42–43, 46, 106–8
tropics, 79, 89–91, 104, 150
Tsoongia axillariflora Merrill, 84
tsubakki (tzumacky), 11–13, 12*fig.*
Tutcher, J., 86, 165
type specimen, 7, 80–81, 112, 159

U

United States of America, 24–26, 36, 170, 176; and Boxer Indemnity payments, 145, 148; Chinese students in, 38, 40, 85, 86, 145; Department of Agriculture, 175
Universal Encyclopedia, The (San cai tu hui), 16–18, 17*fig.*, 19
universities, 3, 29, 38; Aurora University (Zhendan Daxue), 130, 160; University of Copenhagen, 151; Cornell, 38, 73, 145; Guanghua University, 89; Imperial University of Peking (Jing Shi Daxue), 36, 84; Soochow University, 39; Third National Sun Yat-sen University, 84, 166; Tokyo University, 81; Tsinghua University, 149; Uppsala University, 79; wartime relocation to unoccupied areas, 41, 152–55; Xiamen (Amoy) University, 39; Yale University, 145. *See also under specific universities' names*
University Council (Daxue Yuan), 114

V

varieties: of camellia, 9–10, 15–19, 56; of chrysanthemum, 124; of crops, 78, 93, 154
vascular plants, 5, 105
vegetables, 46, 93, 107–8, 154, 163
vegetation, 39, 149, 166; on Jizu Shan, 97–99; mapping of, 92, 162
Versailles, Treaty of, 39, 96
Verse Primer of Botany, A (Zhiwuxue gelüe), 35–36
Vienna, 73, 80, 96, 166
vines and lianas (*teng*), 46, 103
von Handel-Mazzetti, Heinrich, 94–98

W

Wang Hanchen, 80, 82
Wang Jie, 120
Wang Xiangjin, 53
Wang Zhan, 175, 177–78
war and warfare, 27, 75, 137, 153–55
warlords, 27–28, 37, 91, 168; in Sichuan (Liu Wenhui and Liu Xiang), 153
West China Academy of Sciences (Zhongguo Xibu Kexueyuan), 151, 153–54
Western: botanical illustration, 30*fig.*, 131, 134–36; sciences, 7, 28–29, 33, 47–49, 58–61, 179–80; scientists collaborating with Chinese scientists, 39, 63, 177, 178; Western powers and China, 28, 31–36, 39, 51, 145. *See also* botanists; imperialism; plant hunters
Williamson, Alexander, 32, 46, 109
Wilson, Ernest Henry, 3, 81, 85
Wissen und Wissenschaft (Xueyi She), 146

wood sciences, 151, 156, 168
Wu Jiaxu (pen name Wu Bingxin), 44–45, 76, 110–11, 117, 143
Wu Qijun, 77, 129, 133, 179. See also *Research on the Illustrations, Realities, and Names of Plants* (Zhiwu mingshi tukao)
wu wei (non-willful action), 44
Wu Zhonglun, 80, 82, 175
Wuchang, 34; Wuchang Normal College, 113
Wuxi, 63, 65; Wuxi Third Normal College (Wuxi Di San Shifan Xuexiao), 134

X

Xishuangbanna Tropical Botanical Garden, 91
Xie Zhaozhe, 10, 78; *Fivefold Miscellany* (Wu za zu), 78
Xu Shou, 63, 65, 141
Xu Xiake, 93–99
Xue Jiru, 175, 178–79
Xunzi (Xun Kuang), 100, 103, 180

Y

Yan'an, 154–55
Yan Fu, 49, 66, 71
Yaquan Journal (Yaquan zazhi), 37, 142
Ye Lan, 35, 50
Ye Qizhen, 37
yin and yang, 43, 101, 119
Yongle emperor, 57
Yongzheng emperor, 30, 63
Yu Dejun, 10, 22–25, 27, 91–92, 151; *Garden Camellias of Yunnan, The*, 10, 23, 24*fig.*, 179
Yu Heqin (Yu Heyin), 54, 83, 142

Yuan dynasty, 57
Yuan Shikai, 37
Yunnan, 80–81, 156, 181; Cai Xitao's expedition to, 87–88; camellias in, 9–10, 18, 179–80; tobacco in, 156; Western collectors in, 23, 75, 92; Yunnan Institute of Agriculture, Forestry, and Botany (Yunnan Nonglin Zhiwu Yanjiusuo), 10, 41, 85, 137, 171; Yunnan Province Bureau of Education, 41, 155–56
Yunnan Scholarly and Critical Weekly (Yunnan xueshu piping chu zhoukan), 100

Z

Zeller, Suzanne, 75
Zeng Guofan, 64
Zeng Zhaolun, 145
Zhang Jingyue, 40
Zhang Zongxu, 113
Zhao Chengzhang, 82
Zhejiang, 85, 92; Zhejiang Agricultural College, 167; Zhejiang University, 84
Zheng Qiao, 119
Zheng Wanjun, 173–75
Zhiwuxue. See *Elements of Botany*
zhizhi zai gewu (the extension of wisdom and the investigation of things), 47
Zhong Guanguang (Zhong Xianchang, K. K. Tsoong), 4, 82–84, 83*fig.*, 105, 158, 166; aligning traditional and scientific nomenclature, 68, 115–16, 180–81; founder of the Shanghai Scientific Instruments Factory, 82, 142; in Guangxi and Yunnan, 39–40, 84, 92–99
Zhou dynasty, 46, 105–6, 130

Zhu Xiao, 121
Zhuang Zi, 43
ziran: translation of "nature," 42–43, 45, 54; *ziranzhi li* (self-generation's pattern) 47, 101

zoology, 49–50, 147, 159–60, 205n1
Zou Bingwen, 73
Zou Yigui, 21–22
Zuo Jinglie, 81

CULTURE, PLACE, AND NATURE
Studies in Anthropology and Environment

Ordering the Myriad Things: From Traditional Knowledge to Scientific Botany in China, by Nicholas K. Menzies

Misreading the Bengal Delta: Climate Change, Development, and Livelihoods in Coastal Bangladesh, by Camelia Dewan

Timber and Forestry in Qing China: Sustaining the Market, by Meng Zhang

Consuming Ivory: Mercantile Legacies of East Africa and New England, by Alexandra C. Kelly

Mapping Water in Dominica: Enslavement and Environment under Colonialism, by Mark W. Hauser

Mountains of Blame: Climate and Culpability in the Philippine Uplands, by Will Smith

Sacred Cows and Chicken Manchurian: The Everyday Politics of Eating Meat in India, by James Staples

Gardens of Gold: Place-Making in Papua New Guinea, by Jamon Alex Halvaksz

Shifting Livelihoods: Gold Mining and Subsistence in the Chocó, Colombia, by Daniel Tubb

Disturbed Forests, Fragmented Memories: Jarai and Other Lives in the Cambodian Highlands, by Jonathan Padwe

The Snow Leopard and the Goat: Politics of Conservation in the Western Himalayas, by Shafqat Hussain

Roses from Kenya: Labor, Environment, and the Global Trade in Cut Flowers, by Megan A. Styles

Working with the Ancestors: Mana *and Place in the Marquesas Islands,* by Emily C. Donaldson

Living with Oil and Coal: Resource Politics and Militarization in Northeast India, by Dolly Kikon

Caring for Glaciers: Land, Animals, and Humanity in the Himalayas, by Karine Gagné

Organic Sovereignties: Struggles over Farming in an Age of Free Trade, by Guntra A. Aistara

The Nature of Whiteness: Race, Animals, and Nation in Zimbabwe, by Yuka Suzuki

Forests Are Gold: Trees, People, and Environmental Rule in Vietnam, by Pamela D. McElwee

Conjuring Property: Speculation and Environmental Futures in the Brazilian Amazon, by Jeremy M. Campbell

Andean Waterways: Resource Politics in Highland Peru, by Mattias Borg Rasmussen

Puer Tea: Ancient Caravans and Urban Chic, by Jinghong Zhang

Enclosed: Conservation, Cattle, and Commerce among the Q'eqchi' Maya Lowlanders, by Liza Grandia

Forests of Identity: Society, Ethnicity, and Stereotypes in the Congo River Basin, by Stephanie Rupp

Tahiti Beyond the Postcard: Power, Place, and Everyday Life, by Miriam Kahn

Wild Sardinia: Indigeneity and the Global Dreamtimes of Environmentalism, by Tracey Heatherington

Nature Protests: The End of Ecology in Slovakia, by Edward Snajdr

Forest Guardians, Forest Destroyers: The Politics of Environmental Knowledge in Northern Thailand, by Tim Forsyth and Andrew Walker

Being and Place among the Tlingit, by Thomas F. Thornton

Tropics and the Traveling Gaze: India, Landscape, and Science, 1800–1856, by David Arnold

Ecological Nationalisms: Nature, Livelihood, and Identities in South Asia, edited by Gunnel Cederlöf and K. Sivaramakrishnan

From Enslavement to Environmentalism: Politics on a Southern African Frontier, by David McDermott Hughes

Border Landscapes: The Politics of Akha Land Use in China and Thailand, by Janet C. Sturgeon

Property and Politics in Sabah, Malaysia: Native Struggles over Land Rights, by Amity A. Doolittle

The Earth's Blanket: Traditional Teachings for Sustainable Living, by Nancy Turner

The Kuhls of Kangra: Community-Managed Irrigation in the Western Himalaya, by Mark Baker

Lightning Source UK Ltd.
Milton Keynes UK
UKHW010950280122
397775UK00002B/110